NON-TIMBER FOREST PRODUCTS OF INDIA

NON-TIMBER FOREST PRODUCTS OF INDIA

S.P. Rawat

2008
Gene-Tech Books
New Delhi - 110 002

ISBN 81-89729-86-1
ISBN 978-81-89729-86-8

Published by : **GENE-TECH BOOKS**
4762-63/23, Ansari Road, Darya Ganj,
NEW DELHI - 110 002
Phone: 41562849
e-mail: genetechbooks@yahoo.co.in

Laser Typesetting: **Twinkle Graphics**
Delhi - 110 088

Printed at : **Chawla Offset Printers**
Delhi - 110 052

PRINTED IN INDIA

PREFACE

Since the beginning of the history of forestry various efforts has been made to classify forests throughout the world basing on the timber yielding plant species they produce. This is because timber has been taken in to account as major forest products. But in recent years it is accepted that plant and animal species associated with a forest other than timber yielding species are equally valuable and sometimes much more important than wood products of forests. Scientists/Forester classify these products as Non-timber Forest Products (NTFP). Policy makers often assume tropical forests have no economic value unless they are logged or formed. But the reality is something else. Out of 25,000 medicinal plants of the world about 12,000 species have so far been botanically identified in India of which about 5000 are of higher plant groups. Some of them are the rarest of the rare. All these are used in various transitional systems and the modern allopathic system of medicines in India.

On the diversified uses point of view it can be mentioned that there are almost 4000 edible species; 1000 species are used in various social, cultural and religious performances; 500 species have todder value; 60 species have been recorded as insectivorous plants; 400 plants yield gums, resins and dyes and 150 species are the source of essential oils used in perfumery industries. Most of these species comes under NTFP. Most of these plants are verged in the threatened category. Any geographical area with greater number of overall bio-diversity and maximum number of endemic species are considered as "Hot Spots" area. We are fortunate to have two of them. One is Eastern Himalayas and other is Western Ghats. But the forest cover is decreasing day by day due to population rise, expansion of industries and introduction of exotic species for plantations. So it is an urgent need to record NTFP for the documentation of the species but also to save from further destruction to save economy, environment and society.

In the present book we have documented few of NTFP, its usefullness and suggested how it will help us to improve our economy.

S.P. Rawat

CONTENTS

CHAPTER 1

BIRTH OF NTFP

Non-timber forest products (NTFPs) have been defined as "all goods of biological origin other than wood in all its forms, as well as services derived from forest or any land under similar use."

There are diverse ranges of ways the people from different parts of the world exploit their surrounding forests. Forests provide food, medicines, household equipments and building materials, raw materials for proceeding enterprises and materials for agriculture and other equipments. While the extent of forest resources use varies by household even in particular region, it is clear from the existing reports that forest products are widely consumed.

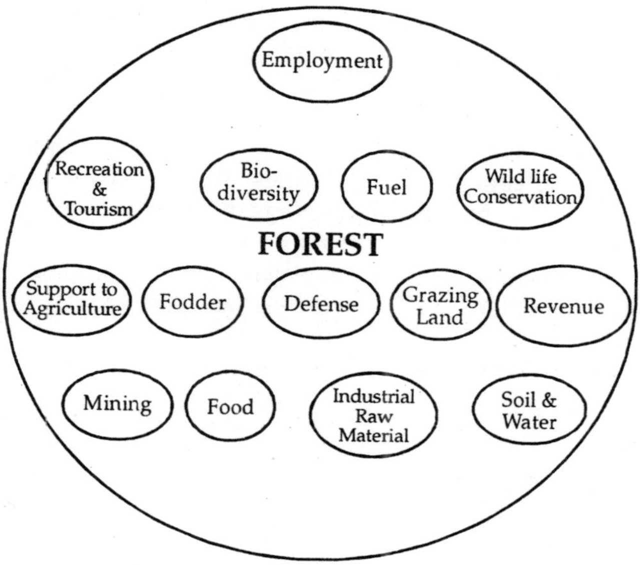

Fig. 1

Therefore, NTFPs can also be defined as all goods derived from forest and their biomass for commercial and industrial use, which can be sustainably extracted from a forest ecosystem in quantities and in such ways, that it does not downgrade the plant communities' basic reproductive functions.

Various functions of forests can be detailed as follows :

1. **Regulatory function** : maintenance of the quality of the soil, air and water; protection of river catchments, stabilization of climate;
2. **Supportive function** : living environment (including land for agriculture and cattle rearing), tourism (including eco-tourism);
3. **Production function :** forest products (timber and non-timber);
4. **Information and reservoir function** : ecological processes, biodiversity, genetic material, social, cultural values and intrinsic natural value.

Two products, which do count as NTFPs, in this connection are firewood and wood for small-scale construction, this being in contrast to the FAO's definition "Non-Wood Forest Products" (NWFPs). This basis of taking such position is that the same socio-economic factors play a role in the case of firewood and wood for small-scale construction as in the case of the other products mentioned above. In addition, many local communities in India view firewood as one of the most important forest products. Collecting firewood does often have destructive consequences for the forest and savannah landscape (specifically in arid areas), but precisely for that reason it should be given a place in sustainable forest management systems. In some cases, harvesting of other NTFPs can be just as destructive.

Why are non-wood forest products important?

For most of recorded history, people have valued forests not for wood, but for other products. Ancient writings from China, Egypt and India record a wide variety of uses for forest plants, and compilations of botanical knowledge from Western Asia were prized by the ancient Greeks. Whereas wood products have become major international commodities in modern times, NTFPs rank among the oldest traded commodities. Ancient

Egyptians imported gum arabic from Sudan for use in paints and the mummification process. International trade in sandal-wood oil dates back to the twelfth century A.D.

Through the experience of forest communities, forestry professionals have recently rediscovered the great importance of NTFPs (ranging from food, fruits and fibres, dye stuffs, flavours and medicines) for meeting people's needs. In recent years, a growing body of scientific research has suggested that, given certain basic conditions, NTFPs can help communities to meet their needs without destroying the forest resource.

The most commonly cited instances come from Latin America, where the term extractive reserves describes a system where forest is set aside for low-impact use by traditional communities in the area. No single model can suit all conditions, however.

Why have modern science and governments overlooked the importance of this non-wood forest wealth for so long? The answer is threefold. First, most of these products are used mainly for rural subsistence or local markets. They often go unrecorded in official statistics, which focus on nationally traded goods. Second, because modern government administration has divided these products among forestry, agriculture and horticulture, statistics do not recognize even nationally and internationally important non-wood forest commodities as originating from the forest. The divisions between, and the lack of clear definition of, agriculture and forestry have created a large blind spot in the way we reckon our dependence on forests. Finally, modern forestry has favoured timber and large-scale enterprises, and has generally regarded non-wood products as incidental. However, studies show that forests produce many more types of products than wood products particularly in some tropical forests. Small-scale forest-based enterprises in Zimbabwe, which mostly are based on NWFPs, employed 237,000 people in 1991, compared to 16,000 employed in conventional forestry and forest industries for the same year.

Various different interest groups are involved with NTFPs. These include users within local communities (men, women, children, the elderly, elites, etc.), district councils, city dwellers, provincial authorities, national authorities, the market, business enterprises, research institutes and the international commercial market and research institutions.

The harvesting of NTFPs is often referred to as "extraction" in this book. Here extraction means the harvesting of NTFPs from natural forests, in contrast to the "production" of NTFPs, which refers to the cultivation of NTFPs on plantations.

Forests and forest products are linked to household livelihood systems in a variety of different ways. Forest products commonly contribute to meeting food and other basic needs, are a source of income and of inputs into the agricultural system, help households control exposure to risk of various kinds, and constitute an integral part of the habitat and of the social and cultural structure of those living within that environment.

Villagers conceptualize non-timber forest products (benefits) as both concrete and intangible. The inclusion of conceptual as well as physical goods in discussions of NTFPs by farmers suggests that to rural communities forests are a product to be weighed and measured *in toto* and for which physical, extractable non-timber forest products are only a limited subset. The sum of the parts in this case is not worth near as much as the presence of the whole.

Non-timber forest products (NTFPs) have long been used for subsistence by communities living near forests. Indiscriminate felling of forests for timber has caused loss of biodiversity and many serious environmental problems, including damage to ecosystems, weather changes and soil erosion. The other valuable forest resources have been neglected.

Many people living in and near forests are unaware of the potential of that resource for income generation because they lack access to information on processing possibilities. Sustainable exploitation of these resources could improve their livelihood through supplemental income and employment. Tropical rain forests are particularly abundant with plants yielding essential oils, gums, medicines, tannins, and colorants.

It is, therefore, necessary to consider how use or sale of material products such as forest foods are conditioned by this broader context of interrelationships between people and their forest environment. This broader framework is likely to reflect other important attributes of the forest, and the processes whereby people transform the forest into a productive system designed to meet their particular needs.

More recently, NTFPs have received international attention as a result of :

- The growth of green consumerism in developed countries;
- More open international markets;
- Increasing awareness of biodiversity conservation and sustainable and protective use of forest resources;
- Realization, based on a growing number of reports, of the potential of multi-use forest management to generate rural income, and that in many cases NTFP yield more income over time than timber species from the same resource;
- Growing recognition of the need to involve people living near forests for sustainable management of forest resources.

The United Nations Conference on Environment and Development (UNCED) in 1992 brought these factors into bold focus and from that emerged an agenda for implementation. The Agenda 21, Chapter 11 on Combating deforestation, includes the promotion and development of NTFPs through value addition, domestic processing, and promotion of small-scale forest-based enterprises for rural income and employment. Many countries are now preparing plans for implementing Agenda 21, and NTFP programmes can be incorporated. Government commitment, the absence of which had previously posed a major constraint to NTFP development, appears serious.

Only a few developing countries have had the resources to carry out large-scale, commercial processing of NTFPs. Historically, most NTFPs have been exported to industrialized countries where they are processed as final products. However, with inexpensive equipment and small-scale operations, local processing of selected NTFPs can be environmentally and economically viable.

Total present value of non-timber goods and services available from tropical deciduous forests in India varies from a minimum of $ 4034 to a maximum of $ 6662 per hectare, if use, service and existence value are all taken into account. The use value of NTFPs, which accrues to local communities and national economy, is about 40% of this figure. Nevertheless, the use value of NTFPs has significant impact upon income and consumption flows in local economies; in particular, this value may have strong implications for the welfare of communities, which use these products for subsistence. The option and existence values of forests, on the other hand, relate primarily to the role of that forests play in global context.

Option values become significant when accelerated deforestation begins to deplete forest resources. Existence value measures the value of forests as carbon sinks, and the preservers of biodiversity. Thus, implication of any national or international agreement aided at depleting or conserving these values are to be looked in to deeply. These values become highly marketable when international agreements are taken.

In some parts of India joint forest management (JFM) is being practiced which implies that the historical property right of rural communities to forestland are restored in some sense. There is an implicit assumption that rural communities understand the need for sustainable extraction and will not damage the forest if they have a long-term stake in its maintenance and management.

There are some significant differences between the traditional system and modern methods of forest regulation, such as interface forestry. First, human and livestock are much greater than they used to be. Their demand on primary products put pressure on carrying capacity of forest and other natural resources. Second, the link with market economy is much stronger, which means that there is greater demand for forest products, particularly for NTFPs such as medicinal plants. When standard of living increases there is a greater demand for meat and milk, which in turn increases the demand of fodder. There are, therefore many links with the larger economy (including the global economy) that increase the pressure on forests.

India (mainland 6°45′–37°6″ N latitude and 68°7′–97°25″ E longitude) with her 329 million hectares of land mass, represent 2.46% of the total global land mass and is the 7th largest country of the world. Due to its eleven phyto-geographical zones, great variety of physiographic diversity and climate the country is recognized as one of the 12-mega diversity centers and as a "miniature flora of the world". The country harbours more than 45000 plant species which is about 11% of the world flora, so far been botanically identified; of which 17,500 species belongs to flowering plant groups under 4000 genera and 315 families which is about 7% of the described species of the flowering plants in the world. It is estimated that in India about 3000 species yield NTFPs. Nearly 500 million people living in and around forests in the country depend on NTFPs. It has been widely acclaimed that NTFPs possess better employment potential than wood products. This sector accounts for 1.2 million man-years and has potential of additional 2.5 million man-years employment by 2000 AD (ICFRE, 1993-1994). Almost 55% of employment in forestry sector of the country is attributed to NTFP sub-sector. NTFP based small-scale enterprises in our country provide up to 50% of income for 20-30% of rural labour force.

India is a country of vast diversity lying at the juncture of the big-geographic provinces of Afro-Eurasia and the Orient. Because of the country's diversified climatic and physiographic factors, India is blessed with all types of vegetation : tropical, subtropical, temperate, and alpine. Due to its wide-ranging environmental regimes and diverse biological communities, the country is one of the world's top 12 "megadiverse" nations. Of the nearly 425 families of flowering plants in the world, 328 families with 21,000 species occur in India. From this varied emporium, non-timber forest products (NTFPs) are derived from over 3,000 species.

Who uses non-wood forest products?

For most of the world's rural households, NTFPs provide essential food and nutrition, medicine, fodder, fuel, thatch and construction materials, mulch and non-farm income. These products are particularly important in relieving the "hunger periods" in the agricultural cycle, and in smoothing out other seasonal fluctuations. Dealing in NTFPs can provide employment during slack periods of the agricultural cycle, and provide a buffer against risk and household emergencies.

Poor households, in particular, depend on these products for their livelihood because they usually have more access to the forest than to other resources.

For the same reason–greater dependence on open-access forests, for lack of other options women usually rely more than men on NTFPs for household use and income. In many places, women are responsible for the household activities that involve forest-based foods and medicine, as well as fuel-wood. In this respect NTFPs are particularly important to women, addressing their needs for food security and nutrition.

In local, urban, national and international markets, forest foods and medicines contribute substantially to national economic growth.

NTFPs are therefore important to three main groups :

- *rural populations* (the largest group) who have traditionally used these items for livelihood and social and cultural purposes;
- *urban consumers* (a smaller group, but growing faster), who purchase these items;
- *traders,* and product processors, whose numbers in the NTFP sector increase as urban markets for these products grow.

Local participation is important for sustainable management of NTFPs for several reasons : (1) to recognize the full extent of local demands on the forest resource; (2) to fully consider the local knowledge of the resource that has developed over time; (3) to engage nearby communities as stake holders in managing the resource, ensuring their commitment to long-term management goals; (4) to engage the energies of local people in their own economic change, which can include decisions on social and cultural priorities that outsiders do not realize.

What are the challenges in NTFP study?

It is evident that there is a need for increased application of biometric principles in NTFP resource assessment, however it is necessary to first understand the nature of the problem. The following is a short list of the problems encountered by practitioners working with NTFPs :

- Difficulties with traditional forestry designs
- Lack of properly tested sampling designs tailored for NTFPs
- Few NTFP mensuration techniques available–or at least not accessible to practitioners.
- Little cross-disciplinary exchange of ideas or techniques
- Difficulties, conceptual and practical, in the determination of sustainable yields.

The apparent deficiencies in NTFP inventory could be easily solved if it were possible to use traditional forestry methods as these are well described and understood. Unfortunately, this does not appear to be a practical solution as there are certain characteristics of NTFPs that are not easily accommodated in traditional forest inventory. The main problems being :

- Rarity–many NTFPs are rare which means that only a few plots of a conventional inventory designs will contain the species of interest. This results in very inefficient and costly inventories, which often do not produce the quality of data required.
- Clumped distributions–NTFPs often occur in relatively dense patches within the landscape.
- Imperfect detectability–people dealing with trees have rarely come across the problem of searching for an elusive or moving target because trees are generally large and static. Unfortunately, many NTFPs are not so obvious (*e.g.* truffles and epiphytes) and these require that detectability be considered.
- Seasonality–many NTFPs are seasonal but timber accrues slowly over time consequently forest inventory methods do not cope well with seasonality.
- Mobility–animals run away, fruit falls off a tree and rolls down a hill but trees are static.
- Quantification of yield for non-destructive harvesting–most of the methods for determining timber yield from a forest are concerned with the harvesting of entire individuals. For NTFPs often only a small part of the individual is harvested.

The concept of integrated resource management plan (IRMP) encompasses management of NTFPs along with tree species to enhance the food and other non-wood forest products with the object to manage and conserve the floral and faunal resources. With these challenges ahead and demands increasing day-by-day need for better knowledge of NTFPs has enthused many workers to make detailed study of these resources. Only multidisciplinary efforts can make considerable progress in this direction.

CHAPTER 2

CLASSIFICATION OF NTFP

Since the beginning of the history of forestry various efforts has been made to classify forests throughout the world basing on the timber yielding plant species they produce. This is because timber has been taken into account as major forest product since State initiative in forest management emerged. But in recent years it is accepted that plant and animal species associated with a forest other than timber yielding species are equally valuable and sometimes much more important than wood products of the forests. Different authors have classified NTFPs on the basis of the various aspects of the forest products. Sometimes economic and use values in human life or environmental goods and services have been added to it. But all these are artificial classifications.

Before going to discuss the attempts taken up by various organizations and individuals to classify NTFPs, different categories and uses of NTFPs can be mentioned as follows.

NTFPs can be subdivided into various categories, making clear the large variety of products covered by the term NTFP :

- by user purpose (for example for food, medicines, roofing materials, etc.)
- by level of use (self-supporting, commercial)
- by type of NTFP harvested (for example leaf, fruit, stem, exudates, skin).

The fact that some species are used for several purposes means that there may be some overlap in the categorization by user purpose. Nor is it possible to strictly separate NTFPs by their level of use; NTFPs are nowadays used both for self-support and commercially. The third category–the type or part of the NTFP used–is an effective way of clarifying the impact of harvesting on the individual NTFP and on the NTFP species but also on the forest ecosystem. This is therefore a useful basis for sustainable systems for managing NTFPs.

International Union of Forestry Research Organization (IUFRO) in 1990 under its "Forest Decimal Classifications-28" system kept NTFPs under the head "Husbandry of Forest Products other than Wood". This head include nine items in the first phase such as 281-Christmas trees; 282-Bark crops; 283-Fruit and Seed crops; 284-Crops tapped for Exudates; 286-Osier *(Salix viminalis)*; 287-Bamboo and Cane; 288-Miscellaneous products. In the second phase under the head "Minor Forest Products-89" it includes 891-Minor products and Vegetable products; 892-Vegetable products; 893-Animal products; 899-Miscellaneous products.

FAO (1991) used the term non-wood forest products (NWFPs) and subdivided it in to (*a*) Vegetal, (*b*) Faunal non-wood forest products and (*c*) Forestry services. This was suggested by Wickens (1991). In another attempt FAO (1991) classified NTFPs basing on items of human usage of the product *viz.* (1) Fiber products (2) Food products (3) Medicinal plants and Cosmetic plants (4) Animal products and (5) Miscellaneous.

Some efforts have been made in India to classify the NTFPs. In this perspective Forest Research Institute (FRI) of India published Indian Forest Utilization (1972) which has mentioned nine items of NTFPs such as 1. Fibres and flosses; 2. Grasses; 3. Essential oils; 4. Oil seeds; 5. Tans and dyes; 6. Gums, resins and oleoresins; 7. Drugs, spices, poisons and insecticides; 8. Edible products; and 9. Animal, mineral and miscellaneous products. National Commission on Agriculture (1976) has classified the products into : 1. Fibre and flosses; 2. Grasses; 3. Essential oils; 4. Oil seeds; 5. Tans and dyes; 6. Gums, resins and oleoresins, 7. Drugs, spices, poisons and insecticides; 8. Leaves; 9. Edible products; 10. Lac and its products; 11. Other products.

Tewari have suggested twenty-two types of NTFPs such as fuel and fodder biomass, bamboo, canes, grasses, fibre, oils, tannin and dyes, gums, resins, medicinal plants, barks, leaves, flowers, fruits, mushrooms, seeds, mammals, birds, reptiles, fish and insects etc. They are also of the opinion that environmental outputs, some of which are intangible like soil and water conservation, amelioration of micro climates, flood control, conservation of biodiversity, sequestering and storing carbon, nutrient cycling and direct or indirect benefits like religious, scenic and recreational uses, should be included within the domain of non-timber forest products. Gupta and Guleria (1982) have classified non-wood forest products (NWFPs), as they call them, in to eleven groups (Table 1) such as : (*i*) Fibres and Flosses (*ii*) Grasses other than those used for oil extraction (*iii*) Bamboos and canes (*iv*) Essential oils (*v*) Oil seeds (*vi*) Tans and dyes (*vii*) Gums, resins and oleoresins (*viii*) Drugs, spices, poisons and insecticides (*ix*) Edible products (*x*) Lac and *tassar* silk (*xi*) Leaves.

Table 1 : Classification of NTFPs as suggested by Gupta and Guleria (1982)

Resource category	*Plant part*	*Species*
1. Reproductive propagules	a. Fruit b. Nut/seed c. Oil seed	*Garcinia sp., Mangifera sp., Artocarpus sp., Eugenia sp., Syzygium sp., Spondias sp., Parkia sp., Gnetum sp., Shorea sp., Aeurites sp.*, etc.
2. Floral nectar	a. Latex b. Resin c. Floral nectar	*Palaquium sp., Dyera sp., Dipterocarpus sp., Shorea sp., Aquilaria sp. Arenga sp., Nipa sp.*, etc.
3. Vegetative structured	a. Plant part b. Root c. Leaf fiber d. Bark e. Apical bud	*Calamus sp., Bambusa sp. Derris sp. Pandanus sp.*, Members of the family Palmae, (Arecaceae) *Litsea sp., Arenga sp.*, etc.

The Economic Botany Data Collection Standards of Royal Botanical Garden, Kew published a system of classification of these products. This was proposed for plants of economic importance and not for NTFPs specifically. This classification has indicated the following major groups :

1. Food, 2. Food additives, 3. Animal food, 4. Bee plants, 5. Invertebrate food materials, 6. Fuels, 7. Social uses, 8. Vertebrate poisons, 9. Non-vertebrate poisons, 10. Medicines, 11. Environmental uses and 12. Gene sources.

The categorization of NTFPs made by De Beer and McDermott (1996) is given as follows :

1. Edible Plant Products : (*a*) *food,* (*b*) *edible oils,* (*c*) *herbs,* (*d*) *animal feed.*

Many plants and plant products taken from forests are used as food for humans and animals. These include whole plants, leaves, roots, fruits, nuts and mushrooms. User groups can be found at local level, but also at other levels : Brazil nuts and herbs such as cardamom, for example, are traded internationally on a large scale.

2. Edible Animal Products : (*a*) *terrestrial animals,* (*b*) *animal products (eggs, birds' nests, honey),* (*c*) *fish and invertebrate water animals.*

For many people in rural areas, forest animals are an important source of protein. Besides larger animals such as pigs, deer, crocodiles etc., insects are also an important component of diet. In urban areas, larger game animals have often become "luxury" items. Birds' nests are not eaten locally (Borneo, Malaysia, Thailand) but are marketed internationally as a delicacy (for example in China). Honey is a good example of an edible animal product that may be of great significance locally, regionally and internationally. Fresh-water fish, given that they form an integral part of the forest ecosystem, count as NTFPs. Fish and invertebrates, which are dependent for a large part of their life on mangrove forests, can also be classified as NTFPs.

3. Medicinal Products : (*a*) *plant and* (*b*) *animal products.*

This category is only mentioned separately because of the enormous number of products with medicinal properties. In practice, however, it does not constitute a separate group since many plants with one or more medicinal properties are also used as food or as ornamental plants, or are a source of nuts, resin or tannin. Animal products may also have medicinal properties, for example honey or bear's gall. Here too, user groups range from local traditional "healers" to the international market and the pharmaceutical industry.

4. Non-edible Plant Products : (*a*) *rattan,* (*b*) *bamboo (also classified as an edible plant product),* (*c*) *ornamental plants,* (*d*) *chemical components (exudates and extracts),* (*e*) *non-industrial timber,* (*f*) *fibres and leaves.*

This is a large and varied category. Rattan and bamboo are among the most familiar and useful NTFPs specifically from Asia. Besides their local value in "forest gardens" and in traditional ceremonies, ornamental plants are of great value on the international market (botanical gardens, houseplants). The chemical components of plants consist of a large group of exudates (resins, gum, latex) and extracts (essential oils, tannin, paint, aromas). Non-industrial timber means timber in the form of poles for local construction, storage for crops, fencing etc. Timber is also used for tools, instruments, artistic or cultural objects, and as firewood. A great deal of this type of wood is used at local level, but there is also a local and national market for firewood and construction wood.

Fibres and leaves are used for clothing, baskets, mats, roofing materials, etc. Humus from the forest is mixed with manure and used in agriculture.

5. Non-edible Animal Products : (a) *insect products (wax, lacquer, mainly collected),* (*b*) *game products and living animals (pets, trophies, traditional ceremonies, clothing, often traded internationally).*

Beeswax is used both at local and commercial level (cosmetics, batik). The "lacquer" exuded by aphids has a wide variety of uses, varying from high-quality varnish and lacquer to insulation for electricity cables. Animals are hunted not just for food and medicines; a large proportion is hunted to serve as pets, for ceremonial purposes, as trophies, for clothing, and for trade with the city and at international level (for example parrots, butterflies, elephant ivory).

All these systems of classification are not exhaustive ones. There are also some minor classifications of forest-based resources, which are mostly artificial groupings in nature and mostly economic value oriented. Probably for this the term 'product(s)' is used as a suffix with the words 'non-timber forest'. To give more scientific meaning to it, it may appear that taxonomic classifications of NTFPs may be attributed on the basis of (1) morphological similarity (both external and internal) and (2) to some extent phylogenetic relationships. But this classification will appear more mechanical because products of same category may be placed apart and products of different categories may be placed close to each other simply because they are morphologically similar.

Therefore, the most important aspects which has been taken into account by the earlier authors while classifying NTFPs are :

(*i*) The sources from where they are obtained (morphological part of the plant)

(*ii*) The type of the product group they represent and

(*iii*) The economic return obtained from the product.

Considering all these facts in the present case no effort has been taken to propose a new system to classify NTFPs. The products have been mentioned along with the scientific names of the species producing them and has been arranged alphabetically followed by the names of families, phytography, phenology and uses by different ethnic groups. This has been done for easy access to individual species.

CHAPTER 3

WORKING OF NTFP

Management of non-timber forest produces includes some important issues like policy of state towards the resource and the way of developing resource by search for sustainable use. Ostrom has enunciated six design principles that could lead to emergence of robust self-governing organizations that could manage common pool resources and preclude the occurrence of any tragedy due to loss of our common pool resources, which are unsustainably used. These designed principles are :

(*a*) clearly defined boundaries of the resources,

(*b*) congruence appropriation of rules and local traditions,

(*c*) collective choice arrangements by the users themselves,

(*d*) graduated sanctions against offenders,

(*e*) existence of conflict resolution mechanism, and

(*f*) recognition of right to organize.

In India it is now well recognized that self-governing organizations are capable enough to manage their natural resources given the right policy environment.

The 73rd constitutional amendment in recreating the *Panchayati Raj* and vesting powers for the management of the natural resources is a pointer to this direction. Further, in the fifth schedule areas the ownership and management of NTFPs is now vested with the tribal panchayats. The act is referred to the clause (1) of Article 244 of the constitution has been effected from 24th December 1997. Hence, it is possible at least theoretically to organize the tribals to at least into a robust self-governing enterprise for the management of biodiversity both for conservation purpose and for benefit of tribals.

NTFP and International Initiatives

NTFPs have also been on the agenda at international level in recent years, namely within the following contexts :

International Forum on Forests

NTFPs were dealt with as part of the action plans formulated at the final meeting of the IPF *(International Forum on Forests)* March 1998. It was recognized, for example, that national forest and land use programmes should be based to a greater extent on a wider appreciation of forest products. In this context, forest products are taken to include not only timber but also the non-timber and ecological functions of forests. The IPF states that forests are under increasing pressure from the use of forest products (timber and non-timber) and other forest functions. There is also an increasing demand for land for other purposes. This requires more intersectoral decision-making with respect to land use planning. The IFF has also indicated that there may be a positive link between the trade in forest products and sustainable forest management but that there is insufficient information on the opportunities for trade in NTFPs. The IFF intends to implement and monitor the IPF's proposals for action.

Convention on Biological Diversity

The importance of protection of biodiversity at local level is a major focus of attention in the Convention on Biological Diversity (CBD). The CBD emphasizes the importance of equitable distribution of the benefits of the use of biodiversity, including among local and native communities. It encourages member states to develop national legislation "to promote the wider application of the knowledge, innovations and management techniques of local communities, with the approval and involvement of those communities, and with encouragement being given to the equitable distribution of the benefits".

The enormous biodiversity within tropical forests has created an industry, which concerns itself with searching for commercially and scientifically relevant products, such as biochemical material for the pharmaceuticals industry. This is known as bio-prospecting. One important component of the Convention on Biological Diversity (CBD) is to emphasise–in an international context–the importance of ensuring that native communities also benefit from the use made of biodiversity. However, clear mechanisms for achieving this are not available. Much native knowledge is being lost because of socio-cultural and economic changes. Universities and companies, however also use that knowledge, to help them find NTFPs, which are of medical or scientific relevance. Intellectual Property Rights (IPR) does not provide a sufficient framework for protecting native knowledge. There is no legal obligation to pass on even a part of the proceeds of this knowledge to the local community.

Convention on International Trade in Endangered Species

The Convention on International Trade in Endangered Species (CITES) regulates international trade in endangered plant and animal species. Between the Second World War and the 70s, there was a large international trade in animals and animal products. Since 1970, trade has been restricted by national and international legislation. Many countries have ratified CITES. However, observance of CITES legislation leaves much to be desired. In 1995, Interpol estimated that the value of the illegal trade in plants and animals–both alive and dead–(25% of the legal trade) amounted to USD 20 billion. This is more than the combined value of illegal trading in both arms and drugs.

Forest Stewardship Council

The Forest Stewardship Council (FSC) is currently drawing up guidelines for certification of NTFP use and management, alongside the existing guidelines for certification of sustainable timber production.

The importance of NTFP in forest management is often insufficiently understood. As a result, few countries devote specific attention to NTFPs within their forestry policy. Because NTFPs do not have a place in forestry legislation, legislation and regulations on the harvesting and marketing of NTFPs, and the international trade in them remain unclear. In order to allow sustainable development of NTFPs in a regional context, it is necessary that the legislation applicable in various adjacent regions and countries should be coordinated. Such a policy should incorporate guidelines for the sustainable harvesting of NTFPs and for monitoring. Many NTFPs can be accessed and harvested by all ("open-access resources"). Without rights to ownership and use, users are not encouraged to manage natural resources in a sustainable manner. Policy aimed at government control of forest use and management often undermines the authority and effectiveness of local institutions at community level aimed at forest management. If policy does exist which permits local user groups to collect NTFPs, this is often only for use at self-support level. The right to collect NTFPs for commercial purposes is granted only to concession holders (including foreign organisations). If local collectors were to enjoy this economic benefit, they would be more highly motivated to manage communal forests for long-term purposes.

The Beginning of Forest Conservation in India

Share of Man in destruction of forest, is undoubtedly greatest. He not only cuts down the trees to provide himself with building materials, fuel, and medicine, but completely wipes out the forests to acquire land mostly agriculture purpose. In every civilized country, it is the later that has become a regular menace to the very existence of forests and India has been no exception to it so far. History shows that in India, the ruthless and relentless destruction of forests began with settlement of the Chalcolithic people in Punjab, Sind, and Gujrat, in the period extending from 3500 to 3000 BC approximately. They came to this country in successive waves from the ancient provinces of Elam, Akkad, and Sumer in the Euphrates and Tigris Valley, and wiped out the forests, they could lay their hands on, to acquire land, not only for building their towns and villages, but also for pasturage and cultivation.

Aryan migration to India from the West in different batches, under the leadership of their priests, the *rishis* (Hermits), some of whose names have been mentioned in the *Rigveda* as composer of hymns *(mantras).*

Their migration to this country took place between 2000 and 1800 BC, approximately, as suggested by archeological evidence (Chatterjee, 1964). Evidently, they introduced in Indian history, a new chapter named as 'Iron age'. The villages *(grama)* laid out by the Aryans, consisted thatched houses, for residential purpose and cattle-pens, which were suitably constructed for protection of their livestock. These were bone of contention between them and the non-Aryans.

The non-Aryans were the jungle folk living in the hamlets built by them inside the forests. They were still using stone and copper for manufacture of their domestic and agricultural implements, while Aryans were using iron for the same purposes. These forest dwellers were the descendants of proto-Australoids who had come to this country in some unrecorded period of history and are represented in our days by Bihors, Santals, Bhils, Savaras, Mundas, Has, Juangs etc. In *Rigveda* they have been graced with some typical expression, as 'nose less', 'dark-skinned', 'gibberish in speech' (or chattering like the apes) and 'not performing religious sacrifices'. During the period when Xth *mandala* of *Rigveda* was composed, the aboriginal people mentioned above, had already found a place in Aryan society, as members of fourth class or the *sudra* caste. *Astadhyayi* of Panini (II.iv.10) reports that Aryans compelled these people to reside outside their settlements because of their obnoxious profession. It is evident from Pali works that long before the sixth century BC the descendants of the forest dwellers were divided into number of professional castes, depending entirely on the forest products. They are the following :

(1) *Nalakaras,* or Rush-workers engaged in manufacture of baskets and boxes for storing things. They collected canes and bamboos from the forests and had cottage industry totally depended on these primary raw materials.

(2) *Usukaras,* or the Fletchers specialized in the manufactures of bows and arrows from canes and bamboos, tipped with iron.

(3) Chammakaras, or Leather-workers

(4) *Tachchhakas,* or Carpenters

(5) *Pasikas* or Trappers and hunters and

(6) *Dantakaras* or the Ivory-workers.

Literary evidence tends to show that the Aryans near about BC 1800 commenced a systematic study of India flora. To utilize the forest wealth in best possible manner, the Aryans not only classified trees keeping in view their character and usefulness, but also their medicinal properties.

During that period whole northern India except Rajasthan was covered with forest. In Rajasthan a big swamp was in existence, which was fed by Saraswati.

Long back 325 to 273 BC Chanakya in his famous book *"Arthasastra"* formulated guidelines for the promotion of forestry. His guidelines have been duly followed by kings during Maurya period. Probably nowhere in the world such valuable analysis has been presented in the era. He in his writings has emphasized on NTFPs as well. He suggested having a superintendent of forest produce (pp. 107, 108 and 109) to collect timber as well as other forest products by employing guards. The suggestion also included the provision of fixing adequate fines and compensations to be levied from those who cause any damage to productive forests except in calamities.

The NTFPs, which Chanakya clearly mentioned are like *tilaka (Barleria cristata)*; many bulbous roots and fruits under the group of medicines; fibrous plants like *Jasminum grandiflorum (Malati), Calotropis gigantea (Arka), Coix barbata (Gavedhuka)*; rope making materials like *Saccharum munja (Munja), Eleusine indica (Balbaja)*; writing materials as leaves like *Corypha talliera (tall); Borassus flabelliformis (tala)*, Birch *(bhurja)* etc; poisons like *Cyperus rotundus* and many others, which has not been scientifically identified as they are given only in Sanskrit name. Besides plants this list included plant products preparing utensils, minerals available from forests, charcoal, bran and ashes, firewood and fodder.

During 600-700 AD forest was dense in the Himalayan foothills particularly of the present UP, Bihar areas, in Assam, Tripura, in Allahabad-Rajmahal zone, Dandakaranya forest area and also in Deccan as stated by the Chinese pilgrim to India Huen Tasng. He described that the forests were full of fruits and flowers. According to the traveler the whole southern India was full of forests intervened with small habitations.

During Moghul period (16th century and 17th century) abundance of forest was also recorded in many parts of India. The Moghul policy on forests was one of indifference. They had no regard for them. The emphasis was on agriculture and reclaiming forest was supported by the State and even incentive was given.

In the days of Tipu Sultan, teak was considered as a royal tree. In the neighboring states like Cochin and Travancore same system existed. In the early part of 19th century British imposed royalty on teak trees. Special officer was appointed who is acquainted with the language and habits of the people and having knowledge of the forest. British then needed teak badly for shipbuilding in view of growing shortage of oak in England. In 1806 Captain Watson of the Police Department was appointed as the Conservator of Forests in India (for Madras Presidency) and under a proclamation issued in 1807, immense power were given to him. Thus, the policing on Indian forest resources by British rulers started. But this system did not exist for a long period due to the resistance from the traders. In 1823, the conservatorship was abolished. But again in 1931, the Indian Navy Board again recommended for establishment conservatorships in various provinces due to shortage in supply of excellent quality of teak for navy. It was then decided that revenue department would be best owed with the duty of forest conservation. Thus, it could be seen that the conservator was basically bestowed with the task of extracting forest resources rather than conserving them.

India was the first country in the British Empire to adopt a forest policy. The need for regulation of forest areas arose out of the demands to supply timber for infrastructural development across British India as indicated by the following extract from an 1862 Dispatch from the Government of India to the Secretary of State : But so long as the supply of timber in the country was generally sufficient for the public works in hand, the question of forest management did not present itself to the Government as one calling for earnest consideration. Latterly, however, while the supply of timber has been steadily diminishing from want of proper conservation, the demand both for state and private purposes has been rapidly increasing and the enormous requirements of the different railways for sleepers has especially brought the matter into very prominent notice and has now made the subject of forest conservancy an important administrative question'.

These policy debates were fuelled by various reports including an influential policy document arising from a report on the improvement of Indian agriculture, written by Dr. Volcker; this had a major impact on the direction of Indian forestry. Interestingly, Volcker's conclusions were that forest policy should 'serve agricultural interests more directly'. In reply to this observation in a 'Review of Forest Administration in British India' the Government stated that : 'The sole object with which State forests are to be administered is the public benefit and the cardinal principle to be observed is that the rights and privileges of individuals must be limited, otherwise than for their benefit, only in such degree as is absolutely necessary to secure that advantage.'

The Government defined four broad categories of forest :

1. Forests the preservation of which is essential on climatic or physical grounds
2. Forests, which afford a supply of valuable timbers for commercial purposes
3. Minor forests
4. Pasture lands

In the detailed definition of these categories it becomes obvious that commercial interests were not to override the interests of local people :

'The needs of communities dwelling on the margins of forest tracts consist mainly in small timber for building, wood for fuel, leaves for manure and for fodder, thorns for fencing, grass and grazing for their cattle and edible forest products for their own consumption. Every reasonable facility should be afforded to the people concerned for the full and easy satisfaction of these needs ... It should be distinctly understood that considerations of forest income are to be subordinated to that satisfaction.'

The uneven development of forest management in different provinces in India was, however, made uniform to some extent by issuing of the Charter of Indian Forestry on 3rd August 1855 by Dalhousie, the then governor general of India. This is said to be the first step towards scientific conservation of forests in India.

The first formal forest policy of modern India was issued in 1894. Prior to that necessity of a vigorous forest policy was strongly felt by British. In the early days of British occupation in India, the then administration paid scant attention to the part played by forests in every day lives of local population and also the immense influence of forests had on general well-being of the country. Even as necessity to the people and a source of revenue, the forest was not considered very significant.

During the period the general perception towards forests were :

(*a*) Except *sal* and teak forests there was not much enthusiasm for revenue earnings from the forests.

(*b*) There was not much difficulty in acquiring forest resources by the people for their simple needs as there were laws to allow them to do so.

(*c*) Trade of forest products including timber and the wood based industries were still in its primitive state and not at all lucrative.

(*d*) Both British government and local Zaminders considered forest areas as an obstruction for expansion of agriculture.

This was the period (roughly in the 18th century) when the direct and indirect values of forests, were grossly under estimated as it clearly seen by the provisions of many of the earlier settlements, particularly in Bengal and Punjab, which transferred large forest lands for ever to land holders or to various cultivators who at time had neither a sight on them, nor in many instances appreciated the gift given to them by the region. The value of forests was so underestimated that in Central Provinces and other provinces, efforts were made to sell out right all government owned forests, with only a few tracts being reserved for supplying timber for public works.

In the hills of Kumaun, Uttaranchal, *Van panchayats* were introduced in the 1920s, following a period of agitation against the Colonial government's policy of reserving increasingly larger areas of forest. This system later on was adopted in many parts of the country as a protest to the British Forest Policies. Over time, a fair number of these committee-community forests have regenerated well, leading some observers to speculate that the quality of forest cover in these forests is often equal to or better than in Forest Department reserved forests that are situated near villages. Since the *van panchayat* probably represents one of the largest and most diverse experiments in devolved common property management ever developed in collaboration with the State, the possibility of learning from them about conditions which contribute to the success of community forest management, or to its demise is enhanced.

Indian Forest Policies and NTFPs

The first formal forest policy of India was issued in 1894. In this policy mention was made regarding the right over NTFPs. Agriculture was given priority over forestry by stating that wherever required forests could be cleared for agriculture. In the forest policy the government took no responsibility of fuel wood and fodder requirement of millions of Indians remaining inside and in the fringe area of forests. No mention and provision were made for forestry research and training neither for timber nor for non-timber. No emphasis was laid on conservation of forests for meeting the need of people at large or environmental needs. This was actually what the British imperialism needed.

The Indian Forest Act, 1927 indirectly regulates collection of NTFPs. The act recognizes two specific categories of forest produce. Section 2(4) (a) consists of a category of items, which are forest produce irrespective of whether they are found outside or are brought from inside a forest. This category primarily includes plant parts like barks, wood and fruits collected for medicinal use. Plants in other category in section 2(4)(b) covers forest produce only if forest is the place of their origin (Ved, 1997). This categorization is so ambiguous that it can be made rarely effective in regulating trade of NTFPs. The convention of international trade on endangered species (CITES) bans commercial trade in an agreed list of currently endangered species included in Appendix I of the CITES, while Appendix II has a list of species which might become endangered in absence of any trade regulations and monitoring. The export import policy (Ex-Im Policy) of India bans any export of items listed in the Appendix I of CITES. With respect to some NTFPs the right of trading has been vested with government controlled corporations. Such NTFPs has been nationalized with a view to eliminate middleman and benefit the primary collector. However, the functioning of such corporations has not been satisfactory and so the very policies are being criticized.

In the national forest policy of 1952, people were longing some relief from independent India so far as use of forest resources by common Indians are concerned. The policy classified forests to four categories. Each and every bit of forestland was brought under government control. Village forests were earmarked to provide fuel wood and yield small timber for agriculture implements and other forest produce for local requirements. But in the same policy it was emphatically mentioned that village communities in the neighborhood of forests should not be allowed to extract resources at any cost. Such use was considered to be at the cost of national interest. It was mentioned that the accident of village being situated close to a forest (functional classification not mentioned) does not prejudice the right of the country as a whole to receive the benefits of a national asset. By this, the right of the forest fringe people leaving in close association with forests for days unknown was totally ignored.

In this policy a reference has been made to "village forests" where some emphasis on use of NTFPs by villagers from this "village forests" has been given. These village forests are also referred to as "fuel forests" and was remarked the job of meeting the need of small timber, manure and fodder, agricultural implements, fencing thorns, grazing and edible forest products. It said that the supply of such forest products to the villagers has to be made at non-competitive rates provided the villagers do not trade those items. Thus, the right of the people to use some non-timber forest products from a very restricted patch of forest (village forests) was allowed under some imposed conditions of the Forest Department. The share of NTFPs allowed to the villagers of the forest fringe villages had to be purchased and the Forest Department was given full right to declare any such extraction as illegal whenever they feel necessary. The clause, which is referred to here, is the issue of trade of forest products available in village forests. The Forest Department, which was assigned the duty of checking NTFP trade from village forest, could not, however, check the trading of our non-renewable and valuable forest products under their control.

It was also mentioned in this policy that the management of the village forest should aim at meeting the need of villagers in future also. It was clearly stated that the extraction should not exceed its annual growth. It was stated 'restriction should be imposed in the intervals not only of the existing generation but also of posterity'. Apprehension was expressed against the trend of handing over some management rights of village forests to *panchayats*. It was mentioned that the involvement *panchayat* as 'lack of appropriate safe guard' and 'hazardous undertaking'.

However, the cooperation of *panchayats* in protection, creation of village forests and distribution of forest produce was felt useful in the forest policy. In spite of all the apprehensions and some liberal attitudes towards village forests, no steps were taken by the forest hierarchy in the country to involve *panchayat* in any form in protection or distribution of non-timber forest produces, which the villagers were supposed to do. In very recent years this is being tried in few states. Orissa is being one of the examples of such an experiment. But lack of an institution (Community forestry in Orissa is not a legally accepted institution) *panchayats* has not been very successful to do this. Though, community forestry is not legally recognized in some parts of the states they have very strong social bond. These unrecognized institutions have been mostly undermined in the NTFP-Policy of Orissa.

One important point deserves worth mentioning here. The issue of measuring the annual removal and annual yield of forest produces (both timber and non-timber) has hardly been studies. Though, some effort has been extended to study this aspect of timbers, non-timbers have been constantly neglected. This issue has been dealt with in detail else where in this book.

National Commission on Agriculture (1976) in its report recommended that tribals should be provided small timbers, fuel-wood and other mirror forest products on priority basis. The Commission envisaged a big gap between the demand and supply of fuel-wood of our country. It speculated that by 2000 AD the fuel-wood requirement of the country will rise to 380 million cubic meters and so recommended for large scale afforestation in all degraded lands of the country.

In connection of providing 'rights and concessions' to forest fringe people, their domestic requirement of fuel wood, fodder and other forest produce has also been granted due importance. It needs to be mentioned here that in the history of 41 years of forestry in independent India (till 1988) no such bold steps was taken before in admitting the rights of ecosystem people. It was felt that the essentials of forest management must include provisions of providing sufficient fodder, fuel and pasturage in order to prevent depletion of forests beyond its sustainable limits. This was because minor forest produce provides sustenance to tribal population and to other communities residing in and around forests. The policy emphasized on enhancing productivity of NTFPs as a management strategy for generation of employment and income.

In providing 'rights concessions' to the people, the forest policy after 1894 (the first formal forest policy of India) it was for the first time altered to provide grazing rights to people keeping it within the carrying capacity of forests. Silvicultural research needs to be enhanced to study such ecological aspects.

Under the head 'tribal people and forests' the symbiotic relationship between the tribal people and forests has been desired to be closer in this policy. The association of tribals not only in protection of forests rather in generation and development of forests was also emphasized.

In our country, 1988 forest policy is considered as policy, which is pro-people. In the basic objectives of this policy it has been stated that meeting the requirements of fuel wood, fodder, minor forest produce and small timber of the rural and tribal population is one. It has also been stated that an essential requirement of forest management is to see that NTFPs should be protected, improved and their production should be enhanced with due regard to generation of employment and income. It has been admitted that minor forests produces provide sustenance to tribal population and other communities residing in and around forests.

Table 2 : Government Policy towards Agricultural and Forest Produce : A Comparison

	Agricultural produce	*NTFPs*
Annual fluctuation in production	Generally within 20% of the normal	Could be more than 200%
Who is concerned with increasing productivity	Farmers, seed, fertiliser and pesticide industry, agricultural universities and government.	Almost no one, it is left to nature. On the other hand, government policies reduced diversity and consequently hurt NTFP production.
Government subsidy in procurement and distribution	Food subsidy was Rs. 250 billion in 2002-3. This generally benefits surplus farmers and urban consumers. In addition, other inputs such as fertilizers, water and power are highly subsidised.	There is no system of minimum support price. Inefficient government corporations do get some budget support to write off their losses, but the scale is miniscule compared to food subsidy, and benefits do not percolate down to producers or gatherers.

	Agricultural produce	*NTFPs*
Producers' political influence	Four states : Punjab, Haryana, Uttar Pradesh and Andhra Pradesh; have always exercised a great deal influence over central government. In general, surplus farmers have a strong pressure lobby in all political parties.	Forest dwellers and tribals are politically least important in Indian politics, and are exploited by of bureaucracy, moneylenders and traders. These groups control local power, and benefit from the schemes meant for tribal welfare. Tribals are confined to the sidelines in the state's political life; while they carry heavy weights in their daily lives, they carry little or no weight in the offices, agencies and Assemblies where, without their active or informed consent, their lives are often shaped.
Regions producing marketed surplus	Agricultural surplus regions, with least poverty and high degree of awareness.	Agriculturally deficit regions with dispersed population and a high degree of poverty.
Insurance against loss in production due to natural calamities, such as drought or floods	Postponement of collection of government dues, and often remission.	Despite extreme fluctuation in production, declaration of famine and drought conditions or starting of relief works is not linked to low production of NTFPs, though in many places almost half of forest dwellers' income is derived from forest produce.
Tenure on producing lands	Land under private ownership, with security of access and operation.	NTFPs mainly come from CPRs, including forest lands, where peoples' rights of access are vague and subject to many formal and informal controls.
Controls on movement and storage	No such control on movement within state, and no license required for farmers for storage. Controls on inter-state movement have been lifted in February 2002.	Apart from controls on collection, there are several controls on movement, storage and sale, even within a district. The general impression is that all NTFPs, even occurring on private lands, belong to government and gatherers are only entitled to wages from collection to sale.

An interesting study has been done by Saxena (2003), which shows the apathy of the National as well as the State Governments towards NTFPs in comparison to agricultural produces (Table 2). These changes between the two sets of policies have persisted despite the declaration in the Forest Policy of 1988 that the domestic requirements of fuel-wood, fodder, minor forest produce, and construction timber of tribals should be the first charge on forest produce. With a low productivity of 0.7 cubic meters per ha, India's forests are not good enough to satisfy both the commercial and livelihood needs of the people, and hence it was necessary to give primacy to subsistence needs.

State Control over NTFPs *(including bamboos and canes)*

Most of the important NTFPs (NTFPs with high commercial value) have been nationalized, that is, these can be sold only to government agencies. The nationalization of these NTFP commodities, done in different states during different years from 1960s up to the end of 1970s, presumably with the intention of helping the poor. The state control of NTFPs affected the protection component of the indigenous people opening it to excess exploitation. Traditionally, the collection of NTFPs has been of low intensity and generally sustainable. However, as the economic potential of NTFPs has become apparent, the intensity of collection has increased and more significant infrastructures for trade and processing have developed. This has raised concerns for the sustainability of the resources and the distribution of the benefits derived from them. In reaction to these concerns, a number of state governments have taken over the control of a number of NTFPs.

The degree of control varies from state to state, and so does the number of nationalised items. For instance, tamarind is a free item in Bihar and MP, but not in AP, where GCC has legal monopoly over its purchase and disposal. The right to the procurement of tamarind was leased in Orissa to both the TDCC and some private traders, but unauthorised traders also operate freely. Since March 2000, tamarind is under *panchayat* control in Orissa. In AP, trade in NTFPs is a State monopoly by law, irrespective of land where they occur. GCC has been designated to procure all NTFPs and trade in them. All items other than timber are included. In Maharashtra, the Tribal Development Corporation has a monopoly of purchase in respect of 32 NTFP items. In Madhya Pradesh, *sal* seeds, gums, *harra* seeds and *tendu* are nationalised, and the rest are free. Resin, which is the main output from pine forests of the Uttaranchal hills, is also nationalised. The Government of Kerala has created a monopoly for 120 notified NTFPs items. The Scheduled Tribes and forest dwellers have no right to make any direct sale to outside parties. They have to sell these to cooperative societies which auction the products gathered by the tribals. The open market price was much higher than the government price. Thus in Kerala, government monopoly was not only inefficient but also exploiting the tribals. The Rajasthan Scheduled Tribe Area Development Cooperative Corporation Ltd. Udaipur has a monopoly over designated NTFPs. It buys *tholi musli,* a medicinal herb, at Rs. 250-400 per kg, although tribals could easily get from Rs. 500-1000 in the open market. Similarly, the Corporation pays only Rs. 18 per kg for honey as against the market price of Rs. 50 per kg. Thus nationalization has not been of any help to the gatherers.

The women living in desert area of Santalpur Taluka of Banaskantha district, Gujarat survive mainly on gathering gum from the Babul trees planted by the Forest department. The Forest Department insists on licenses for gum collection, and since the women had no licenses, they were in the past collecting gum 'illegally' and selling to private traders. After joining SEWA, a voluntary agency of international repute, they formed self-help groups and demanded licenses, so that they could 'legally' sell the gum to the Forest Corporation. The rates for gum are fixed by the Forest Corporation, and to the women's dismay, their legality has resulted in getting poorer rates from the Forest Corporation than what they could get from the open market. The tragedy is that the women can get a better rate for gum in the open market, but the Forest Corporation will not allow the gum pickers to enter the open market, and they have to sell their gum for 1/2-2/3 of the market price. Despite peoples' enthusiasm for JFM in West Bengal, almost nothing was done to improve the marketing framework for NTFPs. The World Bank Implementation Completion Report (1998a) observed, 'Unfortunately, no specific, clearly stated comprehensive objectives for strengthening and development of forest products marketing to contribute to the overall objectives of the project were presented. The outcome of the project in terms of the development and strengthening of forest product marketing remained meagre'.

The state regulations bringing certain NTFPs under monopoly trade are summarized in Table 3.

Table 3 : State Trading Regulations Promulgated by State Govt.

State	*Regulations*	*Implications*
Andhra Pradesh	Andhra Pradesh Minor Forest Produce (Regulation of Trade) Act, 1971 Andhra Pradesh scheduled areas	Trade in NTFPs is declared state monopoly whether ownership is with government or not.
Bihar	Bihar Kendu Leaves (Control of Trade) Act, 1973 Bihar Forest Produce (Regulation of Trade) Act, 1984	Bihar State Forest Development Corporation operates as state government agent for collection and marketing of *kendu* leaves, *sal* seed, *mahua* (*Madhuca indica*)
Gujrat	Gujrat Minor Forest Produce (Regulation of Trade) Act, 1979	Minor forest products identified include *timru* leaves (*tendu* leaves), *mahua* flowers, fruits, seeds and gum
Himachal Pradesh	Himachal Pradesh Resin and Resin Produce (Regulation of Trade) Act, 1981	Resin, bamboo and *Acacia catechu* (*khair*) collection through Himachal Pradesh Forest Development Corporation Ltd.
Madhya Pradesh	Madhya Pradesh Vanopaj (Vyapar Viniyam) Adhiniyam 1969	Items under monopoly include *tendu* leaves, *sal* seed, *harra* and gums; Madhya Pradesh Minor Forest Produce (Trade and Development) Federation acts as agent of state government
Rajasthan	Rajasthan Tendu Leaves Act, 1974	Rajasthan Tribal Areas Development Federation collects and markets NTFPs.
Orissa	Orissa Forest Produce (Control of Trade) Act, 1981 Orissa Kendu Leaves (Control of Trade) Act, 1981	Collection and sale of NTFPs are monopolized by Forest Department and leased to Tribal Development Cooperative Society which in turn delegates to an individual; collection and trade of leaves are handled by Forest Corporation

Some of the explicit objectives for state monopoly of NTFP trade 1996 are :

- to prevent unscrupulous intermediaries and their agents from exploiting NTFP collectors;
- to ensure fair wages to collectors;
- to enhance revenue for the state;
- to ensure quality;
- to maximize the collection of produce.

Nationalization that lacked real motive and skill, had its own set of disadvantages such as (*i*) It reduced number of buyers (*ii*) It chocked the free flow of goods (*iii*) It delayed the payment to the gatherers etc. These deficiencies resulted in entrance of contractors in to the field from the back door. The contractors obviously operated with higher margins of profit required to cover uncertain and delayed payment by government agencies. All these undesired externalities reduced collection and incomes of gatherers (who are mostly tribals) to a considerable extent.

The Forest Act of 1927 is a Central Act and various provisions of the Act are same for all the states of India, though in some states the rules made under the act are more stringent. NTFPs like *tendu* leaves, *mahua* flowers, *sal* and other seeds, gum, honey, resin, skins, and roots are mostly marketed by Forest Development Corporations (FD Corpns.) of different states. The FD Corporations pay royalties to the forest department for such sales. Section 51A of the Act states that the state governments can make rules to provide for the establishment and regulation by licenses in the manufacture or preparation of :

(*i*) *katha* out of *khair* trees;

(*ii*) resin, turpentine and other products out of resin; etc.

The state governments are also given rights to notify through Official Gazette any other forest product to be included in the 'nationalized' list. The Section 69 of the Act says that any forest produce can be included under the list of nationalized NTFPs in case of any dispute though it has not been declared as nationalized previously.

To acquire a monopoly in the procurement, transport, and sale of NTFPs Uttar Pradesh Govt. issued an Act in 1976 on certain NTFPs with retrospective effect from 04/10/1975. Gujrat Govt. enacted the Gujrat Minor Forest Produce Trade Nationalisation Act in 1979. Here collectors are termed as growers. The term grower was used deliberately to include the products (Items nationalized as NTFPs) grown in their own lands. In Orissa during October 2001, the government came up with an order that it is in the process of changing the legal framework to enable the *Gram Panchayats* to fix up prices of NTFPs. The order also said that during the transitional period, the District Collector, along with other district level officials, would fix up the prices of NTFPs. The order at least gives a signal to the district authorities and *Panchayat* bodies on their role in marketing of NTFPs. Hopefully, this may create conducive environment for larger debates on the process of price fixation, ensuring fair prices to the primary procurers. Very recently the forest department of government of Orissa brought out a Resolution on the control and regulation of removal of *sal* leaves from the degraded *sal* forests of the state. The Resolution is under severe criticism throughout the state. The resolution provides certain prescription for collection of *sal* leaves as the forest department feels that the removal of leaves for making leaf plates or other purposes are harming the health of the forests. *Sal* leaf is an NTFP item, which provides subsistence to thousands of forest dwellers during odd seasons in the one hand and its unregulated exploitation makes the *sal* trees stunted in the other hand. The Resolution has been issued to make both the ends meet. The restrictions imposed by the department include restricting collection of the NTFPs to certain seasons and many other practices, which the department thinks detrimental to the forest health. The Resolution debars collectors to collect the leaf from trees below 3 meters of height. Plucking is allowed up to 2 meters of a tree only. The amount of leaves to be plucked from a forest patch is again has to be decided by the forest officials. Plucking of leaves from sanctuaries and national parks has been totally banned. The *sal* leaf procurers have to be registered with the forest department and any person found possessing more than 5 kg of leaf (green or dry not mentioned) would be liable to be punished. Most of these restrictions are not scientifically tested but aims at imposing heavy toll on the lives of thousands of forest dependent people depending on this NTFP resource.

Many states have created Forest Development Corporations (FD Corpn) to carry out the collection sale and marketing of nationalized NTFPs. Some states have legalized monopoly rights for their procurement. Others collect certain nationalized NTFPs (mainly *tendu* leaves) through the forest department and leave the collection of other NTFPs to cooperatives of tribals (LAMPS) on a monopoly basis. For *tendu* leaves, the forest department divides the forest areas into units, auction them, and appoint purchase agents. But all these arrangements do not work always making the whole thing a mess.

The first NTFP brought under state control was *tendu (Diospyros melanoxylon)* leaves, used to wrap traditional cigarettes *(bidi)*. This tree species is found in abundance in tropical deciduous forests, on wastelands to some extent and even on private holdings. Madhya Pradesh monopolized *tendu* collection in 1964, followed by Maharashtra (1969), Andhra Pradesh (1971), Bihar (1973), Gujrat (1979) and Orissa (1981). The monopolization of *tendu* was rapidly followed by similar procedures for other economically important NTFPs, including *sal* seed *(Shorea robusta)*, gums and myrobalan *(Terminalia chebula* and *Terminalia bellerica)*.

Before Madhya Pradesh adopted a cooperative structure for *tendu* leaf trade in 1988, the collection of leaves as per official records ranged from 6 to 7 million standard bags. This was reduced to around 4 million standard bags per year after 1989. The reduction did not result from a lack of resources, but rather from the rejection of leaves that would previously have been collected but were not of high enough quality for the cooperatives. However, local manufacturers of *bidi* have been known to buy additional *tendu* leaves directly from collectors.

In Madhya Pradesh, collectors share in profits through a bonus plan at the end of each season. In Orissa, which is also rich in NTFPs, the collectors get only wages for collection; the bulk of the profit goes to the Forest Development Corporation, which has been given monopoly rights by the state government. As in Madhya Pradesh, collection of *tendu* leaves is being limited by a desire to collect only the best produce.

Broomstick grass grows wild in most hilly tracts of the Raygada district of Orissa. Tribal women, through JFM forest protection committees, have protected this grass from grazing and fire and have obtained income by selling it. However, since the trade in broomstick grass has come under state control its collection and sale have been much reduced. While the women get 1.5 to 3 rupees per kilogram of broomstick grass, the company holding the monopoly (Utkal Forest Products Ltd.) is making profits of as much as 600 per cent. Reduced collection has also been observed for *T. chebula,* gums and *mahua (Madhuca indica)* flowers since these were taken under state monopoly. However, collection of *Buchanania lanzan* and *Chlorophytum tuberosum (safed musli)* has increased, perhaps to the point of unsustainability, in response to high commercial demand.

Perhaps most widely traded NTFPs partially under state control in India is bamboos and canes (claimed to be timber by some people). Its increased demand in industries brought it under state control. There is a constant demand to include bamboos and canes within timbers, as non-timbers are largely neglected both by management institutions and research institutions of India. The artisans associated with bamboos and canes are at the bottom line of social hierarchy. The socio-economic situation of workers in this sector, except in the Northeast, is also quite bad. A survey conducted in 1981 shows that 43 per cent of workers in this sector are women. A problem of all the bamboo artisans in the present day is that of raw material shortage. P. Sainath reporting for Times of India (March 15, 1994) on the *Koya* tribe of Malkangiri district of Orissa wrote, "Bamboo is the socio-economic oxygen for the *Koyas.* An oxygen which is being denied to them thanks the forest laws that have remove their access to the material while granting access to major corporate seeking huge quantities of bamboo for paper (via the Orissa forest corporation). Ironically, *Koyas* are hired sometimes to do the felling on a causal basis, (but) seldom get the government rate of Rs. 25 a day. With the denial of access, indebtness unlike the bamboo forests appear to be growing in central *Koya* pockets".

Due to diversion of bamboos to industries in Bhandara and Chandrapur districts of Maharashtra, for instance, the threat of being rendered jobless confronted over 70,000 *Burads,* a tribe who traditionally make mats and baskets from cane bamboo. They previously used to cut bamboo freely from the forests, until the British government introduced a pass system and levied a fee for this privilege. In the 1960s, the government introduced a system of rationing 150 bamboos per month to each cardholder, at Rs. 90 per bundle. This was later hiked to Rs. 175. Simultaneously, bamboo-growing areas were being leased to paper mills, to enable them to extract the bamboos. The situation deteriorated on account of mismanagement of forest resources. Bamboo prices rose rapidly, and even areas not leased out to the paper mills were being exploited by contractors for extracting bamboo, thereby creating shortfalls in the quotas for bamboo workers. Ultimately, the supply to bamboo workers from forest depots ceased altogether. On March 15, 1982, the *Burads* around the Kohmara depot began a relay fast to press for a resumption in the supply of bamboo. The agitation carried on for days and snowballed in to a *"bamboo-roko"* and *"rasta-roko'* movement several Burads were *lathi charged* and arrested in the process. Finally, the government agreed to supply each of the workers of Kohmara depot 25 bamboos per month.

Similar restrictions have been imposed by various state forest departments on cane workers, who used to cut cane directly from the forests for their requirements. A case in point is Agumbe village in Shimoga district of Karnataka, where landless cane workers have been deprived of their livelihood. Cane was once abundantly available in the nearby forests. However, in the mid 70, the forest department imposed restrictions and a fee on cane cutting, and subsequently, a total ban. Thus, the forest department alleged, was necessary, because the cane workers had over-exploited the forest, and fast becoming extinct. But curiously, plywood manufacturers were not prevented from forest extraction. They felled trees in such large numbers that the support required by the cane trees to spread and grow virtually vanished. Road construction in heavily forested areas had also resulted in the uprooting of cane.

Invariably, throughout the country the traditional rights of people over NTFPs were reduced by nationalization or imposing restrictions to such an extent that people, mostly tribal, suffered a lot. These restrictions led to several conflicts and unrest. Forest bureaucracy of our country remained unaware of these instabilities.

It appears that the objectives of bringing NTFPs under state monopoly to reduce exploitation of tribal people and other forest-dependent communities and to promote sustainable management of NTFPs–are not being realized. State control over the trade of NTFPs has often resulted in compounded problems of restricted access to resources and non-remunerative returns to the collectors.

NTFPs are one of the keys to successful joint forest management, but if local people who are engaged in the arduous task of collecting NTFPs are not able to get fair wages even when the trade is handled by government-appointed agencies, JFM may not be a viable tool in the achievement of sustainable forest management.

Understanding Sustainable Management of NTFPs

Sustainability can be interpreted in a variety of ways :

1. Ecological sustainability (maintenance of biodiversity);
2. Economic sustainability (maintenance of economic benefits);
3. Social sustainability (equitable access and distribution of proceeds);
4. Institutional sustainability (maintenance of various management institutions at a variety of levels : government, NGOs, local communities).

Management systems may be set up on the basis of all these different aspects of sustainability. The management and use of NTFPs will therefore be examined with these various aspects in mind.

To be sustainable, harvest levels need to be based on a sound knowledge of the ecology, distribution and abundance of the resource species. Such information can be obtained from a number of sources including indigenous or local peoples' knowledge as well as formal scientific enquiry. Resource assessment of NTFPs in the tropics is relatively new and has received little formal study; consequently methodologies have been developed by individual researchers in response to local circumstances and the peculiarities of the resources under study. Sociologists stress that at the community level that methods need to be devised that can be carried out by the community, are sensitive to informal local knowledge and responsive to community information and management needs. It is argued that the main benefit of participation in such exercises is often awareness building. The assessment methods used can therefore be simple and need not have a biometric basis.

A sustainable management practice needs efforts in three spatial scales; the macro or national scale, the local or community scale and at the level of the individual species. The results of these deliberations are summarized in Table 4.

Table 4 : NTFP quantification needs as expressed by participants at the Rome workshop

Scale	Research priorities.
Species	Sampling, assessment, monitoring and analysis.
Community	Participatory methods. Interface between local and scientific knowledge.
National	Integration of data from a wide range of sources. Integration of NTFPs in large-scale multi-purpose resource inventory protocols.

It is generally agreed that there is a need for a better understanding of how to design biometric sampling schemes, suitable mensuration techniques, effective monitoring strategies and for analyses including the determination of sustainable yields for individual species. At the community level (village) these technical problems are compounded by the need for protocols suitable for use by the community. Local communities also possess their own knowledge and experience of NTFP ecology and management. A key issue in promoting community management of these resources is the integration of local and scientific knowledge.

At the macro scale all of these issues are further compounded by the need to incorporate NTFPs into multi-purpose inventory designs for national-scale strategic assessments. Lastly, given that resources for commissioning NTFP assessments are always limited it is necessary to find ways to integrate knowledge across all scales. The challenge is to develop efficient single and multi-species inventory and data analysis procedures at a range of scales from local to national level without alienating the people who will benefit from the data collected and who should be given the opportunity to participate in the proceedings.

The Concept of Sustainability

Sustainability is a concept on which social and natural scientists, and philosophers have expressed views from time to time. It meant conservation of resources to Epicures, population management to Malthus, carrying capacity to biologists, maintaining biodiversity to ecologists, maintaining cultural diversity but with social harmony to philosophers, the inevitable stationary state (*i.e.*, level of well-being or living) to John Stuart Mill and so on. In most of these concerns, a weight is certainly given to inter-generational equity. Meanwhile, economists talk of prosperity, income or well being for the present and future with the usual emphasis on the discount rate, which itself is not universally accepted. This focus on inter-generational equity has attracted intellectuals and policy makers, even when issues regarding intra-generational equity are not yet completely taken care of.

The term "sustainable development" was brought in to common use by the "World Commission on Environment and Development" (the Brutiand Commission) in its 1987 report- *Our Common Future*. Gro Brutland (1987) elaborated the concept of Sustainable Development in the following terms :

- It requires elimination of poverty and deprivation;
- It requires conservation and enhancement of the resources base;
- It requires a broadening of the concept of development so that it should cover economic growth along with social and cultural development; and
- It requires unification of ecology and economics at all levels.

Sustainability can be considered in three conceivable dimensions :

- Ecological dimension (avoiding any damage to the environment)
- Economic or production dimension (where a constant flow of goods can be assured)
- Socio-cultural dimension (preserving all socio-cultural diversities).

Some interesting works has been done on sustainability of NTFPs in recent years. Here sustainability has been considered in terms of productivity and extraction of NTFP resources. Peters (1996) has shown that yields of fruit and nuts per unit area in species-poor forest in Amazonia are greater than in species rich forests. Godway and Bawa (1993) pointed out in terms of hypothesis that the costs of extracting NTFPs will increase with increasing species richness in tropical forests. According to them this will happen because of the assumption that the relative abundance of many species will decrease with an increase in species richness. Chambers *et al.* (1989) argue that the potential yields of NTFPs are far greater than the realized yields, but they do not address the issue of sustainability.

Sankar (1998) has studied the impact of extraction of NTFPs on floristic diversity and population structure in the thorny scrub forest of Biligiri Rangan Hills. They have carried out their experiment in the Biligiri Rangaswamy Temple (BRT) wild life sanctuary. According to them six species of NTFPs are most commonly extracted from BRT scrub for subsistence and income generation by an aboriginal tribe, the Solia. Although, the forest has been provided protection from indiscriminate felling since 1978 under a wild life sanctuary cover, changes in vegetation structure are still occurring as a consequence of anthropogenic pressure in the form of extraction of fuel wood and other NTFPs. The results indicate that large woody species are being replaced by small woody species. The population structure has also been reported to be increasingly skewed to the smaller size classes due to extraction-bound mortality of the individuals particularly those >7cm dbh and above.

The data suggests that the community itself may be a manifestation of long persisting anthropogenic pressure in the form of extraction of a variety of forest products, and it currently represents a downward transition from a deciduous forest to a shrub thicket. The authors have suggested some management interventions to check this gradual degradation of forest community. The first suggestion was to divide the forest area into different 'compartments' extracting a few compartments at a given time, and allowing the forest to recover following extraction. The second one was to reduce the level of harvest of NTFPs followed by value addition so that the total income of the local community is increased rather than reduced.

Ganeshaiah (1998) examines the response of forest tree species with different dispersal modes to anthropogenic pressure in a dry deciduous forest of South India. The study was conducted in Biligiri Rangan Hills of India. The result of this study suggested that the population of animal dispersal species than those of wind or passively dispersed are more vulnerable to human disturbance. In fact wind dispersed species seem to be facilitated by human disturbances. They inferred that the forest under continuous and persistent anthropogenic pressure might come to be dominated by the species with abiotic mode of dispersal. Such changes in community structure in turn might have negative effect on populations of pollinators and seed disperses thus leading to lower level of bio-diversity.

Hunting gathering is the most primitive habit of human society. Forest dweller used to depend on gathering of resources from the forests. In the present day also gathering of NTFPs play an important role in the subsistence of tribal communities. In the modern days NTFP extraction from forests is being done not only for sustenance of life of forest dwellers but also for commercial use of these resources. People living away from forests also feel the need of NTFPs and so various types of NTFPs are being traded in a large quantity. The tragedy of the whole story is that neither the government machinery nor the persons engaged in collecting NTFPs from forests as middleman exactly known for what they are being used at the sink. This so called 'trade secret' extras a pressure on our ecosystems as because the traders prefer to collect them from their natural habitats without going through the rigorous process of cultivating them.

Different workers have reported adverse effect of extraction on sustainability of NTFPs. Godoy (1993) comments that extraction of plants and animals from forest by rural populations helps conservation because foraging does not disturb the forest as much a logging. Leaving aside the injudicious extraction of forest plants and animals by recent harvesters, there is evidence that even indigenous far from markets can deplete forest goods. As evidence of Paleo-Indian hunters in North America shows (Martin, 1984), population growth or technological change can lead people deplete natural resources. Secure property rights, low population density, customary rule of use and simple technology seem to encourage judicious use. Extraction may be a viable activity only in areas with low population densities (Homma, 1996). The authors also suggest that sustainable extraction of wild plant and animals will be affected by (*a*) the harvester's access to modern technologies for extraction, processing and transport goods, (*b*) the availability of inexpensive industrial substitutes, and (*c*) the intended end use of the product in household, or market place.

Fernandes (1988) have compiled extensive data on many NTFP harvested in indigenous societies in Orissa. These data indicate a decline in yields. Chanber, Saxena examined government records of yield of NTFP and noted a decrease for several products in recent years. They attribute this decline to changes in market structure and profit incentives from the nationalization of trade in NTFP.

Other explanations of reduced yields in India include deforestation, which reduces the area that can be harvested, and growing human pressure on forests, which may cause sustainable resource use.

Assessing Impact of Extraction on Sustainability of Forests : A Case Study

The forest of the region is dominated by coppice sal *(Shorea robusta)* and the major associates are *Combretum roxburghii, Holarrhena antidysenterica, Bassia latifolia* and exotic species like *Lantana camara* etc. During October maximum number of species was encountered (104) and their number fell to only 38 during April. The unique management decision of the FPC is that they use only six species as fuel-wood.

Table 5 : Annual productivity of NTFPs of fuel-wood producing (mean and range of six observations) species at Bhagabatichak (in terms of biomass).

S.No.	Name of the Species	Biomass (Kg/Ha.)	
		Fenced	Unfenced
1.	*Combretum roxburghii*	5346.37 (3899.43-597.37)	2933.77 (21.55-37-3148.66)
2.	*Hollahena antidysenterica*	182.50 (99.31-208.45)	431-73 (345.91-529-54)
3.	*Ehretia laevis*	94.08 (77.34-104.93)	22.76 (11.91-45.05)
4.	*Lantana camara*	98.00 (88.15-109.67)	8.02 (5.06-10.63)
5.	*Antidesma acidum*	82.00 (65.31-98.55)	190.40 (156.65-219.31)
6.	*Helicsteris isora*	9.99 (6.32-12.46)	24.35 (13.67-29.12)
	TOTAL	**5812.95**	**3611.25**

**Figures in the parentheses give range of minimum and maximum NTFP biomass.*

These species are C. *roxburghii, H. antidysenterica, L. camara, Ehretia laevis, Alntidesma acidum,* and *Helicteris isora.* In protected condition, the annual biomass production of these six species was 5812.95 kg/ha (Table 5).

Within this C. *roxburghii* only contributed 5346.38-kg/ha *i.e.* 91.9% of total fuel-wood. In unprotected condition from where there is regular exploitation of fuel-wood, the annual productivity is 3611.25 kg/ha. This is actually the residual biomass left after exploitation. The annual fuel-wood extraction is estimated to be 262676.44 kg from this forest. The total forest area being 63 ha per hectare extraction is 4169.47 kg/ha. The actual productivity in unprotected condition, therefore, is the sum of the residual biomes and the extracted biomass per hectare, which is estimated to be 7780.72 kg per ha. In the unprotected condition *C. roxburghii* alone produces 2933.77 kg/ha *i.e.* 81.24% of the total residual biomass. It needs to be mentioned here that from the door-to-door survey it is revealed that the families of Bhagatichak village only consume 177185.6 kg fuel-wood annually *i.e.* only 67.5% of the total fuel-wood extracted. The people of neighboring localities collect rest amount.

Major plant species of Ethnobotanical importance of this region are *Hemidesmus indicus, Ichnocarpus frutescene, Smilax macrophylla* and *Andrographis paniculata* (all are of medicinal use). *Basella alba* and *Dioscorea elata* are used as food items by forest community (Table 6).

Table 6 : Ethnobotanical uses of plants extracted from the forest other than fuel-wood.

Name of the species	*Habit*	*Annual/ Perennial*	*Parts used*	*Uses*
Andrographis paniculata	Herb	Annual	Whole plant	Against acidity, worms and general debility of children.
Bassela alba	Herb	Perennial	Whole plant	As leafy vegetable in general and against piles.
Dioscorea elata	Herb	Annual	Tuber	As a food item.
Hemidesmus indicus	Herb	Annual	Root	Against eruption of tongue in children, tonsils, and constitutional disorder of children.

Ichnocarpus frutescence	Shrub	Perennial	Root and latex	Remittent fever, nervous debility and latex is used in sores of nail.
Smilax macrophylla	Shrub	Annual	Leaves	Against unwanted pregnancy.

In all these cases the whole plant has to be extracted during harvest used as in some cases root or tuber is used and in other cases the whole plant is used. Comparing the species wise biomass in protected and unprotected conditions (Table 5) except *H. antidysenterica, A. acidum* and *H. isora* all other species produce less in unprotected condition whereas these two species produce more. In case of species with ethnobotanical importance (Table 7) it is found that in most of the cases protected condition has more biomes than unprotected ones (Except *S. macrophylla* and *D. elata).*

Table 7 : Annual productivity of NTFPs of Ethnobotanical importance (mean and range of six observations) species at Bhagabatichak (in terms of biomass).

S.No.	*Name of the species*	*Biomass*	
		Fenced	*Unfenced*
1.	*Hemidesmus indicus*	8.96 (4.32-10.19)	8.20 (3.87-11.21)
2.	*Ichnocarpous frutiscens*	16.66 (12.99-22.17)	14.50 (9.32-17.12)
3.	*Smilax macrophylla*	9.36 (5.26-11.32)	11.61 (8.00-14.23)
4.	*Andrographis aniculata*	8.28 (5.32-9.46)	3.70 (1.12-4.55)
5.	*Bassia latifolia*	226.30 (188.30-294.32)	140.40 (113.19-159-29)
6.	*Dioscorea elata*	0.98 (0-1.28)	4.99 (2.18-6.49)
7.	*Bassela alba*	13.50 (7.34-19.36)	0 —
	Total	**284.00**	**183.40**

**Figures in the parentheses give range of minimum and maximum NTFP biomass.*

The way whole plant is extracted in most of the cases, leave no scope of regeneration of plants vegetatively. Of course, some of the community members demonstrated that they keep a part of tuber in the soil for propagation of the species while collecting the tuber in tuber yielding species. Collecting species like *A. paniculata* before its seed dispersal season makes the harvest unsustainable. *B. latifolia* also being a non-coppicing tree species yield of less in unprotected condition.

Major plant species of ethnomedicinal importance of this region are *Hemidesmus indicus, Ichnocarpus frutescens, Smilax macrophylla,* and *Andrographis paniculata.* The whole plant of *Bassela alba* and the rhizome of *Dioscorea elata* is used as vegetable. *H. indicus,* (8.96 and 8.20 kg/ha), *I. frutescens* (16.66 and 14.50 kg/ha) and *A. paniculata* (8.28 and 3.70 kg/ha) produce slightly less in unfenced condition but *S. macrophylla* (9.36 and 11.31 respectively) produce slightly more in unfenced condition. *B. latifolia* is a multipurpose tree species of this region. It is heavily exploited from the unfenced plots (226.30 and 140.40 respectively) as grown-up saplings are marketed and used to make agricultural implements.

Sal (Shorea robusta) leaf productivity estimated from litter traps is 6799.15 kg/ha *i.e.* 428346.45 kg/annum from the whole forest. Dry *sal* leaves is usually not collected by people in this FPC and are allowed to decompose. Green *sal* leaf collection was estimated to be 38309.85 kg/annum, which is found to be only 9.09% of total leaf litter produced. These leaves are used for plate making and are sold in the market. Thus, the rest *i.e.* almost 90.91% of leaves are allowed to decompose and recycled as nutrient after litter-fall.

Direct measurement of productivity has a set of difficulties such as it is always influenced by interaction with the tree layer (canopy coverage etc.) and effect of extraction. Lele (1993) pointed out that these effects would typically require conducting controlled experiments that simulate the manner and magnitude of extraction while allowing for the measurement of extracted material. While conducting such measurements, in this case the experiment has been designed to take care of the magnitude of extraction through controlled experiments.

The study reveals some important facts regarding the impacts of forest management by local community on the productivity of the forest. The results from the present study indicate that higher productivity in terms of biomass (in the unfenced plots) of fuel-wood species is encountered even when local people under FPC management exploit forest patch within the management norms framed by them. Earmarking six species to be collected as fuel-wood, restriction on collection of dry litter are some scientifically prudent management practices as they on the one hand serve the purpose of forest fringe people and on the other hand does not exert much pressure on the forest ecosystem. In the past forests have not been managed for NTFPs by foresters and also very little is known about their management science. The question of yield of only 13 species of NTFPs (as stated in Table 5 and Table 7) and their extraction induces us to think that the forests, which has traditionally been valued as timber, provides a multitude of products which features in people's day to day lives if properly managed. From table-5 it is found that the above ground biomass produced by these species is 5812.95 kg/ha. In unfenced condition the actual above ground productivity is calculated to be 7779.83 kg/ha. Therefore, in unfenced condition an excess production of biomass (1966.86 kg/ha *i.e.* 33.84% higher than the protected one) by the six fuel-wood species is noticed. Moreover, it is also noticed that in fenced condition the contribution of biomass of *C. roxburghii* (The dominant species in terms of biomass produced) in fenced condition (91.9%) much more than that in unfenced condition (81.24%). From this it appears that *C. roxburghii* possess better competing ability in an undisturbed condition than a disturbed ones.

Here lies the perception of the community to keep high NTFP resource potentiality of the forest; they manage, in a more effective way, learning from their experiences. From Table 4 it appears that NTFPs have immense use value to the forest community. It needs to be remembered that non-timber forest products are closer to their lives in comparison to timbers (except house construction and agricultural implement requirements). Toledo (1992) calculated the quantum and number of useful products used by indigenous people and pointed out that the uses derived from non-wood parts of the plants, counter the notion that the tropical forests are primarily important sources for timber. From the tables it is mostly found that in case of the species where the whole plant or the under ground part is extracted yields less in unprotected condition.

This is because of the pattern of extraction. It is also found that *B. latifolia,* though; being a multipurpose tree species is extracted mercilessly. The tribal of the locality has an emotional and cultural attachment with the species. The Forest Department a concept of 'Kath' (good quality timber) and *'Kukath'* (bad quality timber) as a management strategy to prevent indiscriminate felling of *Shorea robusta* which is a *Kath* and exempting all other species from being preserved. This encourages unwise extraction of species other than *S. robusta.* The community has to be properly enthused to protect this species from being lost.

Some Management Lessons

1. Harvesting NTFPs can have a major influence on the species concerned and/or on the forest ecosystem. An understanding of ecological strategies of plants and animals is therefore essential; this understanding needs to form the basis for the construction of forest and NTFP management systems. Native management systems are indeed often based on such knowledge.
2. There is a major need for the development of a short, simple, participatory method–based on native knowledge–which can generate the knowledge necessary for taking management decisions. Such a method should be applied at the level of the management unit concerned.
3. In natural forests with a relatively low level of biodiversity (for example flood plains in the Amazon), NTFP management is simpler and productivity higher. In forests with a relatively high level of biodiversity, the potential of NTFP use for economic purposes will depend in part on the presence of NTFPs, which can allow for a high production level both from the point of view of quantity and quality.
4. In order to increase production of NTFPs in natural forest, NTFP species, which only occur in low densities, can be planted ("enrichment planting"). This reduces the impact of harvesting on the NTFP species and increases the daily income of those harvesting it. Many native forest management systems include such enrichment techniques.
5. If there is a major demand for an NTFP from which a relatively large proportion of the biomass is harvested,

cultivation on plantations is the best management approach. In the case of lower demand and a relatively smaller proportion of the biomass, sustainable management in natural forests is an option.

The Impact of Over Harvesting

The impact of harvesting NTFPs depends on the type and/or part, which are harvested. Repeated and intensive use may have a negative effect on the NTFP population and even on the species.

Generally speaking, harvesting nuts, fruits and seeds (for example *Emblica officinales)* does not need to have a negative influence on either the tree or the species. However, constant harvesting of fruits may reduce the number of fruits to such an extent that the number of seedlings (some of which will in any case die naturally) will be too low for the species to survive successfully.

In particular, commercial extraction of fruits, nuts and seeds may also have an impact on the genetic composition of the species. Continual harvesting of large, tasty fruits from the same population will ultimately result in a population of trees bearing small and less tasty fruits and these will consequently have a lower economic value.

Intensive harvesting can reduce the number of fruits to such an extent that fruit-eating animals (such as rats) are disadvantaged. These animals will then leave the area concerned in search of forest areas where more fruits are available. The animals may also spread (or carry the pollen of) other species of plants, which can ultimately have a major influence on the overall composition of the forest.

The Importance of Sustainable Management Systems

The experience described above makes clear that it is essential to understand the ecology of plants and animals. Native and/or traditional forest management systems are often based on such an understanding. It is assumed that native management systems are only sustainable when the population and market pressure is low. Additional studies are therefore necessary of the relationship between extraction and ecological sustainability.

However, these are still few and far between. Most human interactions with the forest have an impact on biodiversity of the plant and animal species within the forest. Although some societies have long coexisted with plants and animals in the forest, the effect of harvesting forest products reverberates throughout the entire ecosystem. An evaluation system must be implemented at the onset of harvesting and marketing for each species. Local communities must share in the responsibility of monitoring the impact of harvesting, for the sake of sustainability.

Besides long-term scientific studies, it is important to be able to assess the sustainability of harvesting in the short term. Monitoring of the harvest also needs to be carried out at local management level. One example of such a quick method of assessment is "Rapid Vulnerability Assessment" (RVA, given below). This is a rapid method of assessing whether a particular plant species is sensitive to use by humans and therefore in need of management and monitoring. The advantage of the method is that it allows a management framework to be created within a relatively short time without first needing to determine precisely the maximum allowable level of harvesting. The sustainable extraction level can then be determined more precisely in the course of management activities.

RAPID VULNERABILITY ASSESSMENT

Rapid Vulnerability Assessment (RVA) is a method, which is capable–in a rapid, participatory manner–of assessing which NTFPs will be sensitive in use and which management measures can best be applied. The plant species which will be sensitive are :

- species which are slow growing, long living and which produce relatively little (depending on the life form);
- species with a highly specific preference as regards their living environment (living environment preference);
- species which occur infrequently and which are not widely distributed (occurrence and distribution);
- species which are slow growing (rate of growth);
- species which do not grow again or which do not increase their rate of growth after being harvested (reaction to harvesting).

Other factors which influence sensitivity and which are being investigated by means of RVA are :

- the part of the plant which is used : using leaves has the least impact; using branches, the trunk and the bark has more impact; using the whole plant naturally has the greatest impact;
- demand (quantity and frequency of harvesting) for the NTFP;
- seasonal NTFP production : demand may be lower if the NTFP availability is seasonal;
- commercialisation can increase demand and therefore the likelihood of non-sustainable use;
- the availability of substitutes will reduce the sensitivity of an NTFP occurring in its natural environment.
- The core of RVA is integration of native knowledge with scientific knowledge (Wild & Mutebi 1996).

Forest Type in Relation to NTFP Harvest

It is often stated that NTFPs are primarily harvested from secondary forests and wasteland, and the role of NTFPs in primary forests is said to be minimal. One reason for greater use of NTFPs from secondary forests (apart from the fact that there are an increasing number of secondary forests) is that these are located closer to the living environment of the NTFP collector. In general, such forests have a lower diversity of species than primary forests and a greater density of certain NTFPs. Less time is therefore needed to find the NTFPs, and it consequently takes less time to harvest a given quantity. The collector therefore gets a higher return on the time he/she invests. There are situations, however, in which important self-support products in fact involve plant and animal species, which are dependent on undisturbed forest.

Instead of differentiating between primary and secondary forest, the potential of NTFP use and management should be seen in relation to the density of occurrence of NTFPs and their distance to the living environment of the NTFP collector. There is little potential for commercialization of NTFPs in :

- forests with a high level of diversity (except for NTFPs, which produce a large amount per quantity of biomass harvested and
- remote forests where there is only a minimum level of social organization and infrastructure.

There is potential for intensifying NTFP use in forest fringes, buffer zones of protected areas and nature reserves, inhabited forests with the necessary infrastructure, and flood plains. Here, NTFP exploitation can provide additional income, thus preventing more forest being reclaimed.

Production of NTFPs in Plantations and in Agro-forestry Systems

Cultivating NTFPs on plantations or as part of agro-forestry systems is often seen as a strategy for reducing the pressure on the natural forest and the NTFPs it contains, for example in the case of increased demand resulting from commercialization. Commercialization in fact has a negative influence on the forest and the NTFP if harvesting the NTFP is already destructive in itself.

However, domestication followed by a decrease in productivity may also lead to lower prices and consequently to a reduction in the value of the NTFP occurring in natural forest. This may then reduce motivation with respect to conservation. Maintaining NTFPs in natural forests is therefore of continuing importance for increasing the value of the forest. Moreover, natural forests are an essential genetic pool, which remains necessary even in the case of domestication. A third method of "domestication" is to introduce systems for managing NTFPs in rainforests, which lead to higher production. In natural palm forests in Brazil, for example, those species of tree are felled which compete with the particular palm that produces palm hearts. This probably results in a reduction in biodiversity.

The decision to introduce NTFP management in natural forests or to cultivate NTFPs in agro forestry systems or plantations is dependent (within the limits of ecological possibility) on :

- the demand for the NTFP concerned (extent of commercialization) and

- the relative quantity of biomass harvested.

If both of these are low, little or no negative impact on the forest is to be expected, so that no additional management is necessary. If demand is high and a relatively large quantity of biomass is harvested, this may have a major influence on both the NTFP and the forest. In such cases, domestication is the best management option. Between these extremes, a sustainable management system for NTFPs in natural forests is desirable.

CHAPTER 4

NTFP : A TREASURE HOUSE

In many parts of the world, NTFPs still play an important biological and social role in local food systems. They can contribute substantially to nutrition, either as part of the family diet or as a means to achieve household food security. They can also improve health through the prevention and treatment of diseases. Poor households residing in and around forest areas particularly landless people, women and children depend to a greater or lesser extent on the exploitation of common property forest resources in their everyday life or in periods of crisis. In every day life, other than food and medicinal uses, NTFPs are used as building materials, preparing agricultural equipments, in rituals, as fibres etc. During the periods of crisis many unconventional food materials are used, which include fruits, nuts, storage roots, leafy vegetables etc.

Since NTFPs are essentially part of local subsistence economies, they have not received the required attention in development planning and in nutrition improvement programmes of the population group, which depends on them. As a result, their potential contribution to human welfare remains unrealized. Although one should not expect forest foods to ensure food self-sufficiency of the local population, they can nevertheless constitute an important element of sustainable diets, *i.e.* balanced diets based on local foods, which can be obtained in a sustainable way. Such resources, which are often seen as relics of the past, should actually be considered as under exploited opportunities for the future.

Another feature that needs to be recognized at this stage is the way that different NTFP activities are linked as components of livelihood strategies that households employ. The concept of food security for rural households in developing countries encompasses all factors affecting a household's access to an adequate year round supply of food. Thus, it is concerned not just with the household's production of food crops, but also with the availability of income to the household with which to purchase food, where this is necessary.

In examining household use of NTFPs, we therefore need to identify their effectiveness in both providing gathered foods that contribute to food self-sufficiency, and saleable products that could supplement income needed to purchase food. In doing so it is also necessary to consider whether, and if so how, income-generating activities based on forest products affect other aspects of a household's capacity to contribute to its food self-sufficiency.

Increased commercialization of forest food products could lead to over-exploitation of the resource, or to diversion of food needed for household consumption from local use to sale on the markets. Households may sell not only what is surplus to their requirements but also food needed in the household. Changes in use of agricultural or grazing land to favour production of NTFPs for sale could reduce the amount of land available to the household for producing its basic food crops. Introduction of time consuming NTFP gathering or processing activities could be at the expense of time women need to cook and look after their children. There are many different kinds of food gathered from forests. Forests also provide the habitat for many commonly consumed wild animals and fish. Forest foods may also be smoked, dried or fermented, making them available over extended periods of time.

NTFPs as Diet : For the majority of rural people, forest foods add variety to diets, improve palatability, and provide essential vitamins, minerals, protein and calories. The quantities consumed may not be great in comparison to the main food staples, but they often form an essential part of otherwise bland and nutritionally poor diets. Diet diversity is an extremely important element of nutritional well-being, in part because more vitamins and minerals are consumed, and also because it improves the taste of staple foods. The most common supplementary foods are leaves and wild animals, both of which are generally added to sauces and soups, which accompany staple foods. Forest and farm tree products are also valued throughout the year as snack foods. They are commonly eaten on the job : while working in fields, while herding and while gathering fuel wood. Forest fruits and nuts are the most common snack foods, especially for children.

Some species are noteworthy as particularly rich sources of vitamins, minerals, proteins and fats. For example, many forest fruits and leaves are good sources of Vitamin A, shortage of which is a common cause of blindness in many developing countries. Forest foods can thus be used to combat nutrient related health problems. The food value of some important NTFPs used in different parts of India are as follows :

FOOD VALUE OF SOME NTFPS

1. ***Aegle marmelos*** : Analysis of the fruit gave the following values :

(*i*)	Moisture	–	61.5g/100g
(*ii*)	Protein	–	1.8g/100g
(*iii*)	Fat	–	0.3g/100g
(*iv*)	Carbohydrate	–	31.8g/100g
(*v*)	Calcium	–	85.0g/100g
(*vi*)	Phosphorus	–	50.0g/100g
(*vii*)	Iron	–	0.6g/100g
(*viii*)	Vit–B_1 (Thiamin)	–	0.13g/100g
(*ix*)	Vit–B_{12} (Riboflavin)	–	1.2g/100g
(*x*)	Vit–B_5 (Niacin)	–	1.1g/100g
(*xi*)	Vit–C	–	8 mg/100g

Its Calorific value is 137 *Kcal/100 g.*

2. ***Amorphophallus campanulatus*** : The elephant-foot yam is used chiefly as a vegetable. It is nutritious and in food value compares well with other vegetables. Analysis of the corms gave the following values :

Moisture – 78.7%

(*i*)	Carbohydrate	–	18.4%
(*ii*)	Protein	–	1.2%
(*iii*)	Fat	–	0.1%
(*iv*)	Fibre	–	0.8%
(*v*)	Starch	–	17.7%

(*vi*) Oxalic acid – 1.3%
(*vii*) Minerals – 0.8%
(*viii*) Ca – 50mg/100g
(*ix*) P – 34mg/100g
(*x*) Fe – 0.6mg/100g
(*xi*) Vit.A – 34 *i.u.*
(*xii*) Vit. B_1 – 0.06mg/100g
(*xiii*) Vit. B_2 – 0.07mg/100g
(*xiv*) Vit. P – 0.7mg/100g
(*xv*) Carotene – 60μg/100g

The calorific value is 79 Kcal/100g.

3. ***Anacardium occidentale*** **:** Analysis of cashew kernel gave the following values :

(*i*) Moisture – 5.9g/100g
(*ii*) Protein – 21.2g/100g
(*iii*) Fat – 46.9g/100g
(*iv*) Carbohydrates – 22.3g/100g
(*v*) Minerals – 2.4g/100g
(*vi*) Calcium – 50.0mg/100g
(*vii*) Phosphorus – 450.0mg/100g
(*viii*) Iron – 5mg/100g
(*ix*) Vit–B_2 (Riboflavin) – 0.16mg/100g
(*x*) Vit–B_5 (Niacin) – 1.2mg/100g
(*xi*) Vit–C – 1.2mg/100m

Its calorific value is 596 Kcal/100g.

4. ***Ananas cosmosus*** **:** Analysis of the fruit gave the following values–(Per 100 gm).

(*i*) Moisture – 87.8g/100g
(*ii*) Protein – 0.4g/100g
(*iii*) Fat – 0.1g/100g

(*iv*) Carbohydrates – 10.8g/100g

(*v*) Mineral – 0.4g/100g

(*vi*) Calcium – 20.0g/100g

(*vii*) Phosphorus – 9.0mg/100g

(*viii*) Iron – 1.2mg/100g

(*ix*) Vit-B_1 (Thiamin) – 0.2mg/100g

(*x*) Vit-B_1 (Riboflavin) – 0.12mg/100g

(*xi*) Vit-B_5 (Niacin) – 0.1mg/100g

(*xii*) Vit-C – 39mg/100g

Its calorific value 46 cal/100g.

5. ***Ananas racemosus*** : The dried, debarked tuberous roots yield a yellowish white powder, which contains–(Per 100 gm).

(*i*) Moisture – 11.4g/100m

(*ii*) Fat – 0.87g/100g

(*iii*) Protein – 5.44g/100g

(*iv*) Carbohydrates – 46.84g/100g

(With uronic acid and free sugar)

6. ***Borassus flabelifer*** : Composition of some edible products form *B. flabellifer.*

	Moisture %	Protein %	Fat %	Carbo-hydrate %	Mineral %	Ca	P	Energy
Rice, fruit, pulp	77.20	0.70	0.20	20.70	0.70	9	33	87
Candy	0.70	0.20	0.40	98.76	0.30	58.7	5.4	396
Palm Jaggary *(Talerqur)*	3.2-8.9	1.4-1.49	Vit B_1–10.7–29.9 mg/100g. Vit C–7.3–33 mg/100g.					

7. ***Buchanania lanzan*** : The kernels have a pleasant, sub acidic flavour, are nutritious and eaten raw of roasted. The kernel contains :

(*i*) Moisture – 3.0 g/100g

(*ii*) Protein – 19.0g/100g

(iii)	Fat	–	59.1g/100g
(iv)	Carbohydrates	–	12.1g/100g
(v)	Minerals	–	3.0g/100g
(vi)	Calcium	–	279mg/100g
(vii)	Phosphorus	–	528mg/100g
(viii)	Iron	–	8.5mg/100g
(ix)	Vit–B_1	–	0.69mg/100g
(x)	Vit–B_2 (Riboflavin)	–	0.53mg/100g
(xi)	Vit–B_5 (Niacin)	–	1.5mg/100g
(xii)	Vit–C	–	5mg/100m

Its calorific value is 656 kcal/100g.

8. ***Cannabis sativa*** **:** The seeds yield 30-35% of a greenish yellow oil. Analysis of the seed cake gave the following values :

(*i*)	Moisture	–	12.55%
(*ii*)	Carbohydrate	–	16.02%
(*iii*)	Fat	–	8.30%
(*iv*)	Fibre	–	22.90%
(*v*)	Albuminoids	–	32.38%
(*vi*)	Minerals	–	7.85%

9. ***Caryota urens*** **:** The trunk yields a flour from which sago is made. Analysis of the flour gives :

(*i*)	Moisture	–	13.1g/100g
(ii)	Protein	–	2.4g/100g
(iii)	Fat	–	0.3 g/100g
(iv)	Carbohydrates	–	81.7g/100g
(v)	Minerals	–	2.5g/100g
(vi)	Calcium	–	130mg/100g
(vii)	Phosphorus	–	60mg/100g
(viii)	Iron	–	20mg/100g

Its calorific value is 339 Kcal/100g.

10. ***Citrus aurantifolia* :** Analysis of lime gave–(Per 100 gm)

(i)	Moisture	–	66.5g/100g
(ii)	Protein	–	1.8 g/100g
(iii)	Fat	–	0.5g/100g
(iv)	Carbohydrates	–	29.4g/100g
(v)	Minerals	–	1.8g/100g
(vi)	Calcium	–	710mg/100g
(vii)	Phosphorus	–	60mg/100g
(viii)	Iron	–	2.7mg/100g

Its calorific value is 129 *Kcal/100g.*

11. ***Dioscorea bulbifera* :** Analysis of tuber of the species gave the following values–(Per 100 gm)

(i)	Albuminoids	–	7.96–15.68%
(ii)	Fat	–	0.59–1.26%
(iii)	Carbohydrates	–	71.67–85%

12. ***Diospyros montana* :** Analysis of dried and powdered fruits gave the following values–(Per 100 gm)

(i)	Albuminoids	–	16.4%
(ii)	Organic residue	–	16.4%

13. ***Emblica officinalis* :** The fruit pulp contains–(Per 100 gm)

(i)	Moisture	–	81.2%
(ii)	Protein	–	0.5%
(iii)	Fat	–	0.1%
(iv)	Carbohydrates	–	14%
(v)	Calcium	–	0.05mgm/100g
(vi)	Phosphorus	–	0.02mg/100g
(vii)	Iron	–	1.2mg/100g
(viii)	Vit-C	–	600–720mg/100m

(Vitamin C *contains* is *nearly* 20 *times higher than orange.)*

14. ***Euphoria longan*** **:** Longan or Ashphal is allied to Litchi. The fresh fruit gives the following values–(Per 100 gm)

(i) Moisture	–	82.39%
(ii) Protein	–	1.42%
(iiii) Fat	–	0.45%
(iv) Reducing Sugar	–	5.99%
(v) Sucrose	–	2.35%

15. **Honey :** Honey is a thick, syrupy, translucent, pale yellow or yellowish brown to dark brown liquid with a sweet and characteristic taste, which varies with the floral origin or pasturage and environmental conditions. Analysis of a market sample of honey gave :

(i) Moisture	–	20.6%
(ii) Carbohydrate	–	79.5%
(iii) Protein	–	0.3%
(iv) Minerals	–	0.2%
(v) Ca	–	5mg/100g
(vi) P	–	16mg/100g
(v) Fe	–	0.9mg/100g
(vi) Vit. B_2	–	0.04mg/100g
(vii) Vit. P	–	0.2mg/100g
(viii) Vit. C	–	4mg/100g

The calorific value is 319 *Kcal/100g.*

16. ***Juglans regia*** **:** Analysis of the seed kernel–(Per 100 gm)

(i) Moisture	–	4.5%
(ii) Carbohydrate	–	11.0%
(iii) Protein	–	15.6%
(iv) Ether extract	–	64.5%
(v) Vit A	–	30 i.u./100g
(vi) Ascorbic acid	–	3mg/100g
(vii) Mineral matte	–	1.8%

(*viii*) Calcium – 61mg/gm

(*ix*) Iron – 2.35mg/gm

(*x*) Phosphorus – 510mg/gm

(*xi*) Vit. B_1 – 0.33–0.40mg/100g

(*xii*) Vit. B_2 – 0.10–0.16mg/100gm

(*xiii*) Vit. B_6 – 0.87–1.05mg/100 gm

17. ***Lycoperdon giganteum* :** (Puff balls/Putka chattu)

(*i*) Carbohydrate – 1.35%

(*ii*) Protein – 2.2%

(*iii*) Fat – 0.56%

(*iv*) Ash – 0.916%

18. ***Madhuca indica* (Flower) :** The corollas are a rich source of sugars and contain appreciable amounts of vitamins and calcium. Analysis of a sample gave the following values :

(*i*) Moisture – 18.6%

(*ii*) Protein – 4.4%

(*iii*) Fat – 0.5%

(*iv*) Total sugars - 72.9%

(*v*) Fibre – 1.7%

(*vi*) Ash – 2.7%

(*vii*) Ca – 140mg/100g

(*viii*) P – 140mg/100g

(*ix*) Fe – 15mg/100g

(*x*) *Vit.* A – 39i.u./100g

(*xi*) *Vit.* C – 7mg/100g

(*xii*) *Vit.* B_1 – 32μg/100g

(*xiii*) *Vit.* B_2 – 878μg/100g

(*xiv*) *Vit.* P – 5.2mg/00g

19. ***Madhua indica* (Seed kernel) :** Analysis of the seed kernel– (Per 100 gm)

(*i*) Moisture – 5 – 12%

(ii)	Fatty oil	–	51.1%
(iii)	Protein	–	8.0
(iv)	N-free extract	–	27.9

20. ***Momordica cochinchinensis*** : The tender fruits of the plant are esteemed as vegetables. Analysis of fruit gave the following values–(Per 100 gm)

(i)	Moisture	–	84.09%
(ii)	Protein	–	2.61%
(iii)	Fat	–	0.66%
(iv)	Carbohydrate	–	5.69%
(v)	Calcium	–	21mg/100g
(vi)	Phosphorus	–	148mg/100g
(vii)	Iron	–	2.59mg/100g

21. **Mushrooms** : The mushrooms are in general very tasty and easily digestible. The food value of mushrooms is :

(i)	Water	–	Sufficient
(ii)	Carbohydrate	–	5.1-6g/100g
(iii)	Protein	–	3.8g/100g
(iv)	Fat	–	Scanty
(v)	Ca	–	10mg/100g
(vi)	P	–	100mg/100g
(vii)	Fe	–	1.5mg/100g

(vii) Vit. B_1, Vit. B_2, Vit. P, Vit. C is also present.

22. ***Syzygium Cumini*** : Analysis of the edible paste gave the following values–(Per 100 gm)

(i)	Moisture	–	83.7%
(ii)	Protein	–	0.7%
(iii)	Fat	–	0.3%
(iv)	Carbohydrate	–	14.0%
(v)	Calcium	–	15.0mg/100g
(vi)	Phosphorus	–	15.0mg/100g

(vii)	Iron	–	1.2mg/100g
(viii)	Vit. C	–	18mg/100g
(ix)	Vit. B_1	–	0.03mg/100g
(x)	Vit. B_2	–	0.01mg/100g

23. ***Terminalia beleric* :** The fruit is the well–known commercial myrobalan called belliric myrobalan. Analysis of the fleshy fruit pulp gave the following values -(Per 100 gm)

(i)	Moisture	–	6.1%
(ii)	Tannin	–	21.4%
(iii)	Water extractables	–	44.0%
(iv)	Fatty oil	–	38.6%

24. ***Terminalia catappa* :** The fruit comprises kernels, fibrous covering, husk and hard endocarp. The air dried kernels contains–(Per 100 gm)

(i)	Moisture	–	3.51%
(ii)	Fat	–	52.02%
(iii)	Protein	–	25.42%
(iv)	Sugar (glucose)	–	5.98%

25. ***Terminalia chebula* :** The dried fruits constitute one of the most important vegetables tanning materials–(Per 100gm)

(i)	Moisture	–	10%
(ii)	Tannins	–	24.6–32.5%
(iii)	Non-tannins	–	13.9–16.4%
(iv)	Insolubles	–	41.1–50.1%

26. ***Tamarindus indica* :** The fruit pulp is the chief agent for souring curries, sauces, chutneys and certain beverages throughout the greater part of India. The pulp contains–(Per 100 gm)

(i)	Moisture	–	20.9%
(ii)	Protein	–	3.1%
(iii)	Fat	–	0.1%
(iv)	Tartaric acid	–	8-18.0%

(v)	Carbohydrates	–	67.4%
(vi)	Minerals	–	2.9%
(vii)	Calcium	–	170mg/100g
(viii)	Phosphorus	–	110mg/100g
(ix)	Iron	–	10.9mg/100g
(x)	Vit. B_2	–	0.07mg/100g
(x)	Vit. C	–	3.0mg/100g

27. ***Zizyphus mauritiana :*** The fruits are a gong source of Vit–C and sugars and contains appreciable amount of mineral constituents. A typical analysis of the pulp of the fruit is as follows–(Per 100gm)

(i)	Moisture	–	81.6%
(ii)	Protein	–	0.8%
(iii)	Fat	–	0.3%
(iv)	Carbohydrate	–	17.0%
(v)	Calcium	–	4mg/100g
(vi)	Phosphorus	–	9mg/100g
(vii)	Iron	–	1.8mg/100g
(viii)	Vit. C	–	76mg/100g
(ix)	Vit. B_5	–	0.7mg/100g
(x)	Vit. B_2	–	0.02mg/100g
(xi)	Vit. B_1	–	0.02mg/100g

NTFPs as Cash Crops

Major non-timber products of wide commercial acceptance include cocoa, rubber, rattan, coffee, coconut and oil palm.

The cocoa tree *(Theobroma cacao)* has its origins in South America, but is grown throughout the tropics. It is a valued crop, providing a living for many people and frequently planted and nurtured by a farmer and his family. Historically the cocoa tree is a shade lover. More recently there has been some success with strains not requiring shade. For some countries–in West Africa for example–cocoa is a key export product and yet growing more cocoa trees is a threat to the forest, because farmers destroy or thin out forests to create planting space. Foresters and timber men are worried about the loss of productive forest to cocoa. Both sectors are important, as is the forest from which they stem. This same dilemma can be cited for coffee, oil palm, rubber and other cash crops, which replace natural forest without the prospect of natural regeneration, which accompanies selective logging.

Whilst rubber *(Hevea braziliensis)* comes mostly from plantations it can be gathered from wild trees in natural forest. It is not the most efficient form of harvesting but it can be an important part of a local rural economy. Rubber is useful and the tree is a renewable resource. Its production is a legitimate activity. Space has to be found for cash crops in land use planning, especially in the case of rubber trees because when they stop providing latex, they can be felled for timber. Malaysia pioneered the introduction of rubber-wood into world markets.

Rattans find their way into a variety of products including furniture. Rattan is a climbing palm, able to grow more than 100 meters in length, and needs trees for support. It can be harvested from natural forest or grown in plantations of young support trees. Rattans or rotans, are the stems of climbing palms of the genera *Calamus, Daemonorops, Ceratolobus, Calospatha, Plectocomia, Plectocomiopsis* and *Korthalsia,* which together form one-half of the sub-family Lepidocaryoideae, of the family Palmae. Rattans find their chief uses in the furniture and basketry industries, where great strength is essential, and appearance, a secondary consideration. Rattans are used by the native population for so many purposes, for example, for tying and binding. Houses, fences, fish-traps, and even boats, are bound together with rattan, often without the use of a single nail. Ropes for tethering buffaloes, drag-ropes for hauling timber, mooring-ropes, anchor cables, bridge cables, and deer snares are all known uses of the rattan among the natives.

Oil palm *(Elaeis guinensis)* plantations provide vegetable oils in bulk and are created in tropical forest regions after forest clearance. In Malaysia land use planning allows for a balance of permanent forest estate and cash crop plantations. Both have their place in the economy of the country. The mangrove timbers are also used for the production of charcoal, firewood and wood-chips. The latter is used in the manufacture of rayon. The man-grove bark was the traditional source of tannin although the market is now largely replaced by chemical-tannin. The palm, *Nypa fruticans,* produces a diversity of minor products. The leaves are used extensively for roof thatching and mat making. The unfolded leaf sheaths are made into cigarette wrappers, which are cheaper than paper wrappers. The production of nipah sugar from the inflorescence and infructescence sap is still an important traditional activity for communities living in the vicinity. Most of the traditional cakes and the sugar produced are consumed locally. Nipah sugar is also used to produce alcohol for the local market although the production has declined in recent years.

In India oil seeds as NTFPs are collected from *Shorea robusta, Madhuca indica, Mangifera indica, Garcinia indica, Azadirachta indica, Pongamia glabra, Schleichera trijuga, Salvadora oleoides,* S. *persica,* and *Actinodaphne hookeri. Sal (Shorea robusta)* seed cotyledons yield the well-known *sal* butter used for cooking and lighting. It is a useful confectionery fat and can be used in soap making. *Mahua (Madhuca indica)* seed : Almost the entire production of oil from this seed is used in the production of washing soaps. Refined *mahua* oil can be utilized for cooking, confectionery and in chocolate making. Both the *karanj (Pongamia glabra)* seed and oil are poisonous but they possess remarkable medicinal properties. A major part of the *kusum (Schleichera trijuga)* oil produced is utilized by the soap industry . The oil compares favorably with other oils in softness and lathering. Neem *(Azadirachta indica)* seed oil is used in soap and local medicines. Mango *(Mangifera indica)* seed oil is used as a cocoa butter substitute. Khakan *(Salvadora oleoides)* and pisa *(Actinodaphne hookeri)* seed oils are used in making soap.

Essential Oils

In India they occur in about 60 plant families and are frequent or abundant in the Labiatae, Rutaceae, Geraniaceae, Umbellifereae, Asteraceae, Lauraceae, Graminae and Fabaceae families. The important essential oils produced in India are oils of sandalwood, lemon grass, palmarosa, eucalyptus, khus and linaloe. These oils are distinguished from fatty oils because they evaporate or volatilize when they come in contact with air. Grass oils : These are mostly obtained from tropical grasses rich in aromatic essential oil, belonging mostly to the ***Andropogon*** and ***Cymbopogon*** genera. Leaf oils are extracted from *Eucalyptus* oil from ***E. globulus*** **and** ***E. citriodora,*** camphor and camphor oil, cinnamon leaf and bark oils, pine needle oil, mint oil, and wintergreen oil. Root oils are extracted from Costus oil from ***Saussurea lappa*** *(kuth)*, and (*i*) Indian valerian oil from ***Valeriana wallichii*** are used in soap and cosmetics; (*ii*) Pharmaceuticals; (*iii*) Confectionery and aerated water; and (*iv*) *attars,* scented tobacco, *agarbattis,* incense, etc.

The cost of production of essential oils variety depending on the raw materials, facilities for processing, and labour and energy costs. Their production on a large industrial scale requires significant variations in the technologies used. In case of rural small-scale processing these technologies have to be adapted to suit the existing situation at field level.

The international market for essential oils is dominated by a few countries. Some of these countries import oils from developing countries and export them after refinement or blending. The market is quite competitive and protected. Hence the developing countries have to develop strong market strategies for promotion of their products.

Steam distillation is the widely used technique in the extraction of essential oils. In this process, steam is passed through the plant material whereby the constituents that are volatile in steams are carried along with the steam. Steam can be generated in a separate boiler or at the bottom of the still by direct heating. The advantage is that the volatile components can be distilled at temperatures lower than the boiling points of their individual constituents, and that on condensing, the oils, being immiscible in water, forms a layer thereby easing separation. Though the process sounds simple in theory, the actual commercial process for greatest efficiency and quality varies widely, depending on the characteristics of the raw material and the final product.

Production of some expensive essential oils from flowers such as jasmine and rose takes place through a process called enfleurage. This is a technique by which the raw material is kept in contact with fat in order to get the fat saturated with the essential oil. The adsorbed oil is then extracted from the fat using ethanol. The ethanol is finally distilled off to yield the oil. This will contain all compounds that are adsorbed into the fat.

Some parts of plants that contain very minute amounts of essential oils such as flowers are extracted using low boiling solvents such as hexane, pentane or petroleum ether. These extracts will contain all material soluble in the solvent including the essential oil compounds. The process of separation of the essential oil needs an additional step of removal of the solvent in a separate evaporator. This is an expensive step due to the cost of solvents and the cost of production will depend on the efficiency of the recovery of the solvent for re-use. Hence this process has to be done at a central facility using good manufacturing practices. The scale of production will determine the economy of the process.

Oleoresins

The extraction of spices such as ginger, pepper, chili, cardamom using solvents and the removal of these solvents as in the case of concretes yields oleoresins, which contain not only the essential oil compounds but also other flavour principles of the spice such as pungency, waxes and other solvent soluble extractives. Oleoresins therefore are more representative of the spice than the essential oil obtained from it. The extraction process could be a continuous soxhlet extraction process or solvent percolation through a series of extractors.

Downstream Processing of Essential Oils

Essential oils could be further processed or rectified to add value. Rectification of essential oils is sometimes needed depending on their end uses. The process of rectification may consist of one or more of the following :

- removal of moisture, colour and sediments;
- removal of compounds in order to improve the odour characteristics, stability and sustainability;
- isolation of compounds that are highly valued;
- enrichment of the oils by removing or adding other fractions.

This is done by a process of fractionation which could be carried out in some developed rural areas having small-scale processing with back up from national research institutions who would carry out the analysis and develop the fractionation parameters. This needs more training and equipment and may not be possible in certain rural communities or with forest dwellers. It would be better to carry out the fractionation of oils at a central facility, which can afford to invest the funds and personnel required for this activity. As a result of fractionation pure isolates of added value such as citronellal and citral could be produced.

These isolates could further be processed using chemical methods to produce high value aroma chemicals, which have an export market. These aroma chemicals can be used in blending of perfumes and flavours for local industries.

Fibre Boards

Some parts of plants or residues after extraction of the main product could be used to produce other products such as fibre board boxes and hand made paper. Pine needles are one such source of lignocellulose, which could be converted to fibre board for use as packing material.

It is estimated that on an average, the availability of needles of *Pinus roxburghii* per ha is between 2.5-3 t/ha. Even a portion of the needles (15-20 per cent) if collected from easily approachable forests, will be sufficient to meet the requirement of number of small mills. It will help in preventing the forest fires without disturbing the ecology. The areas under *Pinus roxburghii* in the sub-Himalayan region are 1 million hectares. The needles could be collected (April-June) with rakes and baled at the site with mobile baling presses for transportation. The process for the production of fibre boards from pine needle used in the sub-Himalayan region in India is simple and can be adapted by other countries.

The process does not require any binder. The process consists in giving a softening treatment in a rotary digester followed by defibration to get a suitable pulp, in a Hollander type beater. The pulp is blended with pulp made from waste paper, then converted into a sheet in a sheet-former of a single cylinder machine. The wet sheets (2 or 3) from cylinder machine are laminated and hot pressed to get a thickness of 3.2 mm. The boards are given a coating to make them water resistant. The yield of the fiber board is 55-60 per cent based on the moisture free weight of the needles. The board is converted into packing boxes using wooden battens, which are stapled with a machine specially designed for the purpose. A few perforations or slots are given to the board for fruit breathing. Fiber angles could also be used in place of wooden battens for converting the board into a box.

The main effluents of the plant are dissolved matter and small quantities of suspended lignocellulosic fines. These pass through the washing screen. The effluent from the mildly cooked pine needles contains the dissolved matter consisting of mainly colouring matter, resinous substances and some quantity of lignin. About 7-8 cubic meters of liquor will be obtained from each ton of digested needles. The pH of the liquor is around 6 and could be disposed in a close dry channel if available. If not available, some detention tanks may have to be provided, along with an arrangement to remove the suspended particles. The tanks could be emptied during rainy season when the liquor gets diluted with rainwater.

Social Role of NTFPs

The *sal* tree has great significance in the lives of the indigenous populace of the Chota Nagpur plateau. It is the focal point of the festival of *sarhul,* which means the *sal* blossom festival.

Sarhul, which literarily means the *sal*-blossom festival, is a semi-religious festival of the inhabitants of Chota Nagpur. This festival is celebrated by several tribes of the region such as the Oraons, Mundas and the Santhals, the Santhals call it *Baha* and celebrate it in end February-March whereas the Oraons celebrate it a little late in March-April. The festival revolves around the *sal* tree. With much of the forest having been cleared for the purpose of cultivation, these tribals have spared a cluster of *sal* trees to serve as a place for worship. This protected area is known as the *saran,* the sacred grove. The festival of *Sarhul* is celebrated in the month of *Baisakha* and announces the onset of spring. No one in the community is allowed to use the sal tree, its flowers, or fruits in any manner till the Sarhul celebrations. Women cannot use the flowers to decorate themselves; the honey and the flowers cannot be eaten. If anyone breaks these rules; the village priest boycotts the house of the family for a year till the next *Sarhul.*

In early days, the festivities were spread over a period of three days though now in many townships the duration of the festivals has been cut down to two days and in some areas even to one, depending on the number of holidays declared by the state. In the remote village area, the celebrations still have the verve and vigour of earlier times and not much has changed.

Different tribes have a different significance attached to this festival. The tribes practicing cultivation for their livelihood such as the Oraons, celebrate this festival in the month of *Baisakha* (April) before the commencement of the sowing of paddy. The *puja* (prayer) is undertaken with the aim of securing the blessings of the gods and goddesses of nature for a good bumper crop.

Each Oraon family saves at least a handful of the rice blessed by their village priest during the previous Sarhul festival for the following sowing season. This rice is believed have special qualities after having been placed on the *saran-sup* which is a special winnowing basket and the seat of the 'Goddess of the Grove'. This basket is ceremonially hung at the priest's house. This sanctified rice *(asirbadi)* is mixed with a little cow dung and when the Oraon cultivator sows the paddy seed, this *asirbadi* is added to it, with the hope of a good yield.

Besides this, animal sacrifices are made for different reasons. If a piece of wasteland is to be reclaimed for cultivation purposes, a sacrifice has to be made to appease the spirits of such areas. These sacrifices are normally made on Tuesday and Fridays as these are considered to be two auspicious days in an Oraon calendar. These days of festivities give everyone the opportunity to rejoice and make merry. Everybody gets together to eat, drink, sing and dance.

The hunter-gatherer tribes have a special meaning for this festival. They pray for the protection of the inmates of the jungle because of their direct dependency on them. For instance, every part of the *sal* tree has great importance in their lives. The root and the new leaves have medicinal properties; the leaves also serve as a good fertilizer; its timber is used for beams, posts and doors in their house-building; bowls and plates are made out of the leaves of *sal;* marriage invitations are given in the form of folded *sal* leaves with a little bit of turmeric and a few rice grains inside. Liquor is made out of the *Mahua* honey and the flowers are eaten.

The Santhals *Baha* is a three-day long festival, which begins with the day of purification, when the young men of the village gather at the *jaherthan* (the Holy Grove) and make two huts there. One hut is reserved for the Santhal deity *Gosan Era* and the other hut is shared by *Jaher Era,* the deity of the grove, *Monrenko* and *Maran Buru.* Spreading of cowdung by the priest purifies these shrines. Thereafter the priest cleanses a winnowing fan, a basket, a pickaxe, a broom stick and daubs these with oil and *methi* (fenugreek). Later in the day, three young men impersonating the three deities enter the priest's house. *Jaher Era* places the basket on his head and picks up the broom; *Monrenko* takes the bow and arrow and Maran Buru takes the pickaxe. Then all three followed by more men run to the holy grove and *Jaher Era* sweeps the shrine while the other two supervise. In the evening, the three return to the priests house for a meal and they are entertained with songs and rice beer.

The second day is spent in collection of the *sal* and *mahua* flowers by the three impersonating deities. Later these deities are enshrined in their huts and the priest sacrifices fowls in their honour. People assemble here and the priest offers them bunches of *sal* and *mahua* flowers. The priest washes the feet of the deities and water is sprinkled on all gathered there. Thereafter a procession is taken out and everyone joins in. All then return to the village singing and dancing. The priest and his wife remain at the grove.

The third day is solely devoted to merry making. Sometimes the festival takes the shape of a water festival–something similar to Holi of Northern India. People indulge in pursuing their traditional games and sports like shooting and archery. In India many other social roles of NTFPs could not be recorded here. Many of them has been recorded elsewhere and many of them has not yet been recorded. As a whole it is can be concluded that, though, the modalities of the celebrations are changing over the years, one can still observe a large element of belief and faith persisting in various parts of India regarding NTFPs.

Gums and Resins

Gums are translucent, amorphous substances, which are degradation products of the cell wall of woody species. They exude spontaneously from trees and are soluble in water. Resins also are exudates but are soluble in alcohol, not water. Closely related to true gums are gum resins, which are also produced by plants. Since these are a combination of gum and resin, they do not dissolve in water completely. Resins often occur mixed with a high percentage of essential oils known as oleoresins. When oleoresins include some gum, as in the case of exudation from *Boswellia serrata,* they are called gum oleoresins. Gum from ***Acacia nilotica*** (called "gum Arabic") and from other ***Acacia*** species such as ***A. catechu, A. modesta,*** and ***A. senegal*** are collectively categorized as Acacia gums. Gum karaya, or *katira,* from ***Sterculia urens,*** and oleoresin from ***Pinus roxburghii,*** are tapped in significant quantities to be of commercial importance.

Commercial gums enter the market in the form of dried exudates. The varieties having the least color and highest adhesive power and viscosity are the most valuable. The finer grades are used in clarifying liquors, "finishing" silk, and in the preparation of quality watercolours. Intermediate grades are used in confectionery, pharmaceuticals, and printing inks, in sizing and finishing textile fabrics, in dyeing, and in the paint industry. In the cosmetic and pharmaceutical industry, gums serve to emulsify or bind mixtures in creams, lotions and ointments. Many gums add body and bulk to foodstuffs (*e.g.* commercial ice cream).

Resins are used in the manufacture of lacquers and varnishes. Resinous substances can be used for waterproof coatings. Resins dissolve readily in alkali to form soaps. They are used in medicines, for sizing paper, for incense, and in the preparation of sealing wax and other products. In India, ***Pinus roxburghii*** trees yield the highest amounts of oleoresin. Blazes on the tree trunk are cut to facilitate the flow of oleoresin from resin canals. Traditionally, oleoresin is collected by the cup-and-lip method from March to early November.

Important oleoresins are turpentines, balsams, copaiba, and elemi. These are used in perfumery and medicines, for making varnishes, and lacquers, as fixatives, and in scenting soaps.

Gum resins have varied uses. Gamboge is used to colour golden lacquers, as watercolor pigments, and in medicines. Assafoetida is used for flavoring curries and other food products, and also as a drug.

Galbanum is used in medicine. Myrrh is used in incense, perfume, and embalming. Frankincense is used primarily as incense.

Medicinal Plants Products

About 80 per cent of the world population still depends on medicinal plants for their health care. Medicines prepared using traditional methods are still used by the practicing healers in the developing countries. Hence there is a demand for these traditional medicines, which are prepared using wildly growing plant species. Around 20 per cent of the drugs in modern pharmacopoeias are also plant derived, either as pure phytopharmaceuticals extracted from plants or as synthetic derivatives of them. Although the raw materials for the production of phytopharmaceuticals have been produced by crop-wise cultivation of selected varieties of plants, most of those used for traditional medicines have been collected from the forests as NTFPs. In fact this had in certain instances lead to threats of extinction of some valuable species. Hence it is imperative that domestication and cultivation of medicinal plants have to be initiated. In order to encourage cultivation, guarantee of purchase of these plants has to be given. This could only be achieved by introducing small-scale processing units for the production of traditional medicines. These NTFPs in some cases could be sustainably harvested from forests for small-scale processing whilst systematic cultivation has to be introduced to meet the demands for medicines at the national level.

The medicines for internal use prepared in the traditional manner involve simple methods such as hot or cold water extraction, expression of juice after crushing, powdering of dried material, formulation of powder into pastes *via* such a vehicle as water, oil or honey,.and even fermentation after adding a sugar source.

Downstream Processing of Medicinal Plants

Preparation of standard extracts and conversion of them into dosage forms are activities that can be done as a rural based small industry to meet the demands of the local population. This is necessary in areas where traditional medicines are the main form of treatment.

The value of medicinal plants as a source of foreign exchange for developing countries depends on the use of those plants as raw materials in the pharmaceutical industry. These raw materials are used to :

- Isolate pure active compounds (phytopharmaceuticals) for formulation into drugs;
- Isolate intermediates for the production of semi-synthetic drugs;
- Prepare standardised galenicals (extracts, powders, tinctures).

If one is to produce known pure phytopharmaceuticals used in modern medicine such as quinine, reserpine more processing stages and more sophisticated machinery are required. Furthermore safety and pollution aspects have to be considered. As these compounds occur in small quantities in the plant source, large amounts of raw materials have to be processed in order to benefit by economy of scale of production. Hence large-scale central processing is the only solution but costs have to be minimized to be competitive in the market.

Most of these processes and formulations are patent protected. Even transferring technology through contractual agreements and payment will not be of much help unless there is a large local demand for these drugs. Often the drugs so produced are more expensive than world market prices owing to the limitations of the economy of scale of production. Hence any country venturing into such activity should be able to use most of the production within the country.

If the isolation of the pure compound is patent protected or involves complex and sophisticated procedures and equipment, some value of raw materials can be retained by producing concentrates of the plant in the countries of origin. But this will need prior agreements with the buyers as some pharmaceutical firms do not favour the purchase of extracts.

Certain plants are rich sources of intermediates used in the production of drugs. The primary processing of parts of plants containing the intermediates could be carried out in the country of origin thus retaining some value of the resource material. For example : diosgenin (from *Dioscorea* sp.) and hecogenin (from sisal) used in the production of steroids can be commercially produced in the countries of origin where there are steady supplies of sufficient raw materials.

The processed product (galenicals) from the plant could be a standardized fluid/solid extract or a powder or a tincture.

These have to be formulated for incorporation into modern dosage forms. New formulations require some development work, particularly on account of the nature of the processed products. Plant extracts are difficult to granulate; they are sensitive to moisture and prone to microbial contamination. Hence the types of excipients to be used and the processing parameters have to be determined.

The potential of the know-how available with traditional healers and the rural people with regard to the use of plants as curatives, has not yet been fully tapped. Many plants from traditional uses have been useful in the development of new modern drugs. Many have been models for subsequent new drug development. The R&D with respect to new drug development is very expensive and requires long periods of work.

Other Products

Nagas of North East use the seeds of *Chenopodium murale* and the leaves of a wild aubergine (The fruit of the eggplant, *Solanum melongena)* are used to produce yeast. Rice bear, locally brewed, has nutritious value and is used in three forms : as fermented liquor, infused beer mildly fermented. Before drinking, Nagas generally pour a few drops on the ground or blow upon the surface of the drink to drive away the spirits.

Lentinus spp. (a fleshy fungi or a mushroom) is a major edible NTFP of the North Eastern India. The north-eastern part of India, stretching from 80°04'97°25'E longitude and from 21°57' 29°25'N latitude, is bounded by the Eastern Himalayas on the north; Southern China and Burma (now Mynamar) on the east; and West Bengal and Bangladesh on the west and south. Heavy rainfall, high humidity, a moderate temperature range, and a rich but varied vegetation-cover make this region quite conclusive for the growth and proliferation of a variety of macrofungi. Regular forays into the forests of States like Meghalaya, Manipur, Mizoram, Arunachal Pradesh, Nagaland, Sikkim and Tripura were therefore made during the period 1983-1990, which could lead to identification of 95 species of fleshy fungi, majority of which were edible. The survey indicated that the tribal population of the region are very much fond of mushrooms and have special fascination for *Lentinus spp.* Two of them, *viz., Lentinula edodes* and *L. lateritia* were found to be the most popular mushrooms not only in the region but are in good demand by traders coming from the Western parts of India also, and are fetching high prices.

Leaves of *Bauhinia vahlii* locally known as "mahul" in Uttar Pradesh and Madhya Pradesh, *'chihord'* or *'siali'* in West Bengal and Orissa. Leaves are used for making cups and plates and for wrapping food. Leaves vary in size from 20 to 40 centimeters in diameter, and are bilobed at the apex. Collection takes place two to three months after the rains. Leaves are collected by tribals and other forest dwellers and packed and tied with fibers obtained from the same species. No standard practice exists as to the number of leaves in each pack. Average collection per person per day is 5 to 6 kilograms. Green leaf packets are sold in the market without additional processing.

Fibers and flosses are also obtained from different parts of Indian forests in large quantities. The most important fibers coming from the forests of India are from the species like *Agave sisalana, Abroma augusta, Abutilon spp., Ananas cosmosus, Antiaris toxicaria, Boehmeria nivea, Borassus flabellifer, Cannabis sativa, Cordia dichotoma, C. rothii, Girardinia heterophylla, Grewia glab*ra, *G. elastica, G. optiva, Hibiscus spp., Malachra capitata, Marsdenia tenacissima,* M. *volubilis, Oreocnide integrifolia Oreocnide integrifoli, Phormium tenax, Sensivieria roxburghiana, Sesbania bispinosa, Sida rhombifolia, Sterculia foetida, S. urens, S. villosa, Themeda arundinacea, Trema orientalis, Typha elephantina,* and *Urena lobata.* Flosses are obtained from certain wild fruits. Important species are ***Bombax ceiba*** and ***Ceiba pentandra.***

Pectins exist in all higher plants, particularly as structural materials of soft fruits and fleshy roots. Most common commercial sources are citrus peel and apple pomace, both by-products of fruit juice manufacture. From citrus peel (about 30%) pectin is isolated by acid aqueous extraction, followed by filtration, concentration and precipitation with ethanol, methanol or isopropanol. Although most widely used in jams and jellies, pectins also stabilize fermented and acidified milks, fruit yogurts and related foods. They act as protective colloids in cosmetic emulsions and surgical dressings. Pectic substances are extractable from many tropical plants. An interesting source is the *jujuba* tree (*Ziziphus mauritiana./Z. jujuba* Miller), which is highly resistant to drought and salinity. The fruit, like a small plum, has a long post-harvest shelf life, can be eaten fresh, dried or candied and is high in vitamin C (about 75 mg per 100 g); when under ripe it is ideal for jams, jellies and chutneys.

Bio-diesel can be processed from non-edible oilseeds such as *Pongamia pinnata (karanj, honge)*, *Jatropha curcas* (Ratan Jot), *Hevea braziliensis* (rubber), *Madhuca indica* (Mahua) etc. Such oil-bearing trees are abundant in India and there is need for systematic propagation and processing of these oilseeds, which have many benefits. Some of the key issues of National and International importance involved in non-edible oilseed trees are dependence on imported oil, fuel security, increasing rural income and employment, empowering women, sourcing organic fertilizer and increasing renewable energy, promoting clean development mechanisms and sequestering carbon. The term "bio-diesel" is broad and covers all the fuels the earth produces, including fossil fuels and other agro-based fuels like ethanol etc. India is not self-sufficient in petroleum and has to import about two-thirds of its requirement. In the year 2000 India produced 32 million tons of oil and imported 90 million tons, which is 64 per cent of its requirement. The import cost during the current year will be around Rs. 90,000 crores. Development of bio-diesel from non-edible oilseeds from trees is completely decentralized and is largely located in rural areas. It would increase the income of rural populations, as these trees can be grown on land, both common and private, in a decentralized manner. Collection of these oilseeds is mainly done by rural women. After extracting the oil, as much as sixty per cent of the non-edible oilseeds is in cake form and is used as organic fertilizer by the farmers. This is full of nitrogen and is therefore highly beneficial in enriching soil for farming. As synthetic fertilizer is costly and environmentally harmful, the need for natural and organic fertilizer is obvious.

Animal Products

Honey and Beeswax

Honey forms a natural nutritious food for rural people. Also, it is used widely for medicinal purposes. Two species of bees, *Apis dorsata* (rock bee) and *Apis indica* (Indian bee) produce honey. The former is wild in the montane and submontane regions throughout India and is the major source of forest honey. It is a good honey gatherer and a single comb may yield up to 35 kg of honey and 1 kg of wax. The latter is amenable to domestication, but it is not a good honey gatherer. The yield per hive ranges from 3 to 13 kg of honey in the hills and 3 to 8 kg in the plains (Gupta 1994).

About 250 tonnes of rock bee honey and 98 tonnes of Indian bee honey are produced annually in the country. At a price of Rs. 40/kg, the total value of honey produced is Rs 139 million. Beeswax is used in the manufacture of furniture and floor polishes, dressing and waterproofing of leather goods. It is also an ingredient in shoe polish, cosmetics, lipstick and face cream. About 28 tonnes of wax are produced annually, valued at approximately Rs. 1.6 million.

Natural waxes are NTFPs of commercial value used as components of industrial products like candles, varnishes, pharmaceuticals and cosmetics. Some of them are collected, melted and formed into cakes or pieces. Some waxes such as candellila can be obtained by solvent extraction. The cosmetics industry is the largest user of bees wax. Cosmetics containing beeswax include : skin creams (*e.g.* cold cream for facial use), emulsions, make-up foundations, face powders, cheek pomades, hair creams/pomades, lipsticks and eye make-ups (*e.g.*, eyebrow pencils and mascaras). The pharmaceutical industry is the second largest user of beeswax, since it is used as an ingredient in certain ointments, for coating pills, and in some manufacturing processes. Candle manufacturers are the third major users of beeswax.

Other uses include preparation of polishes for furniture, wood and leather, finishes for leather, rexine, wood and paper, lithographic and engraving materials, castings, dental equipment, ornaments and confectioneries. These too can be processed at rural level for income generation. Even with severe competition from synthetic waxes, some specific properties of natural waxes have kept them in demand. The processing and refining of the wax oils are simple but important in order to produce good quality grades.

Lac

Lac from the insect *Laccifer lacca* (commonly known as shellac) is used presently for various purposes in plastics, electrical supplies, adhesives, leather, wood finishing, printing, polish and varnish, ink and other industries. It is also the principal ingredient of sealing wax. Two main strains of the lac insect are recognized : rangeeni and kusumi. The rangeeni crop is raised on several host plants, the most important being *Butea monosperma* and *Zizyphus mauritiana*. The kusumi strain is raised on *Schleichera oleosa*. Both strains produce two crops per year. Other plants that can host the insect are *Samanea saman, Cajanus cajan, Albizzia chinensis, A. odoratissima, Acacia mauritian, Ficus drupica, Combretum quadrangulare, Acacia catechu, A. villosa, Dalbergia chinensis, D. nigerescens* and *Albinia lucidor.*

Lac is collected in two forms, *ari* and *phunki*. The former is cut from the host plant and the latter is collected from the brood lac, after being used for inoculation. The lac is then sold as such or freed from the sticks and then sold. The lac removed from the sticks is known commercially as stick lack. The annual production of stick-lack is about 14 500 to 20 000 tonnes and the price varies from Rs 4500 to 16000/tonne (most of the produce sells at around Rs. 14000/tonne). The total value of the annual production is Rs. 203 million to Rs 280 million. Lac finds a variety of uses in plastics, electrical goods, adhesives, leather, wood finishing, lacquer work, printing, polish and varnish, ink and a number of other industries. It is also the principal ingredient in sealing wax. The lac also yields an edible dye, which is used in colouring soft drinks and foods.

Silk

India produces four kinds of silk : mulberry, *tassar, muga* and *eri*. The silkworm *Bombyx mori* is fed on mulberry leaves cultivated in plantations. Silkworms are also found wild on forest trees, *e.g., Antheraea paphia* which produces the *tassar* silk. *Antheraea paphia* feeds on several trees such as *Anogeissus latifolia, Terminalia tomentosa, T. arjuna, Lagerstroemia parviflora* and *Madhuca indica*. Wild silkworm *Antheraea assamensis* produces muga silk, and another wild silkworm *Philosamia synthia ricini* produces eri silk. The estimated annual production of tassar silk is 130 tonnes. Production of other types of silk exceeds 10,000 tonnes.

Insect Galls

Some of the galls produced by insects are quite beneficial. The gallnuts of commerce, also called Allepo, Mecca, Chinese or Turkey galls, are produced on various species of oaks and other trees by certain Eurasian cynipid wasps. Among the best grades containing more than 50 per cent tannic acid, are those harvested in Iran, Turkey and Syria. These galls, long used in parts of Asia to make dyes and medicines, are now processed for use in tanning, ink and pharmaceutical industries. The chief active principles are the astringent, tannic and garlic acids.

Synthetic Products *Versus* Natural Products from NTFPs

Cheaper synthetic substitutes have always threatened markets for NTFPs. However, the long-term effects of synthetic drugs and the development of resistance by pathogenic parasites to synthetic drugs have weighted the scales on the side of natural NTFPs. More and more R&D is done through joint collaboration to study the medicinal and other uses of plants products. The resurgence of interest on naturals has resulted in more funds being allocated to this type of R&D required to reduce costs of production.

Any threat from synthetic products should be taken as a challenge for vigorous R&D work to improve the economic competitiveness of the product. In the essential oil industry, the survival of many essential oils and flavour industries is largely due to intensive research on breeding new and better oil yielding varieties, improvements on agrotechnology and post-harvest technology, by-product utilization, value added product development, new formulations resulting in improving their competitiveness *vis-a-vis* synthetic substitutes.

Pine oleoresin and its two main products, the rosin and turpentine oil, afford another excellent example of an industry that survived and co-exists with its petrochemical-based synthetic competitors, synthetic resins for surface coating industry and mineral turpentine as a solvent and thinner in paint industry. Through sustained research on chemical modification of natural resin and development of natural resin based derivatives like hydrogenated resins, disproportionated resins, alkyd resins and melamised resins, the natural resin retained a place of its own in the surface coating industry. Similarly, research and development on natural turpentine elevated it from the position of a cheap solvent and thinner to a valuable chemical feedstock for the manufacture of perfumery chemicals and a wide range of pesticides.

Need for Research and Development

Much research and development work is required to tap the full potential of NTFPs. Industries of essential oils, dyes, medicinal plant products established to date have been a result of R&D on naturally occurring plant species. Research has studied only a fraction of the flora. Research needs range from development of superior propagation materials, agrotechnology, to new products and marketing of finished products. Moreover once research leads are found, it also takes time and more development work before being accepted for use as drugs or as additives in other consumer products.

Because industrial research in developed countries focuses on synthetic substitutes as soon as a new useful natural product is discovered, it is vital for developing countries to safeguard the property rights of the original resource with international conventions, while at the same time developing higher-yielding and disease-resistant varieties through genetic improvement. The suggestions that can be put forward for developing R&D in this field are :

- R&D on process and product development. Appropriate technology adaptable at forest or rural locations.
- Development of downstream processing to produce value added products.
- Strengthening of R&D institutes including provision of pilot plant processing facilities for testing viability and for training perconnel.
- Setting up facilities for design and fabrication of process equipment for appropriate technology as well as pilot plants for scale up operations.
- Develop indigenous scientific and technological capabilities in industrial utilization of NTFPs by training overseas.

Employment Generation and Social Benefits

In India, unemployment has always been a concern for planners and policy makers. The forestry sector, with 23 per cent of the country's geographical area, provides 2.3 million person-years of employment. Of this total, 1.6 million person-years is related to NTFPs. Employment generated by various NTFPs is presented in Table 18.

It is estimated that NTFPs are capable of generating 4 million person-years of employment annually, if their full potential were exploited. The government is committed to increasing employment opportunities, so NTFPs should be one of the first items to be considered. A special national drive has been launched to intensify collection, procurement, processing, and marketing of NTFPs.

Most NTFPs currently provide employment during only part of the year because processing of NTFPs is still poorly developed. Improved labor-intensive technologies for processing NTFPs would increase the employment opportunities for longer periods of the year and ensure higher prices for the produce.

The Constitution of India enjoins the state to promote the educational and economic interests of the scheduled castes and scheduled tribes, and to protect them from social injustice and exploitation. It also requires the protection of the forests and wildlife of the country. The National Forest Policy, 1988, also stresses the conservation of the country's natural heritage and the efficient utilisation of all forest produce. The policy suggests that NTFPs, which provide employment to the population residing in and around forests, should be protected, improved, and managed for increased production. It emphasizes the need for research in the conservation and management of forest resources and the application of advanced scientific and technological measures.

Today, all the states of the country have forest corporations dealing with collection, processing and marketing of forest produce. Poor forest laborers, who were previously unorganized, have been brought under the umbrella of various organizations like the Forest Labour Cooperative Societies, the Large-Area Multipurpose Societies (LAMPS), the Tribal Development Corporations (TDC), the Minor Forest Produce Federations (MFPF), and other organizations.

Table 21 : Production and employment generation from non-wood forest products in India

Product	Collection Period	Production (thousand tons)		Employment (thousand person-years)	
		Current	Potential	Current	Potential
Fibers	March-May	2.5	45	4.4	79
Kapok flosses	May-June	3	4.5	10	15
Grasses	Oct-March	350	535	1,200	1,800
Bamboo	Continuous	1,932	4,309	48.3	110
Canes	Continuous	14	21	0.7	1.05
Lemon grass oil	May-June	1.3	1.95	21.7	32, 55
Palmaros oil	Oct-November	0.09	0.135	1.5	2.25
Eucalyptus oil	Continuous	0.14	0.21	2.32	3.48
Cinnamon oil	Continuous	0.003	0.004	0.05	0.07
Sandal wood oil	Continuous	0.15	0.225	1.5	2.25
Deodar oil	Continuous	0.015	0.023	0.15	0.23
Pine oil	Continuous	NA	100	NA	100
Mahua seed	April-June (Northern) Oct-Nov. (Southern)	85	490	28.6	163
Neem seed	May-June	6	418	1	70
Karanj seed	June-Oct.	56	111	19	37
Kusum seed	June-July	30	90	6.7	30
Sal seed	April-June	240	5,504	53	1,123
Kokum seed	May-June	0.5	2	0.167	0.7
Khakan seed	May-June	NA	46.3	0.570	15.3
Nahor seed	May-June	1.7	5.7	NA	1.9
Undi seed	April June & Sept-November	NA	3.8	NA	NA
Babul bark	Continuous	27.4	50	4.57	8.3
Avaram bark	Continuous	30	45	5	7.5
Wattle bark	Continuous	30	45	5	7.5
Myrobalans	Jan-March	100	150	6.6	9.9
Karaya gum	April-June	15	22.5	50	75
Ghatti & babul gums	April-June	2	3	7	10.5

Table 21 – Contd...

Product	Collection Period	Production (thousand tons)		Employment (thousand person-years)	
		Current	Potential	Current	Potential
Resins	March-June	74.2	150	30	60.2
Lac & lac products	Oct-Jan & April-July	22	33	7.3	10.95
Tassar silk	Aug-Dec	0.3	1.9	1.5	9.5
Tendu leaves	April-June	210	300	74.9	107
Sarpagandhaa	Variable	0.6	1	16	42.67
Kuth	October	0.6	1	16	26.67
Cinchona	Variable	1.42	2	23.635	33.335
Edible products	Variable	NA	NA	NA	NA
	Total	3,325.8	12,492.2	1,647.2	3,995.8

CHAPTER 5

NTFP–OIL SEEDS

In the past decade considerable advances have been made in the collection of oil seeds from forest trees and the utilisation of oil in soap industry which consumes about 80 per cent of the current production of the oil. The estimated gross potential of minor oils of forest origin is 4 million tonnes, of which the potential of *Sal* fat alone is around 988,000 tonnes. The production of minor oils in 2004-05 was around 417,000 tonnes.

The socio-economic benefits of the utilisation of forest oil seeds, although well known, need reiteration to emphasize the scope and potential of this vast natural renewable resources.

1. The greatest benefit is in the expansion of total oil supply. The import of nearly 1 million tonnes underlines the gap and tapping the full potential would go a long way in achieving self-sufficiency.
2. The collection of oil seeds is an economic activity, which while adding to national wealth virtually without any investment, generate large scale employment in rural areas. The National Commission on Agriculture has estimated that employment in collection of 233.000 tonnes of minor oil seeds in 1970-71, was 17 million man days. During 1977 *Sal* seed collection alone was estimated around 15,000 tonnes, generating employment of about 6 million man days, on the basis of 25 kg kernel collection per person per day.
3. Export of valuable product to earn foreign exchange, *e.g.*, processed *Sal* fat, kokum, mango-kernal fat.
4. Seed collection is a deterrent for indiscriminate felling of trees thereby ensuring a healthy attitude to forest preservation and development.

5. Considerable proportion of meal, remaining after extraction of fat provides additional feed material for cattle and poultry.

The main constraints in extending the usage of minor oils for industrial and edible purposes are that these oil are generally dark and not amenable for conventional bleaching. They also contain small proportions of non-lipid components of diverse chemical composition. In addition, some of these possess strong odour. A secondary problem is the extensive damage to the seeds during collection and storage.

Kusum Oil ***(Schleichere triguja)*** : The presence of cyanogenic material in this oil has been known for a long time, but the precise structure was established only recently. The oil contains 1.15 per cent nitrogen and yields 1.2 per cent CN^-, expressed as HCN, on saponification. The oil contains only a third as much glycerol as other vegetable oils. Therefore, only a third of fatty acids are combined with glycerol and present as normal glycerides. The other two third of fatty acids appears to be combined with non-glycerol alcohols, containing nitrogen function and termed as cyanolipids. This is shown by TLC, where the oil separates into glycerides (Rf. 0.55) and cyanolipids (Rf. 0.66). Attempts to isolate the alcohol part of cyanolipid by saponification led to intractable product. Acid hydrolysis of the oil, however, gave β-methyl, α-tetronic acid, the hydrolysed and rearranged product of cyano alcohol in 4.6 per cent yield.

The occurrence of cyanolipid in other seeds was reported by Mikolajezak and co-workers of Northern Regional Research Laboratory, Peoria. They assigned structure (I) for the cyanolipid from *Cordia verbenacea* on the basis of spectroscopic and chemical data which included the formation of β-methyl α-tetronic acid acetate by acetic acid-sulphuric acid hydrolysis. Based on our earlier report on the formation of β-methyl α-tetronic acid, they have suggested that cyanolipid from kusum oil may also be related to the cyanolipid from C. *verbenacea.*

We have isolated the major cyanolipid of kusum oil by molecular distillation and preparative thin layer chromatography and confirmed its structure as I, differing only in fatty acids, on the basis of NMR and mass spectral data.

In addition to the major cyanolipid, the presence of a small proportion of a more polar cyanolipid was reported from this laboratory by Vatakencherry. Its spectral characteristics and structure were also reported. Its identify with the cyanolipid (II) isolated from *Koelreuteria paniculata* has been reported by Mikolaczak.

The composition of kusum oil can now be represented as shown below :

38% glyceride
58% cyanolipid
5% isomeric cyanolipid

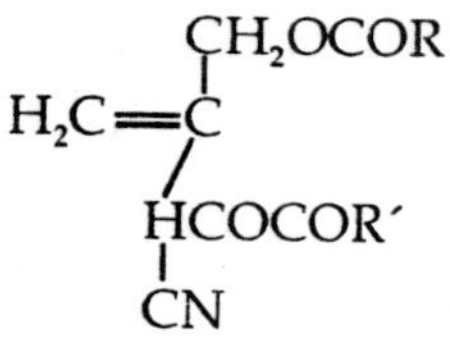

(I)

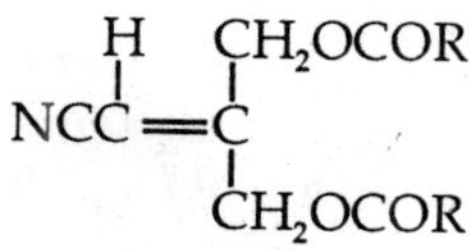

(II)

In order to eliminate the formation of $NaCH_1$ on saponification of kusum oil, we have developed a method based on the selective hydrolysis of the cyanolipids. By controlled hydrolysis, it is possible to hydrolyse the non-glycerol esters leaving the glyceride unaffected. Thus, the oil, when treated with small proportions of zinc oxide, sodium hydrosulphite and water (0.8 parts) at 100-150°C for 3 hours, yields a fatty layer consisting a fatty acids derived by hydrolysis of cyanolipids and clyceride which has survived by hydrolysis. During the reaction the cyano alcohol is hydrolysed to amide and no HCN is produced. The fatty matter after treatment with sulphuric acid to hydrolyse the small amount of zinc soap, is useful for soap making. This process is in commercial operation at Hindustan Lever Limited factory at Calcutta.

Karanja Oil *(Pongamia globra)* : Prof. Seshadri and co-workers have examined the minor constituents of the oil and isolated four furano-flavones, of which karanjin is the principal constituent. They have also isolated pongachromene and two flavanols, demethoxykanugin and kanugin from the roots and stem bark.

Karanja oil gives red colour with alcoholic alkali. However, none of the compounds isolated and characterised from the oil, could give rise to this characteristic colour reaction. Therefore, we re-examined the oil and isolated, by molecular distillation followed by counter-current extraction, a flavanone, isolonchocarpin, besides karanjin, pongamol and demethoxykanugin, the last of which also occurs in the stem bark. Isolonchocarpin gives a red colour with alkali and its presence at 0.01 per cent concentration in the oil is responsible for the characteristic colour reaction of the oil. The same compound is perhaps partly responsible for the darkening of the colour of soap made from karanja oil.

The oil is unique in containing large number of f rano flavones. Furans are believed to be derived from a common C-5 precursor unit, whose familiar variant in nature is a dimethylchromeno group. It is not, therefore, surprising to find the co-occurrence of pongapin, a furan variant and pongachromene, a chromeno analog in the same species. We have now isolated the chromeno analogue of the principal flavone karanjin, by counter-current extraction techniques. We have proposed the name "karanja chromene" for this analogue.

The use of karanja oil for industrial or edible purposes is limited due to the presence of non-lipid minor constituents. Solvent extraction, as a process for improving the oil, would become attractive if some value could be realised for the recovered minor components. There is a scope in this direction as the oil is considered as medicine for skin ailments and expelled cake and leaves of the plant are reported to be useful in pest control.

Neem : The theoretical potential availability of neem oil is around 83,000 tonnes, next only to Sal and Mowrah (171,000 tonnes). The oil is semi hard and used in small proportions in soaps. The main constraints in large scale utilization is the presence of non-lipid components and sulphur compounds, giving characteristic odour. The non-lipid minor constituents, loosely termed as bitters, are present to the extent of 2 per cent. These belong to tetranortriterpenoid family. The sulphur compounds appear to be a mixture and are unstable. The isolation of strongly odoriferous Tiglic acid from neem oil has been reported from this laboratory by Raman.

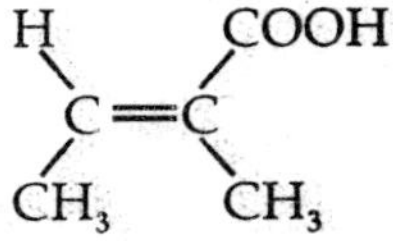

Further work is on hand to decide whether the acid is part of a glyceride or non-glyceride component.

Alcohol extraction removes non-lipid components yielding good quality oil but the economics depend on the value realised for the non-lipid components, which are reported to have potential use in theraputics and as insecticides.

CHAPTER 6

NTFP : MINOR OIL SEEDS

India is primarily an agricultural country and a major producer of oilseeds in the world. But the production of oilseeds has remained static and the consumption has increased considerably. The demand is made good by the import of edible oils from Malaysia, Russia, USA and Canada. In 1977, we have imported nearly 900,000 tonnes of vegetables oils and this trend will continue. India will continue to be a prominent purchaser of vegetable oils in the International market in the years to come. However, the country has vast potential resources of minor oilseeds of trees origin which is properly exploited can wipe out the deficiencies to a certain extent.

The main problems associated with the exploitation of the seeds are :

1. The trees grow in scattered way in many interior forest areas, connected mostly by fair weather roads.
2. The fruit bearing season for most of the trees is short; and organized and planned collection of the seeds is needed.
3. The fruiting season coincides with the monsoon, and unless collection and storage is down appropriately the fat will undergo deterioration.
4. The presence of characteristic chemical constituents and odor, makes the utilization difficult and requires special refining technology, which is available at present.

There are more than 25 varieties of oilseeds available from forest sources, but only some of these are currently exploited for various reasons. The major utilization is in the manufacture of soaps. The use of minor seed oils dates back to mid fifties of the current century by the Khadi and Village Industries Commission who developed their own processes for utilizing these oils in soap. On the other hand, the large scale sector of the soap industry undertook extensive Research and Development work over the years to up grade non-traditional minor oils to make them suitable for soap making. Of the various oils such as Mahuva, Karanja, Neem, Sal, Undi, Khakhan, Tamarind, Dhupa, Kusum and Kokum, only four of them are available in commercial quantities and these are even quoted in the list of oils in the commodities market. The four oils referred above are Mahuva, Karanja, Neem and Sal.

Today the total requirement of vegetable oil for the soap industry is around 370,000 tonnes per annum of which the small scale sector requires 200,000 tonnes per annum, while the large scale sector requires 170,000 tonnes. The bulk of this oil requirement is met through the increased use of minor oils, rice bran oil and hardened castor oil. The problems and prospects of use of some of the minor oilseeds is discussed here.

The minor oil seed and the oil potential is tabulated in Table 1 :

Most of the minor oils cannot be used as such in soap manufacture due to the strong odour, dark color and high content of unsaponifiable matter. Deodorization of these oils makes them ideal from the point of view of odorless soap base. This helps to use lower quantity of perfume in the finished toilet soap, as also stability of the perfume on storage is enhanced. Some of these oils are refined and hydrogenated apart from deodorization, before using in the manufacture of soap. The best examples of these are that of hydrogenated Karanja oil, deodorized neem oil and distilled fatty acids of neem oil. Some of the characteristics of these products are given in Table 2.

Bleaching of these oil improve the color, making the soap base of acceptable quality. The conventional methods of bleaching may not work in all these oils and one has to use either solvent refining or chemical bleaching or other methods.

The minor oils when used by the small scale sector to make the soap by the cold process or by the semi boiled process, leave the glycerine in the soap itself, while in the large scale sector the glycerine could be recovered. The only way to save the valuable glycerine is to split the minor oils into fatty acids and recover the glycerine. Many large scale sectors have facilities for the splitting and distillation of the fatty acids, which enables them to up grade the minor oils and use them in larger quantities in soap. The distilled fatty acids give light colored and odorless fatty acids with lower content of unsaponifiable matter. Some of these oils are even hydrogenated to the required melting point and then split to obtain fatty acids. Karanja oil is an example of this class. Alternatively, one can hydrogenate fatty acids but this requires special s.s. equipment. In fact, distilled fatty acids of Mahuva oil, Karanja Oil, Neem oil and Hydrogenated Karanja oil are commercially available. These fatty acids produce good quality soap which could be used even in toilet soaps. However, there is a problem faced by the small scale sector of soap manufacturers, who are located in far off places, to transport these economically. The quality of the soaps in the small scale sector could be very much improved by the use of these distilled fatty acids. Good quality toilet soaps are made purely from distilled neem, fatty acids, while laundry soaps are made with Karanja fatty acids.

Hydrogenation of fatty acids at high pressures in s.s. autoclaves enables greater utilization of minor oils for the manufacture of commercial stearic acid for use in Rubber, Grease and Metallic stearates. Thus, minor oils which have limited use in their natural state could be upgraded by the fatty acid route and the utilization could be increased.

In the recent past, some of the minor oils have found a place in industrial food products such as chocolates. In this category sal fat is becoming popular as one can suitably process this fat and use in a mixture with cocoa butter as a good extender for cocoa butter in chocolates. Similarly Mahuva oil, Dhupa fat and Kokum butter fall in this category. Application of these oils as an extender for cocoa butter in chocolates will give an impetus to these oils for export as there is a great demand for these products overseas since cocoa butter has become very expensive of late.

Minor oils could be used in various other industrial products like fatty alcohols, fatty amines etc.; after suitable processing. Several laboratories in India are working on different applications for these.

Utilization of the meal is another important aspect of the utilization of minor oilseeds. The sal meal is considered to be toxic to the animals due to the presence of high level of tannins. However, for poultry feed 5 per cent of the defatted sal meal could be used. Recently industrial research organizations have established that the tannins could be reduced considerably, and the meal could be fed to animals without any toxic affect.

Mahuva meal contains saponnins and is used in sugarcane plantations and grass lawns as manure. Attempts have been made to remove the saponnins which will improve the utility of the meal.

Neem meal contains certain sulphur compounds which give insecticidal properties when used as manure.

The Kusum meal is likely to contain the toxic ingredient, namely cyanogenic glucoside and one has to dispose this with caution.

Karanja meal also has a bitter odor and could be used only as a manure. Thus, the meal of minor oil seeds finds application more as a manure than as a feed to animals.

The utilization of some of the important minor oilseeds has been discussed briefly. It is not possible to go into the details in such a short time. However, there is lot of published literature on this subject. Some of the additional information on these minor oils is presented in Appendix.

APPENDIX

Mahuva : (*Madhuca latifolia or M. longifolia*) : Mahuva trees are found in U.P., West Bengal, Orissa, Tamil Nadu, Madhya Pradesh and Maharashtra. The kernal contains 50 per cent of a pale yellow semi solid fat. The fat is obtained from the seeds by screw press, hydraulic press and ghani. The fat gives a hard soap and major quantity goes for soap making. It can be used for edible purposes although the unsaponifiable matter is little high (1–3%). By certain treatment such as solvent crystallization, partial hydrogenation or interesterification one can make a cocoa butter substitute. The meal is used as a manure.

Sal (*Shorea rubusta*) : Sal Forests are concentrated in Northern and Central India in apparently 2 zones–one sub Himalayan region more or less in the north of the Gangetic plain and the other on the South of Ganga. The kernel yields 17-18% of fat and the best method to extract is by the use of solvent after conditioning and flaking the seeds in a particular manner. The fat is dark green in colour and has a characteristics odor. It contains pigments of the chlorophyll type which becomes white on exposure to sunlight. Chemical methods are preferred than the usual physical method for bleaching the fat. Pretreatment with acids such as sulphuric, nitric or phosphoric in conjunction with sodium chlorite in aqueous solution gives a light coloured fat.

Sal fat could be solvent fractionated for use in chocolates as cocoa butter substitute. Equal parts of sal fat and cocoa butter in chocolates serve as a good extender. Its use in soap making is valuable as it has a high titre and low iodine value.

Defatted sal meal contains 10-12% proteins, 50-60% starch and 10-15% tannins in addition to other regular constituents. The presence of tannins renders the crude protein unutilizable and also affects the digestibility of the other materials.

Kusum (*Schleichera trijuga*) : Kusum trees grow from Himalayan Region (south) throughout India. The seed contains 60-65% kernel, and 59-72% fat is obtained commercially in mills and presses. The oil brown in colour contains about 1% of nitrogen, approx. half of this nitrogen appears as sodium cyanide on saponification. Hydrogenation of kusum oil under conventional conditions lead to dark oil with high free fatty acids. There is no drop in iodine value. However, reduction of the oil with Raney Nickel catalyst gives a white solid of m.pt. 56°C and iodine value 4.2.

Kusum oil is unique in that it contains 20-25% of arachidic acid and 8-15% of other saturated fatty acids. It is used in soap making as it gives a hard and granular soap with fair lathering power. However, one has to be careful in making the finished soap free of sodium cyanide. One can use 2-15% of this oil in soap with advantage. The soap from Kusum oil does not become rancid and thus could be used in toilet soap (2-5%) giving good lather.

Khakhan (*Salvadora olecides*) : Khakhan, also known as pilu, trees are found in the dry regions of India, such as Gujarat, Saurashtra, Kutch, Rajasthan, Punjab and Western parts of M.P. The kernel contains 40-42% fat and the light green colour could be easily bleached even on exposure to sunlight. The khakhan fat contains lauric acid, myristic acid and palmitic acid in large quantities which makes it useful as a substitute for coconut oil in soap making. But the availability of this oil is not much. The soap has good lathering properties.

Tamarind (*Tamarindus indica*) : Tamarind trees are largely found in South India. Although some trees are found in North India, but the climatic conditions are unfavourable for the growth. The seeds are removed from the pulp of the fruits. The seed consists of a cream coloured kernel which on extraction with a solvent gives tamarind seed oil. The kernel power is used in the sizing of cotton and jute. The Forest Research Institute at Dehra Dun was the first to invent its use as a sizing material for textile. Tamarind seed oil is pale brown in colour and has characteristics like groundnut oils, and could be used after partial hydrogenation in soap. Of course, the extracted meal could be used as a starch substitute in many applications.

Undi (*Calophyllum inophyllum*) : Undi trees grow abundantly in the State of Orissa, West Bengal, A.P. and West Coast. The yield of the oil from the kernels is about 60% and is green color. The oil on the extraction with alcohol (95%) give a pale yellow oil which could be used for soap making. The refined oil could be bleached and hydrogenated to obtain a light colored fat.

Undi oil is found to have antibacterial activity against germ positive organisms and the soap also has antibacterial properties.

Karanda Oil (*Pongamia glabra*) : Karanja trees is considered to be a native of Western ghats and is found near rivers and sea coast beach and forests. The seed contains 28-38% of an oil containing bitter principles. The oil is used for leather dressing soap making lubrication, illumination and for medicinal purposes. Recent findings indicate the oil is antibacterial. The color and odor of the oil is not easily removed by the conventional methods of purification. De Sa and Kane has suggested alcoholic alkali refining and bleaching. This oil could be hydrogenated and the hardened product is available in commercial quantities for soap manufacture.

The seed cake is unsuitable for feeding livestocks due to bitter taste. Hence only use is as a manure for sugarcane and coffee.

Pisa (*Actinodaphne bookeri*) : The trees are found growing at high altitudes in the Western Ghats in the forests of Nasik, Satara, Dharwar and Karwar. The fruits are available in May and collection time is very short due to monsoon. The seeds have kernel rich in fat as also an outer covering containing oil. The expressed fat is a solid reddish yellow fat with m.p. 42-53°C; iodine value 10 and saponification value 256. Among the indigenous minor oil, Pisa fat is a good substitute for coconut oil in soap making because of the presence of 94% trilaurin. But its availability is meagre and is not used in commerce.

Neem (*Azadirachta indica*) : Neem trees are often found outside forests, planted as avenue trees on the roadsides or growing wild in open waste lands. The seed contains about 20-25% oil (after drying) and was fetching a low price due to the unpleasant odor and dark color. Today the situation has changed as this oil could be refined by alcoholic alkali or solvent refining, and the color could be bleached by chemical methods. Even one can hydrogenate the oil to obtain a hard fat. The deodorized oil could be directly used in soap making and the distilled fatty acids form a good base of toilet soap.

Neem seed meal contains bitter and foul smelling constituents and cannot be used as cattle feed. It is consumed as manure for sugarcane for its supposed insecticidal properties.

Table 1 : Estimated Total Potential of Minor Oilseeds of Tree Origin

Commodity	*Seed Potential (tonnes)*	*Oil Potential (tonnes)*
Mahua	4,90,000	1,71,000
Neem	4,18,000	83,600
Karanja	1,11,000	30,000
Kusum	90,000	30,000
Sal	55,04,000	6,88,000
Khakan	46,300	15,400
Kokum	2,000	800
Nahor	5,700	2,300
Undi	3,800	2,300
	66,70,800	10,23,400

Kokum (*Garcinia indica*) : Kokum trees are found on the coastal belt of Western Ghats, Assam and Meghalaya. The kernel contains 40% white hard and brittle fat. The quantity of the seeds available in the country is meagre and efforts are being made to raise kokum plantations in forest areas. The fat can be obtained by screw press, hydraulic press and later the cake extracted with solvent. It is harder than cocoa butter because of high melting point. It is an edible fat and people use it as

Table 2 : Characteristics of Minor Oils

Tests	*Mahua*	*Karanja*	*Neem*	*Undi*	*Kusum*	*Khakhan*	*Tamarind*	*Dhupa*	*Kokum*	*Sal*
A.V.	5-6	12.2	11.20	30	–	0.71	12.2	2-30	3–5	3–30
S.V.	187–196	185.7	195.6	191–192	187–192	2.33	185.7	187–192	187–192	187–191
I.V.	55–63	99.9	69.2	90	25–27	3.6	94.5	36–43	25–37	38–40
% Unsap.	1–3	3.75	1.80	–	–	0.5	3.75	0.6–2.5	–	0.9–2.2
m.pt. °C.	23–31	–	–	–	39–43	42	–	30–40	39–43	36.5
Titer °C.	38–40	32	39	–	60	–	–	52–55	60	52

substitute for ghee. It could be modified and blended with cocoa butter in chocolate manufacture. It is a good raw material for soap making.

Dhupa (*Veteria Indica*) : Dhupa, commonly known as Malabar fallow is obtained from the kernels of *Veteria indica,* an ever green tree found abundantly in the regions of Karnataka state, Malabar and Travancore. The kernel contains 22-27% fat which is pale yellow in colour, and sometimes dark brown if extracted from damaged seeds. The fat has slight taste and slight odor and would be used as a substitute for cocoa butter. It is used for edible purpose as a cooking medium as also in soap making and candle manufacture. Its use in sizing as substitute for tallow is well known.

CHAPTER 7

NTFP : OIL SEEDS OF RAJASTHAN

Rajasthan has a striking geological feature in the presence of Aravali range of hills which divides Rajasthan into two distinct zones, eastern and western. There appears to be no complete published account of the flora of Rajasthan. The herbaceous flora of the hills and plains is rich during rains and is dominated by species of *Leucas, Indigofera, Euphorabia, Ocimum* and others. In an attempt to screen seeds oils of hitherto uninvestigated species, our plant collection unit toured the eastern sector of Rajasthan and collected samples of seeds from additional species. This survey of herbaceous seed oils was carried out under a PL-480 research scheme of ICAR, New Delhi.

In this study we have selected seed oils of 15 species belonging to seven botanical families. The seeds were collected from different regions of Rajasthan and some of them are very interesting in having unusual fatty acids.

The seed oils were analysed for their oil properties (Table 1) and fatty acid composition (Table 2) using different chromatographic (TLC-Ag^+/TLC, RP/TLC; GLC) and spectrocopic (UV, IR, NMR) techniques. Six seed oils of Leguminosae and one from Tiliaceae family were found to be usual in having very low oil content ranging from 1.6 to 3.5%. Fatty acid compositions of these species are given in Table 1. Out of these, only two species *Indigofera linifolia* and *Cyanopsis psoraloides* are interesting in having a good percentages of nitrogen *i.e.,* 36 and 37%, respectively. Two seed oils of Convolvulaceae family are interesting in having a very high content of unsaturated acids (81.6 and 79.8%). *Sparmacoce hispida* seed oil contains 45% of linoleic acid and has a combined content of 18 : 1, 18 : 2, 18 : 3, unsaturated acids upto 82%. *Celastrus paniculata* seed oil of Celastraceae family seems to be very promising as it contains a very high percentage of oil (56%) and nitrogen 37%. It has a combined content of 10 : 1, 18 : 2, 18 : 3 unsaturated acid up to

Table 1 : Analytical Data on Oils and Oilseeds (Rajasthan State)

S.No.	Source	Seed analysis					Oil Properties		
		Oil content %	Protein content Nx6.25 %	Moisture %	Infrared	Ultraviolet	I.V. [1]	S.V.[11]	Ref. Index 40 ND.
Leguminosae									
1.	*Indigofera linifolia*	1.6	36.2	7.0	Usual	Normal	119	176.0	1.4796
2.	*Indigofera trita*	2.7	24.0	7.6	Usual	Normal	141	238.5	1.4780
3.	*Abrus precatorius*	2.1	25.0	12.8	Usual	Normal	127	125.7	1.4705
4.	*Bauhinial malabarica*	4.0	19.0	8.5	Usual	Normal	120	202.0`	1.4713
5.	*Cyanopsis psoralioides*	3.0	37.5	8.6	Usual	Normal	105	186.0	1.4775
6.	*Cassia saphera*	3.5	23.4	7.1	Usual	Normal	60	169.0	1.4712
Convolvulaceae									
7.	*Ipomea pilosa*	11.0	26.8	10.0	Usual	Normal	132	185.0	1.4689
8.	*Merrima pentaphylla*	9.6	19.6	10.0	Usual	Normal	135	122.0	1.4750
Rubiaceae									
9.	*Spermacoce hispida*	5.7	15.0	11.2	Usual	Normal	134	196.0	1.4590
Celastraceae									
10.	*Celastrus paniculata*	56.0	37.8	8.0	Usual	Normal	129	223.0	1.4708
Tiliaceae									
11.	*Grewia flavscens*	1.5	9.0	11.0	Usual	Normal	124.2	180.2	1.4800

(Contd...)

Table 1 – Contd...

S.No.	Source	Seed analysis					Oil Properties		
		Oil content %	Protein content Nx6.25 %	Moisture %	Infrared	Ultraviolet	I.V. [1]	S.V.[11]	Ref. Index 40 ND.
Compositae									
12.	*Sclerocarpus africana*	17.5	30.0	6.0	Usual	Normal	110.5	182.3	1.4790
Hippocostanaceae									
13.	*Aesculus indica*	4.2	15.0	8.3	Cyanolipid	Normal	85.0	192.0	1.4810
Malvaceae									
14.	*Abutilon indicum*	10.0	21.3	5.3	CPFA+Epoxy	Normal	108.0	153.5	1.4764
Acanthaceae									
15.	*Blepharis edulis*	8.2	8.1	7.2	Hydroxyl	Conj.	70.0	162.0	1.4790
Labiatae									
16.	*Leucas urticifolia*	24.3	10.4	7.5	Allene	–	62.0	176.0	1.4770

* Iodine value ** Saponification value

Table 2 : GLC Analysis of Methyl Esters

S.No.	Source	Fatty acid composition (%)					Others
		16:0	*18:0*	*18:1*	*18:2*	*18:3*	
Leguminosae							
1.	*Indigofera linifolia*	22.7	3.4	25.9	37.6	10.3	–
2.	*Indigofera trita*	18.1	4.7	10.1	55.3	11.7	–
3.	*Abrus precatorius*	9.8	2.2	54.4	7.5	23.3	20:0 (2.8)
4.	*Bauhinia malabarica*	19.6	3.4	19.9	57.1	–	–
5.	*Cyanopsis psoralioides*	14.3	7.9	33.9	8.3	19.9	14:0 (15.7)
6.	*Cassia saphera*	47.1	tr	6.2	19.1	6.9	12:0 (2.0); 24:0 (4.1); 22:0 (8.3); 24:0 (6.2)
Convolvulaceae							
7.	*Ipomea pilosa*	18.2	0.2	24.7	50.7	6.2	–
8.	*Merrima pentaphylla*	18.3	1.9	29.8	34.8	15.2	–
Rubiaceae							
9.	*Spermacoce hispida*	15.5	2.5	28.0	45.2	8.8	–
Celastraceae							
10.	*Celastrus paniculata*	24.1	5.6	16.6	18.3	29.9	10:0 (0.9); 12:0 (2.2) 14:0 (2.4).
Tiliaceae							
11.	*Grewia flavscens*	12.4	7.6	11.2	68.8	–	–

(Contd...)

Table 2 – Contd...

S.No.	Source	Fatty acid composition (%)					Others
		16:0	18:0	18:1	18:2	18:3	
Compositae							
12.	*Sclerocarpus africana*	18.0	12.9	15.4	53.7	–	–
Hippocostanacea							
13.	*Aesculus indica*	TG 13.9	2.3	41.9	23.3	2.6	20:0 (7.0); 20:1 (7.8); 22:0 (1.2).
		NCLF 6.4	1.4	28.6	16.6	tr	20:0 (18.0)' 20:1 (24.0); 22:0 (5.0).
Malvaceae							
14.	*Abutilon indicum*	26.8	4.9	12.8	50.7	–	Malvalic (2.3); Sterculic (0.9); 12, 13-Epoxycleic (1.6).
Acanthaceae							
15.	*Blepharis edulis*	15.6	1.7	65.7	5.9	–	10:0 (1.3); 12.0 (0.9); 14:0 (1.6): Hydroxy acid (7.1).
Labiatae							
16.	*Leucas urticifolia*	14.7	7.0	39.1	*31.6	7.6	*18:2 + Allene

64.8% and the fatty acid composition resembles somewhat with oils presently used for edible purposes. The presence of 5% of lower acids of C_{10} to C_{14} makes it more interesting.

Three species were found to contain some unusual constituents. *Abutilon indicum* seed oil contains HBr–reacting, epoxy and cyclopropene, acids. Their details are given in an other study. *Aesculus indica* (Indian horse chestnut) seed contained 4% oil.

The oil gave positive picrate and prussian blue colour test, indicating the presence of cyanogenetic lipids. The oil was resolved into a triglyceride fraction (80%) and a cyanolipid fraction (20%) by preparative TLC. Cyanolipid fraction contained 2.1% of nitrogen. Its structure was established by a comparison of its IR and TLC with those reported for a corresponding cyanolipid present in the seed oil of *Cardiospermum halicacabum* (*Sapindaceae*). It was observed that the NCLF exactly corresponded to the minor cyanolipid constituent of the standard oil. This suggested that the cyanolipid of *A. indica* is likely to be a diester of 1-cyano-2-hydroxymethylprop-1-ene-3-oil.

Presence of hydroxy acid was established in the seed oil of *Blepharis edulis* by preliminary analysis. IR showed the presence of hydroxy acid. A concentrate of the pure hydroxy acid was obtained by preparative TLC of total methyl esters. IR and NMR spectral analysis showed it to be a unsaturated hydroxy ester. Chromatographic, spectroscopic and GLC retention time of the hydroxy ester and its silyl derivative showed it to be a hydroxy *cis*-12-octadecenoic acid. GLC analyses of silylated methyl ester on DEGS column showed the presence of 7% hydroxy acid.

Leucas urticifolia seed oil of Labiatae family was found to contain an allenic acid. The oil and methyl ester showed a band at 1935 cm^{-1} in its IR spectrum. The percentage of allenic acid was calculated by using preparative argentation TLC separation. It was found that *L. urticifolia* seed oil contains 24% of this acid. Further work to establish its structure is in progress.

These findings are preliminary in nature but will serve as a base for further investigation, ultimately leading to better utilization of the lesser known species.

CHAPTER 8

ABUTILON INDICUM SEED OIL

During our survey of oil-bearing seeds of Uttar Pradesh and Rajasthan, a number of herbaceous species of the genus *Sida* and *Abutilon* belonging to Malvaceae family were found to be of common occurrence in the forest flora. This paper reports the presence of HBr-reactive acids in the glycerides of seed oil of the genus *Abutilon*. The previous work on the seeds of *A. indicum* has not shown the presence of HBr-reactive acids in its oil. In continuation of work on HBr-reactive acids-containing seed oils, an attempt has been made to isolate and characterise these acids in the seed oil of *A. indicum*, a widely grown wild herb. Durbetaki titration, Halphen test, picric acid TLC and IR spectrum clearly showed the occurrence of cyclopropenoid and epoxy fatty acids in this seed oil.

The analytical values of oil and seeds were determined according to the procedure recommended by the AOCS methods and the data are summarised in Table 1.

The characterisation of unusual acids was achieved by first separating the oxygenated acid from the non-oxygenated ones and GII analysis of the methyl esters using the ester of *Sterculia foetide* and *Vernonia anthelmintica* as the reference standards for cyclo-propenoid and epoxy acids, respectively. Quantitative results were preparative TLC separation and GLC analysis.

The oil was acetolysed, saponified, and the liberated acids chromatographed on the layers of silica to isolate the resulting dihydroxy acid. This procedure effectively separates non-oxygenated acids from oxygenated acids. The non-oxygenated fraction contained acids with cyclopropene moiety (1010 cm^{-1}) and the oxygenated fraction showed hydroxyl absorption at 3460 cm^{-1} with no evidence for *trans* unsaturation (965 cm^{-1}) in IR spectrum.

The non-oxygenated acids were esterified and the resulting methyl esters examined for the component acids by a combination of TLC-GLC techniques. The GLC analysis of the silver nitrate-methanol treated methyl esters demonstrated the presence of malvalic and sterculic acids in addition to normal fatty acids by a comparison of relative retention times of similar derivatives of *S. foetida* esters. Qualitative results of direct, reversed phase, and silver-ion TLC supported the findings of GLC analysis.

On hydrogenation the dihydroxy acid absorbed one mole of hydrogen. The original dihydroxy acid when oxidatively cleaved by the von Rudloff procedure gave hexanoic and azelaic acids; the saturated dihydroxy acid have hexanoic and dodecanedioic acids. The cleavage fragments placed the double bond at C_9 – C_{10} position and the hydroxyl at C_{12} and C_{13} positions. Further, the melting points, IR spectra, and the TLC retention characteristics of saturated and unsaturated dihydroxy acids on silica and silica-impregnated with boric acid resemble with those of the saturated and unsaturated dihydroxy acids derived from *V. anthelmintica* seed oil. The acid is, therefore, *threo*-12, 13-dihydroxyoleic acid and the original component of the oil from which dihydroxy acid was obtained is *cis*-12, 13-epoxyoleic (vernolic) acid. The conclusions are depicted in Table 2.

Table 1 : Analytical Data on *Abutilon indicum* Seeds and Oil

Oil content, %	10.0
Unsaponifiable content, %	2.0
Protein content, N × 6.25%	21.3
Moisture, %	5.3
Iodine value (Wujs)	108.0
Saponification value	153.5
Refractive index, n_D^{40}	1.4764
Halphen test	Positive
Picric acid TLC	Positive
Oxirane oxygen, %	0.08
HBr, equiv, 3°C (% epoxyoleic)	1.56
HBr, equiv, 55°C (% cyclopropenoid)	3.42

The cyclopropenoid fatty acids have been the subject of much investigation due to their profound biological effects on animals and cocarcinogenic properties. The co-occurrence of cyclopropenoid and epoxy acids in seed oils is of chemotaxonomic interest and may throw light on the fatty acid biosynthesis.

Table 2 : Component Fatty Acids (% wt.) of Seed Oil

Palmitic	26.8
Stearic	4.9
Oleic	12.8
Linoleic	50.7
Malvalic	2.3
Sterculic	0.9
Vernolic	1.6

CHAPTER 9

FATTY ACIDS OF SEED OILS

(Sida Grewioides and Hibiscus caesius)

In an effort to survey the forest oil seed potential of our country a modest beginning has been made in carrying out a systematic botanico-chemical screening of the seeds of herbaceous plants. During the last decade the application of modern methods has shown that naturally occurring cyclic fatty acids like the cyclopropenoid acids (CPFA) in the seed oils are uncommon. The CPFA that have been identified in oils are mainly malvalic and sterculic acids. The two homologous acids have often been found to occur together in seed oils. Fatty acids in naturally occurring glycerides, plant or animal are almost always found in homologous series with even numbered chains differing in length by multiple of two carbons. In contrast sterculic and malvalic acids differ in chain length by a single carbon; malvalic acid has an odd-numbered (C_{17}) chain. The cyclopropenoid fatty acid have recently been the subject of intensive investigation due to their biological importance.

Results obtained in a continuing screening programme indicated that the seed oils of *S. grewioides* and *H. caesius* are quite unusual. Both the oils responded positive to Halphen test for CPFA and so were undertaken for the estimation and characterisation of individual cyclopropene acids. The analytical data on seeds and oils (Table 1) were determined by AOCS methods.

UV and IR spectra of methyl ester showed no conjugation and *trans* double bond, respectively. The IR spectra and HBr-titration confirmed the presence of cyclopropene acids. Characterisation of the two cyclopropene acids was achieved by GLC analysis of the silver nitrate-treated methyl esters, a method applied for the quantitation of low level cyclopropenoid-containing seed oils. The fatty acid compositions of the seed oils are summarised in the Table 2. The values obtained by GLC are in fair agreement with the values obtained by HBr-titrtion (Table 3).

Table 1 : Analytical Data on Seeds and Oils

Species	Seed analysis			Oil properties		
	Oil content %	Protein content Nx 6.25%	Moisutre %	I.V.[a]	S.V.[a]	Ref. Index n_D^{40}
Sida grewioides	13.8	15.0	8.9	120.4	160	1.4745
Hibiscus caesius	15.0	23.2	3.9	122.0	182	1.4795

Table 2 : Fatty Acid Composition of Seed Oils

Component	Area % by GLC	
	Sida grawioides (non-oxygenated fraction)	*Hibiscus caesius*
16:0	18.8	19.5
18:0	4.3	4.6
18:1	19.5	11.9
18:2	53.9	57.2
18:3	–	tr.
Malvalic	2.1	5.7
Stercuilic	1.3	1.0

Table 3 : Percentages of Hydrogen Bromide-Reactive Acids

Species	GLC analysis			HBr
	Sterculic acid, %	Malvalic acid, %	Epoxy acid, %	
S. foetida	51.2	7.1	–	57.8
H. caesius	1.0	5.7	–	6.5
S. grewioides	1.3	2.1	tr	3.6

The methyl esters of each oil showed the superimposable IR band with the *S. foetide* esters. Although the total methyl esters of *S. grewioides* gave no informative IR spectrum for epoxy group, presumably due to its very low concentration, the oxygenated ester fraction isolated by preparative TLC had a weak band at 848 and 826 cm^{-1}, attributed, to epoxy group. The isolated epoxy ester on analytical TLC gave a red spot by picric acid test. The *Sida* epoxy acid, although present as a trace is under further investigation.

CHAPTER 10

PLANTAGO MAJOR SEED OIL

In our country there are several million tonnes of NTFP minor seed oils of forest origin which are not yet exploited for their use in oil seed industry. The species rich in individual fatty acids have recently been of potential practical interest for oil and fat industry. Previous work on *Plantago major* oil revealed that it contains usual fatty acids. But during the course of our screening of oilseeds, *P. major* oil was found to be a rich source of palmitic acid. A new β-hydroxy olefinic acid has been characterised as a minor constituent of the oil. *P. major* is a perennial herb grown in hilly areas and its seeds are used medicinally as a substitute for Isapaghul (*P. ovata*). Oil and seed characteristics were determined according to AOCs methods, and the results are given in Table 1.

Table 1 : Analytical Data on Seed Oil of *Plantago major*

Oil content	14%
Protein Content (Nx 6.25)	26.2%
Moisture content	3.8%
Iodine value (Wijs)	55.2
Saponification value	257.9
Refractive index at 30°	1.4765

The concentrate of the hydroxy ester was obtained by preparative TLC and subsequently purified by column chromatography. The IR spectrum of pure hydroxy showed no *trans* unsaturation (965 cm^{-1}), but exhibited a wide shall hydroxy band at 3500–3300 cm^{-1}. The NMR spectrun gave signals at 4.65 (2H, -- CH = CH–)m, α 6.4 sharp singlet (3H, $-COOC\underline{H}_3$), α 6.7, (1H, $C\underline{H}$–OH) √7.25 (1H, $C\underline{H}$–OH, disappeared on addition of D_2O), 7.8 (6H, overlapping signals ascribable to allylic protons and protons α-to the ester carbonyl), 8.7 br, (chain methylene), and

9.12t like (3H, terminal methyl). Its acetate derivative gave a strong band at 1235 cm^{-1} (IR for acetate). Disappearance of the hydroxy band confirmed the original acid as hydroxy acid. Its NMR spectrum (acetate derivative) showed no unusual features apart from the two expected but significant signals at 8.13s (3H, $OCOC\underline{H}_3$) and 75.2 (1H, –C$\underline{H}$–OAc). The signal at 77.25 was not observed).

Few reactions were carried out as given in Scheme 1 to determine the absolute identity of the hydroxy acid. On catalytic hydrogenation (Pd/6) the acid (Ib) gave a crystalline compound (IIb), as 9-hydroxy octadecanoic acid mp. 81.5°C, its methyl ester had m.p. 52–54°C. (Ib) on chromic acid oxidation yielded 9-keto-stearic acid m.p. 80–81°C. Methyl 9-hydroxystearate (IIa) was further reduced by hydrogen iodide and phosphorous followed by zinc and HCl to give stearic acid. The positions of the double bond and of the hydroxyl function in the acid (Ib) chain were further established by Von Rudloff oxidation. The acidic fragments after diazomethane treatment, were examined by gas chromatography. GLC analysis using reference compound showed the presence of methyl heptanoate and a compound which could be a C11 lactone methyl ester. All these data were consistent with the presence of the double bond at C11 and the hydroxy group at C9 in the *P. major* fatty acid.

On the basis of the above physical and chemical evidences, the hydroxy acid (Ib) has been characterised as 9-hydroxy-*cis*-11-octadecenoic acid.

The structure of the hydroxy acid (Ib) was further confirmed by the mass spectrum of its methyl ester (Ia). It gave molecular ion peak at m/e 312 ($C_{19}H_{36}O_3$) followed by other significant peaks at m/e 311(M1), 310 (M/e 311-1), 281 (M–OCH_3), 241 (M–C_5H_{11}), 240 (m/e 241-1), 201 (M–C_8H_{15}; $C_{11}H_{21}O_3$), 187 (M–C_9H_{17}; $C_{10}H_{19}O_3$), 185 (M–C_9H_{19}; $C_{10}H_7O_3$), 155 (M/e 187–CH_3OH), base peak, ($C_9H_{15}O_2$), 138, 127, 115 and lower mass peaks.

The fatty acid composition of the *P. major* oil was determined by GLC analysis of the methyl ester as their trimethyl silyl (TMS) derivatives. The GLC analysis (Table-2) showed the presence of hydroxy ester in an amount of 1.5% and high content of palmitate (40.0%), alongwith moderate amounts of stearate, oleate and linoleate.

Table 2 : Fatty Acid Composition of *P. major* Seed Oil as Silylated Methyl Ester

Component	*Area % by GLC*
Caprate	3.5
Laurate	5.0
Myristate	5.1
Palmitate	40.0
Palmitoleate	3.1
Stearate	10.8
Oleate	14.1
Linoleate	10.7
Linolenate	6.2
9-Hydroxy-*cis*-11-octadecenoate	1.5

CHAPTER 11

BORAGINACEAE SEED OILS

The planty family Boraginaceae includes some 100 genera and 2,000 species, mostly herbaceous and grown in forest areas. Recent discovery of the occurrence of cyanolipids as a constituent of seed oils of Sapindaceae led us to search for these unusual lipids in the oils of Boraginaceae species. The present report deals with the identification and characterisation of cyanogenetic constituents in two species of the genus *Heliotropium*. It has earlier been suggested that the seed oils having the little or on C_2O acids would not contain cyanolipids as non-glyceride components. However, recently two more species of the same genus *Heliotropium* analysed in our laboratory have been found to contain only C16 and C18 acids esterified with a known cyanodiol (I). Quantitative tests, elemental analysis, TLC and IR spectra of the oils revealed the presence of nitrogen-containing lipid fraction (NCLF) in the oil constituents.

The seed oils liberated HCN by a standard picrate test and gave an intense prussian blue colour, showing the presence of cyanogen group. Elemental analysis of oils of *H. supinum* and *H. strigosum* revealed 1.86 and 1.97 per cent of nitrogen, respectively. Analytical TLC of each oil on silica gel G plates gave two spots (triglycerides, Rf. 0.71 and NCLF, Rf. 0.53). The oils were fractionated by preparative TLC into triglycerides (88 and 84%) and NCLF (12 and 16%) in *H. strigosum* and *H. supinum*, respectively. The pure NCL from oils gave on elemental analysis 2.0 and 2.2% nitrogen in the two oils.

The structure elucidation of cyanolipids isolated from both the species was done by comparison of its TLC, co-TLC, IR and NMR spectral characteristic with those reported for the corresponding cyanolipid isolated from *Cardiospermum halicacabum, C. canesens* and previously isolated NCLF from the *Heliotropium* species.

The infra-red spectral analysis of the NCLF from the two species revealed a nitrile band at 2230 cm^{-1} of moderate intensity. The NMR spectrum of NCLF (I) in CCl_4 (Ref. TMS) showed signals of usual fatty acids at γ 9.12) (terminal methyl), 8.74 (chain methylene), 8.35 (protons β to carbonyl) 8.05 (protons to double bond), 7.72 (protons α to carbonyl group) and 4.7 (vinyl). The two sets of methylene protons adjacent to the oxygen atoms of the dihydroxy-nitrile moiety gave signals slightly upfiled γ 5.82 (2H) and 5.06 (2H) than reported in literature (γ 5.32 and 5.13). The difference in shielding observed in signals is probably owing to the stereo-chemistry of the acyl methyl groups.

These above comparisons suggested that the cyanogenetic constituents from both the oils were identical and were found a diester of 1-cyano-2-hydroxymethylprop-1-ene-3-ol.

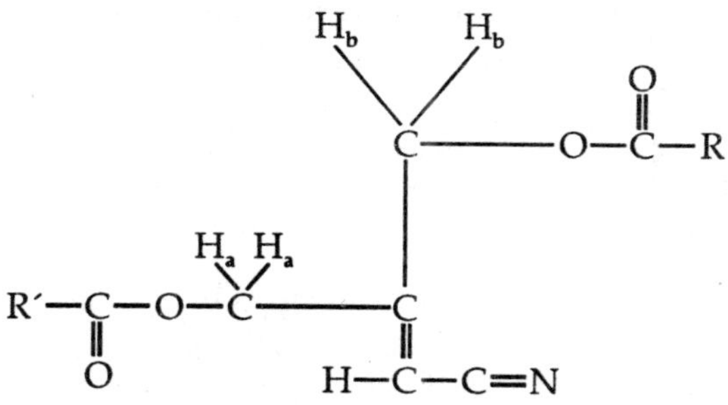

(I)

Diester of 1-cyano-2-hydroxmethylprop-1-ene-3-ol.

The fatty acid composition of triglycerides and cyanolipid fractions was also compared with the help of GLC. In *H. supinum* the triglyceride fraction showed 12:0, 2.2%; 14:0, 3.7%; 16:0, 13.9%, 18:0, 4.5%, 18:1, 36.9; and 18.2, 38.7%; and cyanolipid fraction was found to contain 12:0, 2.7%; 14:0, 6.1%; 16:0, 16.6%; 18:0, 3.9%; 18:1, 32.8%, and 18:2, 37.8%.

In *H. strigosum* the triglyceride fraction showed 12:0, tr. 14:0, 0.3%, 16:0, 8.7%, 18.0, 3.8%; 18:1, 33.0% and 18:2, 53.3%; whereas cyanolipid fraction contained 12:0, 3.3%, 14:0, 8.5%; 16:0, 19.7%; 18.0, 2.6%; 18:1, 30.9%; and 18:2, 34.0%.

From the above results, it appears that the triglycerides as well as NCLF fractions from both the species contain acids of C16 and C18 chain length. C20 acids were absent. The results of the cyanolipid constituents of the two additional *Heliotropium* oils reported in this work lends support that the presence of C20 acids is not a prerequisite for the occurrence of cyanolipids in seed oils.

CHAPTER 12

NTFP - PLANT GUMS

India being a tropical country is fortunate in having a wide variety of plant species. In India, there is also a wide variety of climatic conditions including high rainfall, very low and negligible rain fall; very cool, temperate and warm climate. Thus, we have plants growing in India which are found in cold climates of Europe and North America and those found in Western and South East Asia. Many of these plants, especially those growing in temperate and warm climates, produce plant gums from different parts of the plant. Before discussing plant gums and their applications, it would be appropriate to classify plant gums. Thus, the plant gums can be classified on the basis of source in the following manner.

1. Plant exudates
2. Seel, husk, tuber and root gums
3. Seaweed gums
4. Derived gums

No matter what is the source of the plant carbohydrate gum, certain characteristics are associated with this class of material. One of the most important criterion is that when the gums are lispersed in water, they produce viscosity solutions or pastes which exhibit non-Newtonian flow. The solutions or pastes of gums bring about adhesion between cellulosic surfaces such as paper, cotton yarn and fabric, and wood. The films of the gums are highly hydrophilic in nature and absorb moisture from the atmosphere. The gums are carbohydrate polymers of various structures and, therefore, do not give true solutions in water. In view of this, it would be better to describe these types of materials as plant hydrocolloids. The viscosity and viscosity behaviour (rheology) of different plant gums depend on their chemical and physical structure as well as on the conformation of the polymer in solution.

Because of their characteristic properties, plant gums are used as suspending and dispersing agents, as body builders and as adhesives for various applications.

One sample of leaves from market gave an oil having 60.35% aldehyde content. (Oil of C. *cassia* (leaf) has on an average 75-90 per cent cinnalic aldehyde).

All the samples of oil had characteristic odour of oil from the bark of true cinnanon but lacked in warmth, sweetness and strength to some extent.

Characteristic feature of C. *tamala* (cinnamicaldehyde type) is that the bark is mucillagenous.

Oil samples produced from leaves of both wild growing and cultivated plants were studied. Minor compounds (leas than 1%) were : α-pinene, β-pinene, limonene, β-phellandrene, *p*-cynene, ocimene, γ-terpinene, geraniol, benzyl acetate. Among major components, following were identified.

Component	*Oil from wild plants*	*Leaves of cultivated plants*
Camphor	0.90	3.19
Linalool	15.28	15.67
Borneol	1.07	1.18
– Caryophyllene	7.26	4.00
Cadinene	–	3.06
– Terpineol	1.54	1.77
Benzyl cinnanate	1.87	1.81
Benzaldehyde	2.00	4.11
Cinnamic aldehyde	55.19	41.20
Eugenol acetate	2.06	12.45
Eugenol and Cinnanyl cinnamate	4.23	13.32

In all, out of 32 components indicated by GLC, 22 components were identified.

Industries based on plant gums have been set up in India after the II World War and although the development has not been commensurate with the vast plant gum resources available in the country, there has been a steady growth. Data on manufacture of plant gums in India are given in Table 1. It may be mentioned that some of the data given are based on approximation as data on production of materials by unorganized sector and on agricultural and forest products are not readily available.

Table 1 : Production of Industrial Plant Hydrocolloids

Hydrocolloid	Production (Tonnes)
1. Maize Starch	1,60,000
2. Guar Gum	70,000 *
3. Isabgul Seeds & Husk	12,000 *
4. Tamarind Kernel Powder	10,000 *
5. Gum karaya	7,000 *
6. Sodium Alginate	600
7. Will Onion Gum	250
8. Agar Agar	36
9. Gum Arabic	15 *
10. Carboxymethyl cellulose (CMC)	6,000

Data on production of two important hydrocolloids, namely tapioca starch and gum ghatti could not be located.

* The figures given are approximations based on export.

Plant Exudates

These are produced by plants in the form of tears or nodules spontaneously or by making incissions in the bark of the tree trunk. In response to the injury to the plant, a viscous material comes out at the point of injury to seal off the wound. The viscous, clear solution gradually hardens to form a transparent brittle nodule which can be removed by hand. It is generally believed that the production of gum by the plant is a part of the inbuilt mechanism for warding off attack by micro-organism at the injured portion. This aspect is utilised commercialy and for obtaining large quantities of gum from a tree. A systematic programme of making repeated incisions on successive seasons is launched for maximum production. In India, the major plants which give gum exudates are *Sterculia urens* (roxburgh) (gum karaya), *Acacia arabica* (gum arabic) and *Anogeissus latifolia* (gum Ghatti or Indian gum). Many other trees of Indian origin produce gums such as *jeol, neem, ketha* and *bael.* In fact, most of the trees when deliberately injured at the trunk by an incision will produce some quantity of gum. However, since the amount of gum produced is not much, they are not of commercial importance. In this discussion, only gums of commercial value will be taken.

Gum Karaya

Gum karaya is obtained from *Sterculia* trees in the form of large, irregular, brown to white tears. A mature tree produces 1 to 4.5 kg of gum per season and an average tree can be tapped 5 times during its life time. India is the only country which produces gum karaya, majority of which is exported to the United States. The export figures of gum karaya for the last several years and given in Table 2. It is observed that by exporting this commodity foreign exchange to the tune of about 5 crores is earned. At present, gum karaya in the form of tears or lumps fetches a price of Rs. 14 per kg from exports. However, if it is pulverised to the desired particle size and properly graded it has a potential to realise a price of Rs. 100 per kg.

Table 2 : Exports of Gum Karaya

Year	*Quantity (in Tonnes)*	*Value (Rs. in Crores)*
1967-68	5023	1.78
1968-69	5337	2.67
1969-70	4006	3.01
1970-71	5666	4.09
1971-72	5787	2.93
1973-74	5657	4.43
1974-75	6188	5.70
1975-76	4169	3.12
1976-77	5276	4.48

Chemically, gum karaya is a polyuronida containing residues of D–galacturonic acid, D–glucuronic acid, D–galactose and L–rhamnose. The gum is unique in having 8% acetyle group. Gum karaya absorbs water rapidly to form a viscous mucilage which is short-bodied and gel-like in character. It has low solubility in water. On leacetylation of gum karaya mucilage a coherent ropy basis is obtained.

The major use of gum karaya is as a bulk laxative in view of its ability to form a mucilaginous gel on contact with water. For this use, the gum is ground to a granule size of 8 to 30 mesh. These granules have a capacity to absorb water and swell to 60 to 100 times their original volume. The gum is neither digested nor absorbed by the body.

Based on its property of absorbing large quantities of water, gum karaya is used as a denture adhesive for a comfortable and tighter fit of the plate. The gum is particularly useful for this application because it is unusually resistant to bacterial and enzymic break down.

The gum has also been used in a limited way as a wet-end additive in paper manufacture in conjunction with starches. However, due to the high cost of the gum, it does not appear to be an economic proposition.

Potential New Uses of Gum Karaya

Dextrinization (heating with small quantities of acidic catalysts) of starch is known to bring about degradation of the component polymers followed by recombination of the fragments. As a result of this a more branched polymer than the starting material is produced. In view of the availability in large quantities of gum karaya in our country, a study was undertaken in this laboratory to bring about a copolymerization of gum karaya with maize starch through the process of dextrinization. Evidence of such copolymerization was obtained by chemical analysis. The products obtained by codextrinization of gum karaya with starch were evaluated as adhesives. The results of this study are given in Table 3. It is observed that 9:1 starch: gum karaya dextrins display high adhesive strength. These codextrins were prepared on a pilot plant scale and were evaluated as adhesives for making paper drums. The paper drums were found to be as good as those prepared from PVA.

In view of the high content of uronic acid and the highly branched character of the gum karaya structure, it may be possible to use this gum after suitable chemical modification as a thickener for printing fabrics with reactive dyes. For this application, a given condition is that the thickener should not possess easily accessible hydroxyl groups for reaction with reactive dyes. At present, sodium alginate (Rs. 60 to 100 per kg) is the thickener of choice for this purpose.

Table 3 : Adhesive Strength* (in pounds per sq. in.) of Starch, Karaya and Starch-Gum Dextrins

Time in hrs.	*Karaya gum 153° 0.07% HCl*	*Starch 153° 0.07% HCl*	*S:K (9:1) 153° 0.07% HCl*	*S:K (9:1) 173° 0.07% HCl*	*TS:K (9:1) 153° 0.116% HCl*	*S:Ghatti (9:1) 153° 0.07% HCl*	*S:Guar (9:1) 0.08%*
0	9	4	4	4	4	4	42
2	–	35	73	62	57	64	53
4	–	33	62	51	48	55	51
6	–	44	59	40	48	35	40
8	11	40	55	35	46	35	37

* 25%, w/w concentrations of the dextrins was used for the tests.

Under the conditions used, British gum had an adhesive strength of 13 lbs.

Abbreviations : S = Maize starch; K = Gum karaya; TS = Tapioca starch; Ghatti = Gum ghatti and Guar = Guar Gum.

Gum Arabic

In India, gum arabic is obtained from Babul (*Acacia arabica*) trees. India's contribution to the world production of gum arabic is small since 85% of the world supply of gum arabic is obtained from Sudan. Gum arabic is a very valuable gum and is used in confectionary, bakery products, in pharmaceutical preparations, especially as a suspending and emulsifying agent, and as a domulcent. It is particularly useful as an adhesive for envelopes, lables and stamps. Large quantities of gum arabic are imported in the country. The import figures are given in Table 4.

Gum arabic is a highly branched polyuronide whose structure has been a subject of study by several workers. It is composed of D-glucoronic acid, D-galactose, L–arabinose and L–rhamnose. Gum arabic is the most soluble of gums and workable paste of 40% concentration can be made. This is attributed to the highly branched structure of the gum arabic molecule.

In view of the large quantities of import of gum arabic it would be desirable to develop substitutes from indigenous gums by chemical and physical modifications. Gum ghatti, which is closely related to gum arabic in structure, may be the raw material of choice for such modification.

Table 4 : Import of Gum Arabic

Year	*Quantity (Tonnes)*	*Value (Rs. in Lacs)*
1967-68	3571	54.5
1968-69	3294	50.0
1969-70	3972	61.0
1970-71	4448	72.9
1971-72	3698	70.0
1973-74	3226	92.0
1974-75	916	30.7
1975-76	491	31.3
1976-77	1193	50.4

Gum Ghatti

Gum ghatti is obtained from *Anogeissus latifolia* tree which grows in India and to a lesser extent in Ceylon. The gum is obtained in the form of rounded tears which on fracture give a glass-like surface. The tears have light to dark-brown colour and which sometimes contain air-borne impurities. The export of gum ghatti is less than that of gum karaya and it is reported that from 1967 through 1971 between 450 and 600 tunnes of gum ghatti per year were imported by the United States at a price of Rs. 5 to 11 per kg.

Gum ghatti, like other plant exudates, is a polyuronide containing residues of D-glucuronic acid, L–arabinose, D-galactose, D-mannose and D-xylose. The proportion of uronic acid in gum ghatti is much less than that in gum karaya. The dispersion of gum ghatti in water is less viscous than gum karaya but more viscous than that of gum arabic. The adhesive strength of gum ghatti dispersions is similar to gum arabic, but dispersions of as high a concentration as gum arabic cannot be prepared.

The main use of gum ghatti is as an emulsifying agent for pharmaceutical preparations. In U.S.A. gum ghatti is used for making stable pharmaceutical preparations of oil-soluble vitamins. It is reported that by using gum ghatti the fluid loss in oil-well drilling muds can be prevented. This is true also when the drilling mud contains a high concentration of salts.

Experiments on codextrinization of gum ghatti with starch similar to those carried out with gum karaya were carried out in this laboratory and the products were evaluated as adhesives. The results in Table 3 show that the starch–ghatti codextrin has excellent adhesive properties. This aspect needs further investigation.

Seed, Husk, Tuber and Root Gums

The gums from seed, seed husks, tubers and roots constitute the largest class of gum. These gums include various carbohydrate polymer types, namely, starches, glucomannans, galactomannans, and complex acidic polysaccharides.

Starches

In India, the main source of starch is maize. Maize starch is produced to the tune of 160,000 tonnes per year and in its native form or after modification is used in many industries such as textiles, paper and food.

Another important source of starch in this country is tapioca root. Tapioca starch is largely produced by the unorganised sector around Salem and to a small extent by the organised sector. The major portion of tapioca starch produced is utilised as food after conversion into sago pearls.

Starch is a reserve carbohydrate and is very widely distributed in the plant kingdom. Industries based on maize and tapioca starches are well established and, therefore, these starches will not be discussed further. Here, the emphasis is laid on alternative sources of starch.

Sal or sakhu (*Shorae robusta*) constitutes about 13.3% of the total forest area. Sal is grown extensively in Orissa, Bihar, Madhya Pradesh, U.P., West Bengal and other parts of north-east India. Sal tree, apart from giving valuable constructional and domestic wood, produces seeds which are of commercial importance. Lately the oil from sal seeds has attracted much attention as a substitute for cocoa butter in the manufacture of chocolates and similar products. The proximate analysis of sal seeds is given in Table 5. The nitrogen-free extract is mainly carbohydrate in nature and about 50% of the sal kernels is made up of starch. It is estimated that the total annual production of sal seeds is about 200,000 tonnes (1966-67 figures). If the isolable quantity of starch is considered as 50% of the total weight of sal seeds the potential availability of sal seed starch works out to 100,000 tonnes, which is approximately 2/3rd of the total production of maize starch.

Table 5 : Proximate Analysis of Sal Seeds

Moisture	5.23%
Protein	6.16%
Ether extract	16.77%
Crude fibre	4.81%
N-Free extract	63.25%
Ash	3.78%

The extraction of oil from sal seeds has already been started on a commercial scale. The oil cake, which contains starch, protein and residual oil, is used as a cattle feed. It is apparent that a relatively minor constituent (oil) is extracted and sold at an attractive price while the major component is sold in the impure form at a relatively low price. If a process can be developed whereby the two major constituents of sal seed *i.e.,* starch and oil are extracted without affecting the quality and purity of either sal seeds could become a raw material for a multi-product industry like maize starch.

Mango Starch

Mango (*Mangifera indica*) is the most important fruit crop of India. It is estimated that from a total of 3.3 million tonnes of mangoes produced in India over 1 million tonnes of mango seeds are available. Mango seed kernels contain 50 to 58% of starch which is of good quality. Thus, there is a potential of producing 500,000 tonnes of mango starch. However, it has to be recognised that mango seeds are subject to microbial attack especially during the rainy season and that it is difficult to collect and transport mango seeds from widely dispersed mango producing areas in the country for processing. A project was taken up at this laboratory to work out a practical process for the isolation of starch from mango kernels and to find out the viability of manufacturing starch from mango kernels. Although pure starch could be isolated from mango kernels the process developed was lengthy and involved treatments with chemicals besides repeated centrifugation. In view of the difficulty of collecting mango, seeds and a practical process of starch isolation, manufacture of starch from mango kernels does not appear to be a viable proposition. Be that as it may, considering the vast quantities of good quality starch present in mango seeds and the fact that most of it is wasted by decomposition and biological degradation, further serious efforts should be made to utilise this valuable raw material. We have looked at mango kernels only from the point of isolating starch. It is not impossible that if mango kernels are straight away hydrolysed to intermediate products or to its ultimate unit, *viz.*, glucose and thence to ethyl alcohol, the proposition of utilisation of mango kernels as a chemical raw material may be still worthwhile. If mango kernel can be profitably utilised, uses can be found for the seed coat.

SEED GUMS

Guar Gum

Guar gum is obtained from the endosperm of gaur (*Cyanopsis tetragonoloba*) seeds. Apart from starches, guar gum is the major seed-based industry of carbohydrates. The production and export of guar gum has been steadily increasing over the last 10 years with occasional fall. Thus, with a modest export of 5,000 tonnes in 1967, the export in 1976-77 touched 45,000 tonnes, earning foreign exchange to the tune of Rs. 15 crores (see Table 6). It may be noted that almost all gum exported is in the form of unmodified raw gum which is further processed in advanced countries for specific and uses.

Gaur gum is a unique carbohydrate polymer (galactomannan) inasmuch as it gives very high viscosity at low concentrations. In fact, as compared to other cabohydrate polymers, it gives the highest viscosity at equivalent concentrations, because of its paste characteristics, guar gum is used in large quantities in the textile industry as a sizing and printing gum, in the food applications as a body builder and for suppressing crystallization of such preparations as icecream and sherbets. It prevents water loss from materials and is used for conditioning oil well-drilling muds. Guar gum is used as a wet-end additives in paper industry. Guar gum is also used as an aid in settling, filtering and flotation separations. In many of the speciality uses, guar gum has to be blended or physically and chemically modified. There is vast scope of tailor-making products from guar gum for specific and uses. This laboratory has already carried out considerable work on preparation and evaluation of derivatives of guar gum for various end uses. One of the products based on ATIRA's technical know-how is being manufactured and used as a substitute for sodium alginate for printing fabrics with reactive dyes. In another project, work has been undertaken to develop modified guar gum for use in slurry/gel explosives.

Table 6 : Export of Guar/Gum Flour

Year	*Quantity (Tonnes)*	*Value (Rs. in crores)*
1967-68	5,000	0.8
1968-69	6,000	1.0
1969-70	7,000	1.6
1970-71	23,000	5.4
1971-72	22,000	3.8
1973-74	26,000	7.9
1974-75	37,000	17.6
1975-76	23,000	10.9
1976-77	45,000	15.1

Guar gum is being manufactured by approximately 75 units, half a dozen of which are large units. Thus, a majority of the manufacturers are in the unorganized sector. The units are located in Haryana (Bhiwani), Rajasthan (around Jodhpur), Gujarat (in and around Ahmedabad) and in Bombay. Because of large quantities of exports, the guar gum industry is flourishing. It may, however, be noted that it modified guar gum products of standard quality are exported, the earnings in terms of foreign exchange would be many times more than what are being realised at present. Considering that guar can be grown in semi-arid areas, there is a vast potential for increasing the acreage under guar, thus providing employment and subsistence to people in under developed regions.

There are several other shrubs like *Cassia tora* and *Cassia absus*, which grow wildly and produce seeds containing galactomannans similar to that present in guar seeds. There is considerable potential of growing such weeds and isolating gum products from them for different applications.

Psyllium Seed and Husk Gum

Isabgul, which is the husk of the seed from *Plantago ovata* is a product of importance especially for use as a bulk laxative. A similar gum is also obtained from psyllium seeds. The plant is grown almost exclusively in Gujarat, in the districts of Baroda and Palanpur. Because of the high quality of laxative gum, psyllium seed and husk are exported in large quantities. The data for the last 10 years on the export of psyllium seeds and psyllium husk are given in Tables 7 and 8. It is observed that export earnings to the tune of Rs. 16.5 crores are obtained from these materials. Perhaps, the foreign exchange earnings can be augmented by compounding psyllium seeds husks with other ingredients to form bulk laxative tablets instead of just exporting the raw psyllium gum.

Table 7 : Export of Isabgul (Psyllium seed)

Year	*Quantity (Tonnes)*	*Value (Rs. in Lacs)*
1967-68	181	4.47
1968-69	562	10.67
1969-70	574	10.87
1970-71	800	20.31
1971-72	1072	27.37
1973-74	1003	34.13
1974-75	2225	83.02
1975-76	1850	101.94
1976-77	1905	158.80

Chemically psyllium gum is a polyuronide containing residues of D-galacturonic acid, D-xylose and L-rhamnose.

Table 8 : Export of Isabgul Husk (Psyllium Husk)

Year	*Quantity (Tonnes)*	*Value (Rs. in Crores)*
1967-68	1788	1.42
1968-69	3065	1.83
1969-70	1691	1.02
1970-71	2883	2.55
1971-72	5211	3.63
1973-74	3239	2.99
1974-75	5146	6.90
1975-76	2396	4.60
1976-77	7979	14.94

Tamarind Seeds

The fruits of tamarind (*Tamarindus indica*) tree upon ripening give a dark brown hard seed. It is estimated that about 140,000 tonnes of tamarind seeds are available annually in India from which 90,000 tonnes a year of tamarind kernels can be obtained by a process of roasting followed by grinding and winnowing. Tamarind kernel powder (TKP) obtained by the above process when dispersed in water gives a smooth and viscous paste. TKP has been a subject of considerable research mainly at ATIRA. Its application in sizing of cotton yarn was investigated with a veiw to substituting starch by this material. Whilst TKP could be used for sizing of coarse yarns, due to its inherent drawbacks, it could not be employed for high count yarns and close construction fabrics. TKP was, therefore, chemically modified, so that low viscosity products could be obtained. These products were extensively evaluated in the labortory and in mills as sizing agent. With controlled modification and proper sizing recipe, its performance was found to be equivalent to maize starch, the common sizing agent for textiles. Based on the technical know-how of ATIRA, low viscosity tamarind kernel powder is being manufactured in India and is being used in the textile mills.

Various other derivatives of TKP are being prepared and they are being evaluated for various end uses. A more detailed discussion on modifiction of TKP is taken up in another paper from this laboratory.

TKP has also become an earner of foreign exchange and the data on the exports are given in Table 9.

Table 9 : Export of Tamarind Seed and Seed Powder

Year	*Quantity (Tonnes)*	*Value (Rs. in Lacs)*
1967-68	1575	12.37
1968-69	2118	24.42
1969-70	1859	14.54
1970-71	2683	21.80
1971-72	3029	26.68
1973-74	3182	42.69
1974-75	4806	88.76
1975-76	2503	38.30
1976-77	3910	52.72

Wild Onion Gum

As name indicates, this gum is extracted from wild onions (*Urginae indica kunth,* family Liliaceae) which grow in sandy land and are found in the hills of lower Himalaya, Rajasthan and in the salt range of Punjab. The plant can be cultivated almost anywhere on Indian plains. However, its importance as a raw material rests in the fact that it can be grown in areas where normally nothing can grow, such as in the hilly regions of Rajasthan.

The bulbs of wild onions are similar to ordinary onion except that they are white and more fleshy and produce irritation on touching.

The onion bulbs are cut into small pieces and after drying pulverized to a creamy white powder. This powder contains largely a glucomannan in addition to small quantities of a cardiac glycoside. It is also called Indian squill because of its resemblance to European squill (*Urginea scilla*). The polysaccharide of wild onion gum and its physico-chemical properties have been examined recently. Dispersion of wild onion gum in water gives a viscous and adhesive paste which thins down on keeping. The adhesive powder of wild onion gum is good to excellent and because of this property it is used as a gum in textile sizing. The production of wild onion gum per year is about 250 tonnes. Considering the large quantities of wild onions that are available and if necessary further quantities grown, further studies on this gum and development of useful products may be worthwhile and remunerative.

Seaweed Gum

The main seaweed gum of commerce in India is sodium alginate, manufactured from *Sargassum* seaweeds harvested around the south coast near Madurai. Sodium alginate is a unique polymer inasmuch as it is entirely composed of uronic acids, *viz.*, D-mannuronic acid and L-guluronic acid. It gives viscous pastes of required thixotropic properties for printing. It is especially employed for printing reactive dyes on textiles because of all the carbohydrate polymer thickeners available and tested so far, this is the least reactive towards reactive dyes. Due to the limited availability of *Sargassum* seaweeds around the Indian sea coasts and the high cost of extraction of sodium alginate, attempts, have been made to develop substitutes for this product for printing applications. No product so far has been developed which is as good as sodium alginate, but it will be worthwhile to continue search in this direction considering the pay off if a suitable substitute based on Indian gums can be found. The work at ATIRA on modified guar gum in this direction has been mentioned earlier.

Derived Gums

In this class are gums which are derived from native carbohydrate polymers by physical and/or chemical modifications. In certain cases the native polymer does not give a gummy solution. An example of this class is sodium carboxymethyl cellulose. Cellulose as such is insoluble in water, but when partially carboxymethylated, it disperses in water to give a high viscosity paste having an anionic charge. Because of unlimited resources of cellulose and standard method of carboxymethylation, large quantities of carboxymethyl cellulose of required viscosity grades are produced. They find application in textile, paper, oil-well drilling, food pharmaceutical and mining industries. Similarly, cellulose can be converted into methyl cellulose, which absorbs large quantities of water. This product has been used in slimming diets because it gives a feeling of satiety without much caloric intake.

The extent to which chemical modification can change the character of a carbohydrate polymer is illustrated by the recent development by the U.S. Department of Agriculture of 'Super Slurper,' which absorbs 2000 times its weight of water. Many uses such as retention of water by sand, soil and seeds for agriculture; absorption of body fluds for surgical sponges, bandages and disposable diapers have been mentioned for this product. Super slurper chemically belongs to the family of hydrolysed starch–polyacrylonitrile graft copolymers. If indigenous gums, which have several times more water absorbing capacity than maize starch, are converted into polyacrylonitrile copolymers with relatively less modification, products similar to or better than super slurper may be possible.

Derived gums have the greatest future in this country in view of the availability of large amounts of agricultural and forest product wastes. These, in their native form, may be of little use except as fuel, but when processed and chemically tailored to suit the specific needs of an end application, may be very useful. We have the whole field open to us and it is for the enterprising to enter this almost virgin field for exploitation for national good.

Concluding Remarks

India is rich in plants containing or capable of producing carbohydrate polymers of a wide variety. The exploitation of potential carbohydrate polymer resources in India has so far been very limited. Carbohydrate polymer resources should also be looked upon as renewable sources of energy through production of chemicals and semi-synthetic polymers. Through systematic studies many new carbohydrate polymers of commerce can be discovered. In addition, physical and chemical modifications of the known carbohydrate polymers can generate polymers of required properties for specific end uses.

CHAPTER 13

GUM GUGGUL

Gum guggul has an important place in the Indian system of medicine and is claimed by Ayurvedic physicians to be highly effective in the treatment of rheumatoid arthritis, obesity, neurological disorders, leprosy, urinary disorders and skin diseases. It is also used in the form of a lotion for the treatment of indolent ulcers, chronic tonsilitis and ulcerated throat. The plan gum resin occupies an important position in the Indian mythology as its smoke is considered to be an air purifier and dispeller of the evil. In ancient times, all the Yagnas and Havanas were performed with guggul and this practice is still continuing even to-day.

Recent pharmacological studies have shown that some of its constituents do possess anti-inflmmatory, antirheumatic and hypocholestramic activity. The compound named Z-guggul sterone isolated from the gum is highly effective in lowering the cholesterol level.

Gum guggul is the else gum resin obtained from the plant *Commiphora mukul* Hook. ex Stock (Fams : Burseraceae) from the incisions made on the bark of the tree. The term guggul is often confused with the resinous products obtained from the other plant of this genus namely C. *agollocha* Engl., C. *stocksiana* Engl., C. *berryi* (Aen). Engl. and from *Boswellia serrata,* but the commercial product is obtained only from C. *mukul.*

Occurrence and Dsitribution

Commiphora mukul Hook. ex Stock occurs in the arid rocky tracts of North-Western India. It is a small tree varying from 1-4 metres in height and often branched near the base. The major resin producing areas are Gujarat and Rajasthan while some minor quantities come from Berar and Mysore. A survey for the collection of guggul indicated that these trees are distributed all over Rajasthan especially in the western regions. The guggul bearing areas in Rajasthan are given in Table 1. It was found out during our survey that the resin production in Rajasthan has considerably come down in recent years. On investigations with the local people and the forest departments, it was learnt that the cause of scarcity was the faulty method of tapping which the collectors applied to maximise the yield of oleo gum resin. To increase the secretion the villagers were applying a paste made of horse urine, oleo gum resin and copper sulphate around the incision. This method no doubt increased the secretion for sometime but makes it unfit and ultimately the tree dies. This injurious effect may be caused by the copper sulphate present in the paste. The forest authorities are keeping a strict watch on the people to discourage such practice.

However, the main supply of guggul comes from Gujarat State which is mostly localised in Kutch Division. The major gum producing areas in Kutch under the control of forest department are Bhuj, Nakatrana and Anjar. Table 2 shows all the important guggul producing areas in Gujarat.

An estimate of total producing of guggul is not possible because the collectors do not report to the forest authorities the actual production from each range. The forest department is mainly concerned with the revenue obtained from contracts. In Kutch the trees which "die" and stop producing guggul are burnt by the contractors to produce coal, which is mixed with cowdung and sold in the market at Bhuj; thereby earning more money. However, recently some restrictions have been imposed by the forest department authorities on burning of the trees as they presume that "dead" tree may start producing guggul after some years as it happened in Rajasthan.

Tapping and Collection

Winter is the best season for collecting the oleo gum resin of the tree *C. mukul*. The tapping starts in November/December and collection continues upto May or June. Trees not less than five years old are given circular cuts at a uniform distance of 30 cm. from each other. The workers usually give deep incisions to get maximum exudate in a season but this method kills the tree.

The latex oozes out through the cut as a yellow fluid which hardens in big lumps to give the oleo gum resin. The pieces of the oleo gum resin are often pale yellow, redish brown or dull green in colour with a bitter aromatic taste. After the completion of first collection the subsequent collections are carried out from the tree every 10-15 days during the season. The yield of the oleo gum resin can be increased, without affecting the health of the plant, by spraying the tree with a 50% solution of sulphuric acid or 50% hydrochloric acid as recommended by the Central Arid Zone Research Institute, Jodhpur.

Tapping and collection of the resin is done by men and women both. In Gujarat, it is done by Samma and Kanhi tribes in Kuth and Bheel tribes in Rajasthan. Each worker collects about 1–1½ kg. of the dry material in a day and a mature tree yields 250–500 g. of the resin in one collecting season. It is considered by the forest department in Kutch that for higher yields, the tree should not be tapped continuously, but be given a period of rest.

Marketing

The collectors bring three grades of guggul to the market :

(*i*) the first grade is collected from the trunk and thick branches, and is free from any adhering material;

(*ii*) the second grade is dull in colour and is contaminated with dust and bark; and

(*iii*) the third grade is highly inferior in quality and is collected from the ground.

The final grading of the gum is done by cleaning the material and the inferior grade is improved by sprinkling caster oil over the drug which makes the appearance shining. The wholesale price of guggul is Rs. 20-25 per kg. The Bhuj market meets practically the entire requirement of the country. Bombay, Delhi, Calcutta and Amritsar receive guggul from the Bhuj market alone. It is estimated that about 400-500 metric tons of the oleo gum resin is being exported every year from Bhuj to other Indian markets.

Chemical Composition

The chemical composition of the oleo gum resin is highly

complex and has been investigated by several workers. It is found to contain 0.37% essential oil comprising of myecene, dimyrcene and polymyrcene. Some other components isolated and characterised are Z-guggulsterone, E-guggulsterone, guggulsterol-I, guggulsterol-II, a diturpene alcohol, sesamin and chloesterol.

Cultivation

A systematic cultivation of this important medicinal plant has not been tried seriously. Only recently, some plants were raised in the Regional Research Laboratory, Jammu Nursery from the stem cuttings brought from Bhuj and these grew well within fifteen days. Hence, it is possible to raise this important plant, and conserve the species which is on the verge of extinction, from stem cuttings and sowing the seeds.

Suggestions

1. The plant *C. mukul* – a source of gum guggul – is an important medicinal plant in the indigenous system of medicine. Since this plant is now almost on the verge of extinction in guggul producing areas, the respective forest departments should take adequate steps for its conservation by excessive cultivation, preservation and the collection of oleo gum resin on scientific lines.
2. As this plant grows wild in the areas which are neglected and uncared for, the cultivation of the species may also be included as one of the rural uplift project. The local inhabitants may technically be guided to cultivate the plant in such areas which will offer them better job opportunities and also increase the production of oleo gum resin which at present is much less than its demand.
3. The propagation may be carried out most scientifically by the technical persons from the stem cuttings of the plant since the seeds are less viable.
4. The tapping should be done under the regular supervision of technical personnel, so that the damage and wastage may be minimised. The tapping should not be too deep and touch the cambial portion to keep the plant healthy for the next tapping.

5. Only healthy and mature plant should be selected for tapping. The tree should not be tapped every year, but be given a rest period of 2-3 years for obtaining high yields.
6. Faulty tapping techniques such as the application of copper sulphate paste, etc. by the villagers should be discouraged in order to conserve the species.
7. Some plant growth hormones may be tried for quick growth of the plant.
8. The marketing should be done through Govt. or Co-operative Societies for maintaining the quality of the oleo gum resin as well as assessing its correct production.

CHAPTER 14

MANUFACTURE OF GUM RESIN AND GUM TURPENTINE

On its way to self sufficiency and prosperity, our country, at the present moment, is passing through a phase of rapid all round development. We have entered an era of reconstruction when our planned efforts aim at exploiting all available resources–to their optimum potentiality. The all round advancement, quite naturally, places great demand on the forest resources of the country, which no doubt provide some of the basic raw materials for a number of important industries.

Forest resources can be regarded as inexhaustible, provided we deal with them scientifically and sympathetically. It is very necessary that these sources should be tapped to yield maximum output on a sustained basis. Our industries based on forest product must also keep pace with the rapid advancement in the industrial field.

Resin industry which uses oleopineresin (OPR) from pine species as starting raw material, is one of the foremost and biggest industries thriving on forest products and has an ever increasing importance in the national economy. It is, therefore, very important to explore all possible ways and means of enhancing production of resin in the country by tapping unexploited pine species. In this article we have described the manufacture of gum rosin and gum turpentine from oleopineresin.

OPR is the exudate of living pine trees, found in the foothills of Himachal Pradesh, Uttaranchal, North East J&K. Figure 1 and 2 shows the pine forest and a tapping of a pine tree respectively. There are four types of pine species grown in India, namely *Pinus roxburghii* (chir pine) , *Pinus wallichiana* (Blue pine), *Pinus khasiya* (Khasi pine) and *Pinus gerardiana* (chilgozapine). Of these, the largest area and extent of distribution is that of chir pine which extends from Kashmir to Eastern Himalayas. Though other pines yield a higher grade of turpentine rich in alpha pinene (Table 1), it is only the chirpine that has been tapped so far in India and provide almost all the resin of commercial value. The reasons for this its higher yield for resin than any other species.

Table 1 : Typical analysis of Turpentine Oil

Fractions	*USA*	*France*	*India* % (*P. roxbudghii*)	*Russia*	*Portu-guese*	*Sweden*	*Japa-nese*
A-pinene	65-75	60	20-25	75	80	80	85
B-pinene	20-30	25-30	5-10	–	15-17	5	10
Carene	–	–	55-65	15	–	15	5
Longifolene	–	–	2-5	–	–	–	–
Other terpenes	5	5-10	3-5	10	3-5	–	–

*P. wallichiana (Blue Pine)	A. pinene :	87.9-96.2%
P. Khasiya (Khasi Pine)	A. pinene :	30-40%
	B-pinene :	60-65%
	Others :	5-10%
P. geradiana (Chilgoza pine)	L-pinene :	73.0%
	B-pinene :	7.0%
	Others :	20.0%

Manufacturing Process

Tins of pine oleoresin collected from forest are put in a hot chamber. The tins are opened and drained in a melter where it is treated with oxalic acid for decolorisation and the charge is slightly warmed.

Warm gum is diluted with 30-35% fresh turpentine oil. Molten gum at 95-105°C is passed through a 4 mesh screen placed at the bottom of the melter to remove bark and other suspended impurities. Final filtration is done by Sparkler type filter. Filtered gum is sprayed into tank containing water, the emulsion is allowed to settle and the clean gum decanted. The water layer is further settled in a muck tank to remove residual gum and wash water is rejected through a pump.

The washed gum is pumped to the distillation tank where at 160-180°C complete turpentine distills off and is collected through the condenser and separator and the gum rosin is drained out through the bottom of the still where again calculated quantity of oxalic acid is added to upgrade it.

The details of the process are shown in Flow Chart (Figure 1).

Uses of Rosin and Turpentine Derivatives

Rosin is one of the important forest products of India. Applications of various rosin derivatives are given in Table 2. Turpentine oil is a typical essential oil obtained during the distillation of pine oleoresin which belongs to the chemical family of terpenes. The different uses of turpentine oil is shown in Table 3.

Table 2 : Derivatives of Rosin

Type of rosin	*Derivatives*	*Uses*
1. Unmodified rosin	Metal salts of sodium, lead, manganese, cobalt zinc, calcium, iron and copper. Esters of methyl alcohol, ethylene, glycol, diethylene glycol, glycerol, pentaerythritol and others polyols. Glycerol and pentaerythritol esters modified with P/F condensates.	Soap sizing, driers, varnish resins and fungicides Plasticizers, tackifiers lacquers, varnishes and Plastic compositions.
2. Hydrogenated rosin	Easters of methyl alcohol, ethylene glycol, glycerol and pentaerythritol	Adhesives, plasticizers, tackifiers
3. Disproportionate rosin	Sodium soap	Emulsifier for synthetic rubber
4. Polymerized rosin	Metal salts of zinc, calcium, esters of glycerol, glycols, pentaery thritol Esters of maleated product, P/F condensates	Varnish resins, printing inks
5. Other products	Methylated rosin, Oxonated rosin, Rosin amine, Rosin amine, Rosin alcohol	Polyesters, adhesive fungicides, insecticides corrosion inhibiters, surfactants plastics.

Complete know-how of all the above mentioned products are available with us.

Table 3 : Major Products and Industrial uses of Turpentine Oil

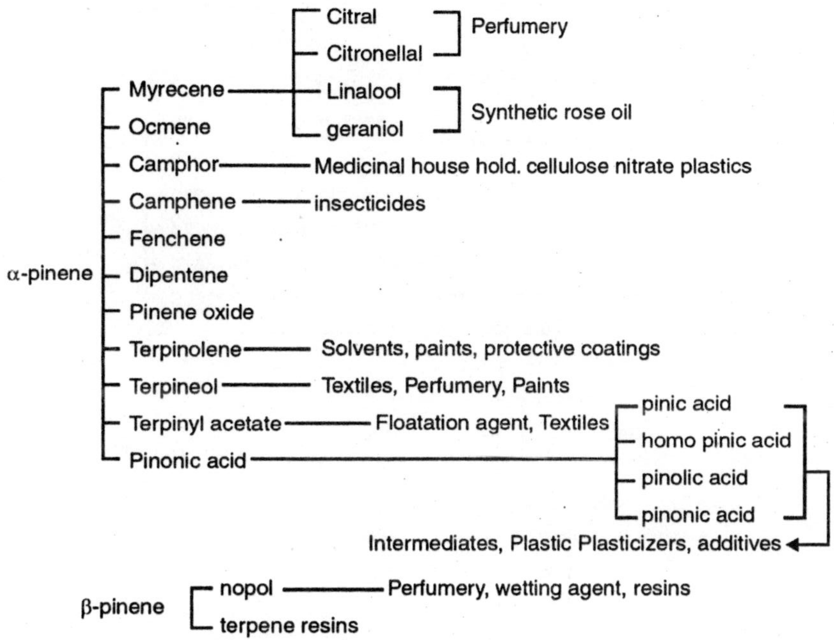

Δ-Carene — p-cymene
- terephhalic acid — Polyesters, resins
- p-toluic acid
- p-cresol — Phenolic resin
- 2-2 dialkyl benzyl hydroperoxide — Catalyst
- carvacrol — Antiseptic, fungicides
- thymol — Anthelmintic
- camphor
- dimethyl styrene — High polymers
- menthol — Medicinal

- dipentene — isoprene — Synthetic rubber
- Synthetic pippermit — Medicine, food

camphene
- 3-isocamphyl cyclohexanol (sandalwood odour) — Perfumery
- isobornyl thiocyano acetate (thanite) — Insecticides
- ethers (cedarwood odour)
- octachloro camphene (Toxaphene) — Pesticides

Longifolene
- Isolongifolene
- musk odour chemicals

(both) — Perfumery

Distilled turpentine oil — Solvent — Paints, Varnishes

Pine tar
- Synthetic rubber
- adhesives
- synthetic resin

ECONOMICS OF THE PROJECT

Plant Capacity and Efficiency

The economics of the project is carried out on the basis on a plant capacity of 5 MT per day or 1500 MT per annum of gum rosin from pine oleoresin. The shift basis or a 300 days working schedule in a year. The average plant efficiency is assumed at 85%.

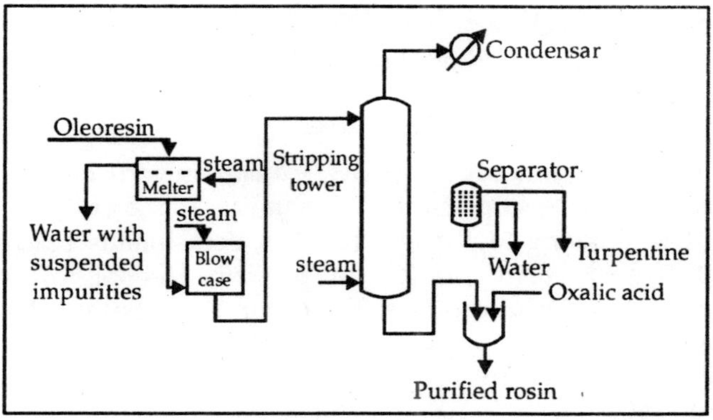

Fig. 1

CHAPTER 15

ESSENTIAL OIL OF DEODAR

Himalayan Cedarwood, *Cedrus deodara* (Roxb.) Loud, (Fam. Pinaceae), has been classified now as a distinct species. Formerly, some taxonomists considered it to be a variety of the "Cedar of Lebnan" *Cedrus libani* Loud.

Deodar, commonly known as *Devadaru* in vernacular, grows extensively on the slopes of the Himalayas. It is a graceful evergreen tree growing to a height of 50 m; occasionally reaching a height of even 75 m and 13.5 m. in girth. Deodar forests are common from Kashmir (alt. 1500–3000 m.) especially is Kishanganga, Kishtwar and Jhelum valley to Garhwal and from bulk of vegetation in that area. It is the most important conifer between 1650 m–2400 m.

Wood of Deodar is oily, fragrant, light (Sp. gravity 0.54; weight 15.91 kg/cft) straight and fairly even grained and medium fine-textured. Sapwood is white, heartwood light yellowish brown, turning brown on exposure. Although Deodar is primarily valued for its timber, its wood is reported to possess diaphoretic, diuretic and carminative properties and is useful in fevers, piles and pulmonary and urinary disorders. Bark is powerfully astringent and useful for fevers, diarrhoea and dysentery. Leaves have mild terebenthate properties. Oleo-resin and dark coloured oil (obtained by destructive distillation of wood) are applied to ulcers and skin diseases. These are valuable in mange in horses and sore feet of cattle.

Deodar timber is one of the most valuable Indian timbers. Primary use is for railway sleepers and for constructional works *e.g.*, beams, floor-boards, posts, door and window frames etc. It is suitable for pattern making, electric poles and battery separators. It is also used for pencil making. Strong odour of deodar wood and its oily nature limits its use for indoor work. While its odour is refreshing in the open, it becomes oppressive in closed atmosphere.

The Essential Oil

Essential oil of Deodar is of comparatively recent origin. It is classical example where due to R & D work in a Research Institute, complete import substitute has been achieved for an item long used in the industry. Himalayan Cedarwood oil is different from that of American cedarwood oil and, as expected, met with some scepticism about its suitability as a substitute for American cedarwood oil; import of which is banned. Oil of deodar is not only in demand in India but is also an export item.

Oil of Deodar was studied in the first and second decade of this century for its composition. However, by late fifties, a beginning had been made for its commercial production and within a few years, production reached about 10 tonnes per year. Current annual estimated production is of the order of about 25,000 kg with prospect of increased production.

Besides Deodar oil, other important oils of this type are :

Common Name	Botanical Name
1. Atlas Cedarwood	*Cedrus atlantica* Manetti
2. American Red Cedarwood or Virginian cedarwood.	*Juniperus virginiana* L.
3. East African cedarwood	*Juniperus procera* Hochst.
4. Chinese cedarwood	*Cupressus funebris*
5. Cedar leaf oil, white cedar leaf oil.	*Thuja occodentalis* L.

Out of the above, American and Atlas cedarwood are the most important. However, oils of East African and Chinese Cedarwood are used and regularly quoted in the trade. Besides these, rectified cedarwood is also regular item of commerce.

Simonsen & Rau indicated possibility of its use in soap perfumery in place of imported oil of cedarwood from Juniperus spp; however, the oil appeared more expensive than the imported oil. Commercial production of the oil started in 1957 in the Drug Research Laboratory at its factory at Srinagar. Its usefulness had been fully established by that time. As expected, the industry faced some difficulty in switching over to deodar oil in place of imported oil and there was some resistance towards banning imports. However, deodar oil established itself gradually and at the present moment import of cedarwood oils from abroad is banned.

Raw Material

Essential oil of Deodar is produced from saw dust, wood shavings and stumps of *Cedrus deodara* by steam distillation. Prior to distillation, however, chips and wood pieces are reduced to smaller size for ease in distillation.

Heartwood and stumps contain maximum oil. There are 2 types of material *i.e.,* saw dust and stumps. So far production of the oil has been from saw dust collected from Lumbering project Srinagar and from the blocks used for pencil making in Jammu. Even though stumps contain more oil, oil is not produced on a large scale from this source.

Oil of good quality can only be obtained if saw dust is not exposed to sunlight, which reduced yield of oil due to resignification. Exposure to moist atmosphere also lowers yield perhaps because of fermentation. Such oils are cloudy, discoloured and lack the typical cedarwood odour.

Yield of oil reported in literature on various cedarwood oils is given in Table 1.

Table 1 : Oil Content in Cedarwoods from various Materials

	Percent oil	*Material used*
1. Atlas Cedarwood (*Cedrus atlantica*)	5.0 – 5.0	Wood
2. Himalayan Cedarwood (*Cedrus deodara*)	1.55 – 5.5	Saw dust and shavings.
3. Port Oxford Cedar Oil (*Chamaecyparis lawsoniana syn. Cupressus lawsoniana*)	1.6	Saw dust and chips.
4. Virginian cedarwood oil (*Juniperus virginiana*)	(*a*) 2 – 2.5 (*b*) 3.5 (*c*) Below 1.0	Fresh chips and saw dust "Virgin cedar". "Sap. cedar".
5. Texan cedarwood *(Juniperus mexicana)*	1.8 – 2.3 3.2 – 3.24	Wood. Shavings & communuted pencil slates. Fresh wood shavings.
6. East African cedarwood *(Juniperus procera)*	(*a*) 1.42 (*b*) 0.28 (*c*) 0.02	Fresh shavings. 2 years old shavings. 7 years old shavings.

Table 2 : Physico–chemical Properties of Oil Samples of Deodar Produced on Studies on Production Conditions

	Punjab	*U.P.*	*H.P.*	*J & K*			*Oil from powdered*
				1	*2*	*3*	*stumps & root*
Specific gravity (30°C)	0.9500	0.9400	0.9529	0.9491	0.9430	0.9441	0.9290–0.9403
Refractive index (30°C)	1.518	1.515	1.520	1.514	1.512	1.513	1.5038–1.5054
Optical rotation	+34.04°	+35.91°	+32.78°	+40.2°	+40.33°	+38.2°	+48°7′
Acid value	1.17	1.76	1.17	1.2	0.66	0.60	0.24–1.11
Sap. value	11.7	6.0	6.0	22.0	15.6	22.1	5.12–21.8
Carbonyl as $C_{15}H_{22}O$(%)	41.0	39.9	33.3	29.5	33.6	31.0	–

Table 3 : Physico–chemical Properties of various Cedarwood Oil Sample

	Cedarus atlantica		*C. libani*		*C. deodara*		*Cupressus lawsoniana*	
	1	*2*	*3*	*4*	*5*	*6*	*7*	*8*
1. Yield of oil %	—	3–5	—	3.5–6.0	—	—	2.3–4.5	1.6 (average)
2. Sp. gravity (15°C)	0.945–0.968	0.938–0.9587	0.939–0.947	0.940–0.947	0.9570	0.9530–0.9756	0.9430–0.9529(30°C)	0.8913 (20°C)
3. Ref. Index 20°C	1.512–1.517	1.5061–1.5132	—	1.5125–1.5134	—	1.515–1.523	1.512–1.520	1.4760
4. Optical rotation	+34° to +62°	+35°10 to +49°14'	+45°to +55°	+68° to +86°	+43°28'	+34°to +53°8	+3504° to +40°.33°	+46°41'
5. Acid number upto 2		0.50–2.16	upto 1	0.70	—	1.6–5.6	0.60 to 1.76	0.19
6. Ester number	3–11	3.75–8.86	—	2.0–3.0	4.84	4.9–20.5	6.0–22.2	19.3
7. Ester value after acetylation	30–46	28.26–40.01	not less than 20	19.80	33.77	30.8–39.2	—	83.3
8. Carbonyls (as $C_{15}H_{22}O$)%	—	—	—	—	—	—	29.5–41.0	—
9. Solubility (90% alcohol)	1 :1–10	1 :1–6	1 :10 (upto)	1 :5–6	1 :1	—	—	—

Table 4 : Physico–chemical Properties of Cedarwood Oil Sample from Juniperus Spp.

	J. virginanana				*J. mexicana*	*J. procera*		*C. lawsoniana*	
	1[4]	*2*[8]	*3a*[21]	*3b*[21]	*4*[8]	*5*[8]	*6*[3]	*I Shimmel & Co.*	*II Thurber & Roll*
Sp. Gravity (15°C)	0.934–0.964	0.949–0.961	0.9623	0.9406	0.945–0.978	0.968–0.988	0.9453	0.8992	0.8913
Opt. rotation	–18° to –42°	–25°27' –37°15'	–25°48′	52°6′	–33°45′ to –62°3′	–16° to –38°3′C	–47°53′ to –63°48′	40°47'	46°41′
Refractive Index (20°C)	1.50–1.51	1.5030–1.5067	1.5050	1.5020	1.5029–1.5084	1.480–1.509	1.499–1.5141	1.47799	1.4760
Acid number	Upto 1.5	—	1.25	0.26	—	1.15	0.56–1.12	—	0.19
Ester number after acetylation	Upto 12	—	2.67	0.87	—	1–9	0.94–5.66	23.3	19.3
Solubility (90°Alc.)	26–68 1.10–20	— 1 :8–10	38.6 1 :1	13.03 1 :5.5	— 1 :8–10	70 1 :0.5–5	— —	80.3 —	83.3 —

Essential Oil Assoc. of USA., (EOA No. 36)	*Characteristics*	*Juniperus virginian*	*Juniper mexicana*
	Colour and Appearance	Colourless to light yellow slightly viscous liquid	Colourless to yellow, slightly viscous liquid
	Specific gravity (14°)	0.945–0.960	0.950–0.960
	Optical rotation	–27° to –45°	–35° to 50°
	Refractive index (20°C)	1.5020–1.5070	1.5040–1.5070
	Solubility in 95% alc.	1 :5	1 :1

Table 5 : Physico-chemical Properties of Oil Samples of Deodar

		Handa et. al.	*Schinnel & Co.*	*Simonsen & Rau**
1.	Specific gravity	0.9212 (25°C)	0.953–0.9756 (15°C)	0.9592 (30°C)
2.	Refractive index	1.516 (20°C)	1.516–1.523 (20°C)	1.5203
3.	Optical rotation	63°18′	34° to 53° 8′	–45.08°
4.	Acid value	0.13	1.6–5.6	1.6
5.	Exter value	—	4.9–20.5	22.1
6.	Ester value after acetylation	26.6	30.8–39.2	42.3

*Oil distilled from wood from a 75 year old tree. Oil yield 2.5%.

It will be observed that yield of oil depends on the type of material distilled; brown coloured heartwood giving maximum oil yield. In the work at Drug Research Laboratory, it was observed that saw dust on the whole, had an oil content between 2-3 per cent. It is not possible to extract full oil in theoretical yields.

Physico–Chemical Properties

Physico–chemical properties of Cedarwood oils from various sources and of different origin are given in Tables 2, 3, 4, 5. Physico-chemical properties of deodar oil samples are summarized below :

Yield	—	1.6 – 5.0
Specific gravity (30°C)	—	0.9290 – 0.9529
Refractive index (30°C)	—	1.5038 – 1.520
Optical rotation	—	+32.78° to +48.70°
Acid value	—	0.24 – 1.76
Sap. value	—	5.12 – 22.10
Carbonyl (%) $C_{15}H_{22}O$	—	29.5 – 41.0
Solubility (%) (90% alcohol)	—	1 : 0.5 – 20.0

Chemical Composition of C. Deodara

Methyl-4-acetyl-cyclohexane 12.5-13.5% cis and trans atlantones, ester of butyric, caproic enanthic and stearic acid, α and β Himachalene, himachalol, allonimachalol, deodarone.

Cedrus atlantica contains Acetone, 1-methyl-4-acetyl-1-cyclohexene, a sesquiterene, d-cadinene, a sesquiterpene alcohol, α-ketone and γ-atlantone.

Distillation

Optimum conditions for the production of oil both from saw dust and stumps have been worked out in RRL, Jammu. Production of oil from saw dust from various localities was studied. Steam distillation was conducted at 20, 35 and 50 lbs/sq. Complete distillation was done *i.e.,* till last traces of oil were obtained. Yield of oil and their physico-chemical properties are given in Table 6.

Table 6 : Yield and Physico–chemical Properties of Oil obtained by different Processes.

	Hydro Distillation	*Stem Distillation Pressure (lbs/sq")*		
		20	*35*	*50*
Oil yield (%)	3.71	4.18	4.41	3.67
Duration of distillation (hours)	20	12	8	8
Specific gravity (30°C)	0.9672	0.9462	0.9470	0.9140
Refractive index (30°C)	1.5172	1.5145	1.5140	1.5126
Optical rotation	40.90	46.20	40.33	47.92
Acid value	0.59	0.59	1.18	1.19
Sap. value	3.27	4.00	15.16	4.00
Sap. value after acetylation	31.80	12.00	18.5	26.40
Carbonyls as $C_{15}H_{22}O$(%)	33.90	33.5	33.6	26.20

Perusal of data would indicate that distillation at 35 lbs/ sq" steam pressure gives optimum quantity of good quality oil.

Agarwal *et. al.*, conducted similar studies with variable quantity of steam used for distillation. Their batch size was also higher. Stumps and roots were cut to size and disintegrated to 4–8 mesh size. Data on recovery of oil is given in Table 7.

Table 7 : Recovery of Oil at Different Rates of Steam Injection

Rate of steam injection (kg/nt)	*Oil Yield (kg)*	*Yield oil on weight of feed (%)*	*Oil recovery on theoretical yield (%)*
1. 3.0	47.33	6.31	63.10
2. 3.5	51.00	6.80	68.00
3. 3.7	51.70	6.90	69.00
4. 4.0	54.00	7.30	73.00
5. 4.5	58.27	7.77	77.70
6. 5.0	55.50	7.40	74.00

It was thus shown that economic cost of production can be achieved if steam is injected @ 4.5 kg/minute for 10-12 hours. Longer hours add to cost of production without giving any advantage. These studies also revealed that stumps and roots of deodar can give oil at lower cost than that obtained from saw dust of timber.

Latest Research Work

A new approach for increased production of Deodar oil involves extraction of the saw dust with a solvent. It was observed that steam distillation of the material can extract only 50 per cent of the available oil (or extractable odouriferous principals). Yield improvement in steam distillation of oil can be made to some extent by prolonging distillation period which becomes uneconomical due to slow recovery of oil with the advancement of distillation. Extraction of saw dust and stumps is done with solvent oil having 60-80°C boiling range. Alcohol also gives good yield but the product so obtained is resinous and deeply coloured. Some data on extraction of stumps is given in Table 8.

Table 8 : Extraction of Deodar Wood Stumps with Solvents product obtained by

	Steam Distillation (40 lb/sq")	*Solvent Oil (60–80°C)*	*Alcohol Extracted*
Size of batch	70 kg	2 kg	–
Duration	8 hrs.	4 hrs (cold)	–
Yield of			
(*i*) Extractive	–	23.8%	38.2%
(*ii*) Oil	7.5%	18.6%	21.7%
Colour	Normal	Light	Dark
	Pale yellow to deep yellow		

Trials were also undertaken in DRL on deodar stumps and saw dust obtained from deodar timber (yellowish heartwood). Data are given in Table 9.

It may be seen from the data that :

1. Yield of oleoresin in stumps is 4 to 5 times more than the oil produced by steam distillation.
2. Steam distilled oil yield from oleoresin is nearly double than produced by direct steam distillation of stumps.
3. In saw dust from yellowish heartwood, the yield of oleoresin is nearly double than the oil produced by direct steam distillation.

Physico-chemical properties of oil produced from oleoresin is within ISI specifications having acceptable odour. Data is given in Table 11.

Table 9 : Extraction of Deodar Stumps and Saw Dust from Timber with N-Hexane

Yield of	*Material used*		
	Stumps	*Saw Dust*	
		Cold Percolation	*Hot Percolation*
1. Oleoresin	21.1%	4.37%	3.63%
2. **Oil yield**			
(*i*) On oleoresin	45.66%	–	–
(*ii*) On wood	11.54%	–	–
3. Oil yield on steam distillation	5.6%	2.04%	1.84%
4. Percentage increase in yield as compared to steam distilled oil :			
(*i*) Oleoresin	34.8	114	97.4
(*ii*) Oil from oleoresin	94.3	–	–

Data on quality of oil sample as examined by GLC is given in Table 10.

Table 10 : Composition of Oil Samples Produced by Steam Distillation and from Oleoresins of Cedar Wood Stumps

Component	*Oil obtained from*			
	Solvent oil extractive	*Alcoholic extractive*	*Steam distillation*	*Market oil*
1.	–	–	–	0.20
2.	–	–	–	0.25
3.	1.10	0.73	0.50	0.33
4.	0.60	0.82	3.70	2.00
5.	0.90	1.00	1.40	1.17
6.	1.00	0.82	0.80	0.55
7.	13.00	11.36	22.60	16.50
8.	34.20	36.55	56.90	65.12
9.	1.70	3.82	1.50	0.17
10.	1.50	1.36	1.00	0.75
11.	1.50	1.64	1.40	1.42
12.	1.80	0.82	1.50	0.17
13.	1.60	1.00	–	0.17
14.	5.00	5.45	–	1.25
15.	9.40	10.00	2.00	2.00
16.	10.80	13.00	2.00	2.50
17.	15.40	11.63	5.00	5.17

It will be observed from the above that while only first two fractions (minor) are absent in the solvent extracted oils, other components are present. High boiling fraction which contributes to the fixative value of the oil are present in higher percentage as compared to market samples. Olfactory properties of all the samples are nearly identical.

Uses of Cedarwood Oils

The oils of cedarwood *i.e.*, Atlas cedarwood and Virginiana cedarwood are used in perfumery, scenting of soaps and as an odour fixative, technical preparations, room sprays, disinfectants and similar products.

Virginiana cedar wood oil is used for clearing microscopic sections and immersion lenses after its selective rectification.

Crude Texan cedarwood oil, besides uses as given above, has wider application in technical preparations *e.g.*, deodorants, insecticides, moth proof clothing bags, floor polishes, janitor's supplies, lubricating greases, etc.

Oil of Himalayan cedarwood has completely replaced the imported American and other cedarwood oils. Besides its normal uses, the oil has also been found to be useful for clearing microscopic sections and with immercian lenses as in the case with American cedarwood oil.

Recent studies conducted at the CDRI Lucknow have indicated that some fractions of deodar oil have potent spasmolytic activity *i.e.*, cedral and deodrol are 1/10th to 1/2 as active as papaverine while isoallohimachalol was nearly equivalent. Himachalol and allohimachalol were the most active agents and 1.5-4 times more active than papaverine.

Table 11 : Physico–chemical Characteristic of Solvent Oil Extractive from Stumps

Characteristics	*Solvent Oil extractive*		*Stem distilled oil*	*IS : 1615–1970*
	Vaccum distilled oil	*Stem distilled oil*		*Steam distilled oil*
1. Colour	Pale Yellow	Pale Yellow	Pale Yellow	Pale yellow to deep yellow
2. Odour	Balsamic	Heavy balasanic	Balasanic	Woody, balsamic
3. Refractive density 30°C/30°C	0.9663	0.9330 (40°C)	0.9487	0.9300–0.9380
4. Optial rotation	—	58.5°	—	42° to 65°
5. Refractive index (30°C)	1.6153	1.5087 (40°C)	1.5104	1.508–1.510
6. Acid value	10.4	0.73	0.4	1 (Max.)
7. Ester value	19.6	11.34	6.3	10–20
8. Ester value after acetylation	58.1	25.8	27.7	20–40
9. Carbonyl content as atlantone $C_{15}H_{22}O$ (Percent by weight)	—	—	—	13 (Min.)

CHAPTER 16

ESSENTIAL OIL OF *Juniperus* Spp.

Genus *Juniperus* (Family Cupressaceae) comprises of about 50 species of hardy or nearly hardy evergreen trees or shrubs native chiefly of the cold or warm temperate regions of Northern Hemisphere with one species extending south to Kenya. Leaves remain green for several years and even after death may remain on the shots for indefinite period. Flowers are unisexual. In some species flowers of both the sexes occur on the same plant. All the species can be propagated through seeds; seeds retain viability for several years when stored in cool dry place. Juniper like open positions and chalky or limestone soils.

Five species are reported to occur in the Himalayan range; *Juniperus communis, J. nacropoda, J. wallichiana* Hook.f.Syn., *J. pseudosabina* Hook, *J. recurva* var. squamata. A few species introduced are : *J. virginiana* Linn (Red cedar, Pencil cedar); *J. procera; J. bermudiana* Linn (Bermuda cedar) and *J. chinensis* Linn (Chinese Junier). According to a recent report, world's second biggest forest of Juniper extending upto 260 sq. km. exists in Pakistan.

Some of the Juniper species (except *Juniperus procera* Hochst commonly known as East African cedarwood and *J. virginiana* whose oils are produced from wood) give useful oil from the leaves and berries. Commercial production of oil is, however, done primarily from the berries of *J. communis*. L. Oil from the berries of *J. oxycedrus* L. was observed to be of academic interest only whereas oil from berries of *J. macropoda* has been recommended as a substitute for imported oil in India.

Juniper berry oil is widely used in essences for the flavouring of beverages and liqueurs, particularly of the gin and sloe gin type. Several European liqueurs, 'Steinhager' among them owe their characteristic flavour chiefly to the presence of juniper berry oil.

The oil has been known for a long time as a diuretic. Because of the local irritating effect on inflammed organs, however, dosage has to be fixed very carefully.

Regarding occurrence and use of various species of *Juniperus* some information is given below.

J. communis Linn is commonly known as Aarar Hindi, Petthri (Punjabi), Havusha (Bombay), Vapusha (Sanskrit), Abhal (Deccan), Chichia (Kumaun), Betar, Pethra (Kashmiri). It grows in Western Himalayas from Kumaun west-wards at an altitude of 3750-4200 m. Its fruits and oil are reported to be diuretic, carminative, stimulant, used in dropsy, gonorrheae, gleet and some cutaneous diseases. *J. excelsa,* Bieb occurs in Himalayas and from West Asia to Greece. *J. Macropoda* Boiss is commonly known as Dhup (Hindi and Garhwali), Dhapi (Nepalese), Luir, Shurgu (Punjabi) occurs in Himalayas from Nepal westwards upto 4200 m. and Baluchistan. Fruits are reported to have some medicinal properties as those of *J. communis, J. recurva* Ham is commonly known as Bettir (Hindi), Tupi (Nepalese), Bettar, Wetyar (Punjabi). Smoke from green wood is emetic producing long continuing vomiting. It occurs in temperate and alpine Himalayas from 2250-4500 m.

Survey of literature reveals production and examination of oils, from berries, leaves and wood of various species of *Juniperus* even though oil acceptable to and used in the industry is from the berries of *J. communis* L. Brief mention of all these types of oils is given in this study.

Oil from Berries

Oils produced from various species and studied by various workers are given below :

(*a*) ***Juniperus communis* L.** : Comprehensive details on this oil are given by Guenther. *J. communis* L., the common juniper is a shrub or tree attaining a height of 3m. or more with trunk diam. of 25-30 cm. It grows wild in many parts of Europe and Asia, including Western Himalayas and North Eastern Asia. It also occurs in North America. It is a variable species having many geographical varieties. Only berries from *J. communis* var. *errecta* Pursh yield good quality oil. This variety thrives on sunny heaths and clearings in pine forests and mountain slopes exposed to sun. It often occurs in thick impenetratable patches covering as many as 70 hectares. *J. communis* is a diocious plant. Both ripe (blue; 2 years old) and unripe berries (green; 1 year old) are found on the same tree. Oil content varies from 0.5-2.0 per cent; blue berries having more oil than green ones. Quantity and yield depends to a great extent on the geographical origin. Of all the producing regions, Italy is the largest producer; other countries being Hungary, Czechoslovakia, Yogoslovia and Austria.

For production of oil, in one method, berries are crushed, immersed in warm water and fermented. After completion of fermentation, the mass is pumped into a still provided with a rectification column and distilled with constant stirring. The distillate consists of water, alcohol and oil. This distillate is rectified till an alcoholic liqueur containing 40-50 per cent alcohol is obtained. Most of the essential oil separates out in the receiver. One ton of berries yield 16-18 litres of alcoholic beverage and 5-6 kg. of an essential oil. This however, cannot be termed as a complete essential oil which is deprived of oxygenated compounds due to the solvent action of alcohol. However, best quality of essential oil is obtained by regular distillation of berries without prior fermentation. Yield of oil in this case varies from 0.8-1.6 per cent.

According to Guenther, pure oils (European), have following physico-chemical characteristics :

1. Specific gravity (25°/25°C) – 0.860 – 0.875
2. Optical rotation – –7°45′ to –13°6′
3. Refractive index (20°C) – 1.4780 – 1.4819
4. Saponification number – 2.8 – 8.4
5. Solubility – Sometimes soluble in 0.5 vol. of 95% alcohol. Usually opalescent to cloudy in 95% alcohol upto 10 vol.

Genuine oil samples having slight dextrorotation have also been reported. Only Norwegian oils have dextrorotation ranging between 33°20′ to 41° (such oils are produced from berries and young shoots as specified by Norwegian Pharmacopoeia).

Solubility of the oil decreases with age; freshly distilled oils are soluble in 90% alcohol.

Recent study on the G.L.C. examination of oil from ripe and unripe berries of *J. communis* in Romania has shown that oil from ripe fruits were richer in hydrocarbons (72.2%) as compared to oil from unripe fruits (53.3%) which on the other hand had higher sesquiterpene content (30%) as against 1.0% in the former. Oil from ripe fruits had higher 1 : 4–cineole content (6.8%) and esters (5.1%) and also higher content of α-pinene, camphene, β-pinene, ρ-menthene, *p*-cymene, limonene oxide, terpinen –4–01 and bornyl acetate. Oil from unripe fruits had higher content of myrcene, phellandrene and nerol.

Oil from berries of *J. communis* L. growing in India was examined by Handa *et al.* Ripe dry berries procured from the Sindh valley on steam distillation yielded 0.77% oil having d^{15}0.9388 and n^{D} 1.488. Besides essential oil, berries contain oxalic acid, resin 8% and Juniperin 0.36%.

(*b*) ***Juniperus macropoda*** **Boiss** : Known as the Himalayan Pencil Cedar and Padam is a tree attaining a height of 12-15 m. with 1.8-2.1 m in girth. It is a native of the arid tract of inner Himalayas from Nepal westwards to end of Western Tibet at altitudes from 1500-4200 m. distributed to Afghanistan, Baluchistan, Iran and Arabia. The plant is highly resiniferous. In Buddhist regions of inner Himalayas. *J. nacropoda* is considered sacred and is frequently planted near Temples; the fruit and wood form part of incense. Dry berries procured from Gurez valley gave 3.3% oil having sweet smell and d^{15}0.8571 and n^{D20} 1.473. The oil was found to closely resemble the imported oil from *J. communis* L.

The oil from the berries of *J. macropoda* was later on studied by Talwar. Oil was obtained by stem distillation of air dried fruits in an yield of 1.5 per cent. Physico-chemical constants of the oil sample were d^{30} 0.8319; n^{D30} 1.4700; $\alpha^{(D)}$ 9.3°, acid value 0.0; sap. value 0.1, sap. value after acetylation 12.1. The oil was found to contain sabinene 25.2%; α-pinene 17.0% cedrene 11.6%; α-terpinene 10.0%, aromadendrene 7.9%, cadinene 6.90%, terpinolene 1.6%, sabinol 1.7% and terpinene –4–01, 1.4%.

Oil sample studied earlier had following properties :

	Tehri Garhwal	*Baluchistan*	*Gurez*
Yield (%)	0.66	1.10–2.04	3.30
Specific gravity (20°C)	0.9006	0.8343–0.0790	0.8531
Refractive index (20°C)	1.4733	1.461–1.468	1.473

Oil from berries of *J. macropoda* is reported to have been used in India as a substitute for oil of Juniper during the World War II.

(*c*) ***J. Foetidissima*** : Oil produced in Azarbiadzan was found to contain β-pinene, myrcene, fenchone, limonene, γ-terpi-nene, ρ-cymol, camphene, phellandrene, α-fenchene, β-bisabolene, cadinene, cedrel and β-bisabalol.

(*d*) ***J. polycarpus*** : Oil from berries produced in Azarbaidzan had cedrol, α–pinene camphene, β-pinene; sabinene, cedrene, –elemene, cedrene, cadinene, α-curcumene, calamenene, calacorene, Δ^3–carene, myrcene, α-phellandrene, γ-terpineol, *p*–cymol, terpinolene, 2–undecanone, thujone, camphor, α-terpineol, linalool and borneol.

Oil from Leaves : Oil from berries along with young shoots is reported to be produced in Norway only. The oil seems to be an item of commerce. Pure leaf oil is not reported to be used in industry. Other such oils reported are :

(*a*) ***J. communis* L. :** Oil from leaves was produced in Romania in an yield of 0.3 per cent containing 56 compounds; 33.7% of which was α-pinene, 4.6% 4–terpineol, 4% 1 : 8-cinene 3.6% *p*–cynol, 3.3% β-phellandrene, 3% γ-terpinene, 27.6% sabinene and 5.5% myrcene.

(*b*) ***J. macropoda* Boiss** : Oil from leaves seems to have been reported and examined in India only. Talwar *et. al.*, obtained oil from air dried leaves in an yield of 1.8% having d^{20} 0.8574; n^{D20} 1.4720, $\alpha^{(D)}$ 58.20, acid value 0.52, ester value 36.46 and ester value after acetylation 46.46. The oil was found to contain mainly camphene 4.47%, sabinene 64.19%, α-pinene 8.03%, terpinen –4–01 7.51% and cadinene 1.62%. The oil was not considered a suitable substitute for imported oil.

Kapahi *et. al.* produced oil from leaves of *J. macropoda* in an yield of 2.0% (on dry weight basis) having d^{20} 0.884 $\alpha^{(D)}$ 44°52; n^{D20} 1.4780, acid number 2.18; ester number 10.40 and ester number after acetylatern 85.6. The oil was soluble in all volumes of 95% alcohol. Oil showed the presence of 37 components out of which limonene 18.37%, methyl heptenone 8.62%, 4–terpineol 15.14%, α-terpineol 2.47% caryophyllene oxide 4.07%, and cedrol 2.31% were identified. According to the authors, the oil possessed light ninty, woody odour with a residual note of vetiver and recommended its use in scenting of soaps, toileteries and deodorants.

(*c*) ***J. oxycedrus*** **L.** : Oil from leaves was obtained by Rutovski and Vinogradova in Crimean Peninsula in an yield of 0.01% having d^{20} 0.9720 $\alpha^{(D)}$ 16°19′, n^{D20} 1.4956, acid number 2.0, ester number 50.68. Oil was soluble in 0.5 vol of 95% alcohol.

Four samples of the essential oil from leaves of *J. oxycedrus* Subsp. *oxycedrus* and *J. oxycedrus* Subsp. *J. macropoda* from Sardinia showed from following characteristics :

n^D 1.4732 – 1.4768; d^{20} 0–8651, 0.8750; $(D)^{20}$ –3.95° to 4.52°, acid value 0.95–2.25, phenols 0.2 – 0.7%. GIC examination of the oil samples showed presence of β-pinene, camphene, α-pinene, Δ^3-carene, myrcene, limonene, γ-terpinene, β-cymene, caryoplyllene and humulene.

Oil from Wood : Oil from wood of Juniperus species is reviewed as under.

(*a*) ***J. macropoda*** **Boiss** : Oil from heart wood studied by Sood was found to contain camphene, β-pinene, limonene, phellandrene and γ-cadinene.

(*b*) ***J. oxycedrus*** **L.** : Huerre, Gattfosse & Igolen and Sontog studied properties of oil produced by stem distillation of wood which are given below :

Oil Samples Studied by

Properties	*Huerre*	*Gattfosse & Igolen*	*Sontag*
1. Specific Gravity (15°C)	0.925–0.927	0.9476	0.9632
2. Optical rotation	–31°25'	21°4′	–28°34′
3. Refractive index (20°C)	—	1.5145	1.5125
4. Acid Number	—	—	0.84
5. Ester Number	—	—	6.31
6. Ester Number after acetylation	—	—	58.92
7. Ester number after formylation	—	—	94.00
8. Ketone content (calc. as $C_{15}H_{24}O$)	—	13.6%	6.54%

The oil contained α-cadinol, cadinene and same unidentified sesquiperpenes.

It will be now clear from the discussion so far that there was no good substitute for the imported oil among the oils described. While working on the aromatic plants of Kashmir, it was found that leaf oil from *J. recurua* var. squamata could be used as a suitable substitute for commercial Juniper oil. This paper discusses details of this study.

Juniperus Recurva **Var. Squamata :** Commonly known as weeping blue juniper, Bhulu Thelu, Bhedara, Watyr (W. Himalaya); Tupi (Nepalese), Chwkboo, (Sikkim) is a prostrate or documbent shrub or a small tree of dropping habit attaining a height of 9-12 m. and is found throughout the temperate and alpine Himalayas and Assam (2100-4500 m. altitude). Leaves awl dull or greyish green; flowers nonoecious or dioecious; fruit ovoid; seed solitary ovoid.

The plant occurs gregariously and has medicinal properties. The wood, leaves and twigs are all aromatic and are burnt as incense. Smoke from the green plant is considered to be emetic.

Literature survey indicated that study has not been carried out on the essential oil from leaves of this species of Juniper.

For the present examination leaves were obtained from plant growing near Khilanmarg. It takes about 7-10 days before the leaves reach the distillation site (Drug Factory, Srinagar). Leaves are mostly green admixed with small quantities of brown and yellow leaves. Leaves are not separated from twigs but are distilled as such.

The area remains snowbound normally from October-November to June-July. Oil content was, therefore, studied during June end to November and data given below :

	Period of harvesting and distillation	*Oil content (Zero-moisture basis)*
1.	June (end)	1.2 – 1.4 per cent
2.	July	0.8 – 1.0 per cent
3.	August	1.0 – 1.5 per cent
4.	September	1.0 – 1.1 per cent

Experimental : The dried leaves on steam distillation terpenic a pale yellow oil (1.1%) having a soft, sweet terpenic odour with the following physico-chemical characteristics.

Composition of the oil was determined by TLC and GLC.

1.	Colour	Pale Yellow
2.	Refractive Index (24°C)	1.4752
3.	Specific gravity (24°C)	0.8827
4.	Optical rotation	+52.0°
5.	Acid value	1.90
6.	Ester value	26.80
7.	Ester value after acetylation	38.54
8.	Phenols	0.3 per cent

TLC of the Oil : 20 × 20 cm. glass plates were coated with a slurry of Silica Gel G, 0.25 mm thick, activated at 110°C for an hour. TLC was performed in two different solvent systems. Spots were visualized and characterised by colour reaction with spray reagents and also by co-TLC. Data is given below :

S.No.	*Component identified*	*HRf value*	
		X_1	X_2
1.	Hydrocarbon	82	84
2.	Citronellal	72	75
3.	Isobornylacetate	65	65
4.	Borneol	48	51
5.	Terpineol	32	34

X_1 – Benzene : Ethylacetate (9 : 1) X_2 – Chloroform : Ethylacetate (98 : 2)

Spray Reagent : 1. Vanillin in Sulphuric acid.
2. Ceric Sulphate-Sulphuric acid.

Spot No. 1, comprising mostly of hydrocarbons was isolated from the oil by preparative TLC and subjected to argentation TLC as described by Gupta *et. al.* Visualization of the spot was done by exposure to iodine vapours. Components were identified by comparison with hRf value of authentic samples.

S.No.	*Compounds identified*	*HRf value*	
		X_1	X_2
1.	α-Pinene	72	75
2.	Camphene	63	70
3.	Limonene	51	57
4.	ρ-Cymene	42	59

X_1 – Benzene Pure X_2 – Benzene : Ethylacetate (8 : 2).

GLC of the Oil : Gas chromatographic examination of the oil sample was carried out on Perkin Elmer Model 3920 having a column 3.6 m long with 2mm. dia m. Other conditions of the operations were as under :

Stationary phase	UCCW 10% on chronosorb (WAWDMCS) 80–100 nesh.
Column Temperature	70–200°C
Detector and injector temperature	250°C
Sample size	6 ul (1 :30 solution in Chloroform).

Peak No.	*Compound*	*Percentage*
1	*2*	*3*
1.	α-Pinene	0.02
2.	Camphene	17.68
3.	u.i.	0.03
4.	Myrcene	8.96
5.	u.i.	2.65
6.	Sabinene	0.46
7.	Limonene	4.57
8.	Terpinene	24.01
9.	u.i.	0.01
10.	u.i.	4.52
11.	u.i.	Traces
12.	u.i.	1.78
13.	Camphor	0.02
14.	u.i.	0.65
15.	u.i.	0.05
16.	Terpineol	7.99
17.	Isobornyl acetate	0.50
18.	Nerol-citronellol citral 'a'	0.48
19.	Geraniol	0.13
20.	Citral 'b'	0.47
21.	u.i.	0.01
22.	Thymol	0.14
23.	u.i.	0.07
24.	Eugenol	0.28
25.	u.i.	0.08
26.	Methyl Eugenol	0.08

(Contd...)

Peak No.	*Compound*	*Percentage*
1	*2*	*3*
27.	u.i.	0.11
28.	u.i.	0.09
29.	Isoeugenol	0.26
30.	u.i.	0.03
31.	u.i.	0.28
32.	u.i.	0.31
33.	u.i.	1.29
34.	u.i.	3.20
35.	α-Cadinene/β-Bisabolene	3.37
36.	u.i.	9.61
37.	u.i.	0.87
38.	u.i.	0.44
39.	u.i.	Traces
40.	u.i.	0.67
41.	u.i.	0.01
42.	u.i.	0.31
43.	u.i.	2.14
44.	u.i.	1.08

*(u.i. – Unidentified).

** In all 44 peaks were obtained. Compounds were identified by comparison with the retention times of authentic samples.

Out of 44 peaks only 17 could be identified. Work on the identification of peaks at 36-44 is under progress.

The oil is now produced regularly on large scale in the Drug Factory of D.P.L. at Srinagar and is now in use in the industry.

CHAPTER 17

OIL OF AGARWOOD

Agarwood (Aloewood or Eaglewood) known as *Sasi* or *Sashi* (Assamese), *Agar* (*Hindi, Tamil*), *Hindigard*, (*Bombay*), *Agru* (*Telegu*) is a source of agar (*Agru*), is the fragrant dark brown resinous portion of wood of a large evergreen tree (*Aquilaria malaccensis* Hook. Syn. *Aquilaria agallocha*, Roxb. (Fam. Thymelaeaceae) which grows mainly in the Khasia and Garo Hills, Goalpara, Darrang, Nowgong, Sibsagar, and the Cachar division of Assam. It is also found in smaller quantities in Sylhet, in the foothills of Bhutan, in Tripura and in the Martban Hills of Burma, Assam, the most important producer in the past, seems to have lost monopoly over this valuable wood.

Healthy wood of *A. agallocha* does not contain agarwood. Unlike aromatic woods of deodar and sandal, in agarwood, dark brown oil bearing resinous irregular patches develop as a result of fungus and/or injury. General indication of presence of agarwood is the diseased appearance of the top and side branches of the tree, welling or depressions and cankers on the bole and appearance of hosts of ants in the fissures on the bole. Wood also assumes a distinctly yellowish tinge when agar formation takes place while the normal wood in the healthy tree is of a pale cream-buff colour. Agar wood is collected by professionals when the tree showing diseased appearance is partially or completely killed. Distillation of agar oil, is confined only to the Sibsagar and Dhansiri valley Forest Division of Assam.

Agarwood is a highly valued item used as incense from ancient times in Egypt, Arabian countries and Far East. Essential oil obtained on distillation is used in high class perfumery compounds, as fixative for its balasmic odour. Agar *attars* and perfumes have been prepared and used in India for a long time. Inspite of its unique position in the field of perfumery industry, agarwood has not been given the attention it deserved, even though work on agarwood was initiated as early as achieved till 1977 on the causes of development of agarwood are still inconclusive.

To recapitulate the work done so far, Tunstall isolated three fungii *i.e.*, (*i*) *Aspergillus* sp.; (*ii*) *Penicillium* sp; and (*iii*) *Fusarium* sp. Out of these *Aspergillus* sp. is reported to be primarily responsible for the production of agar in the trees; 20 healthy trees inoculated with this fungus gave encouraging results with the formation of agar in the infected wood. Later on, Bose isolated a fungus of the type of Manilla. He inoculated 36 trees with this fungus; agar formation was traced in the inoculated region in some of them. Borings taken within 2 cm. radius around the point of inoculation smelt strongly of agar but agar formation did not seem to have spread beyond this radius. Barua's work is equally interesting and informative. He isolated a fungus identical to *Epicoccum granulatum* from the diseased wood. There was some evidence of oil being formed on chips of agarwood when inoculated with this fungus under controlled laboratory conditions. It has also been observed that formation of oil takes place when the tree is in the moribund state due to fungus attack. Experiments conducted by the Silviculturist in the Forest Deptt., Assam, on the inoculation of healthy agar with all the above mentioned fungii have also revealed formation of some oil. Rate of development of agar in the inoculated trees was, however, slow and about 2 cm. per year.

Formation of agar patches was also observed due to mechanical injury alone. It is quite likely that injury was followed by infection by some fungus. Role of any particular fungus was thus not clearly established.

Anatomical studies carried out in the Forest Research Institute, Dehradun have shown that a peculiar type of abnormal tertiary wood is found in the phloem of scented wood specimen only. The abnormal tissue appears to be due to some physiological disturbance resulting from injury by wounding, insect or fungal attack. Agarwood could, therefore, be regarded as the result of either mechanical and/or fungal injury. The insect damage does not appear to play any role in the formation of agarwood in the trees.

Development of agarwood is slow. According to the Silviculturist, Forest Deptt., Assam, as per an old report infection remains confined to the periphery of the wound for as long as 15 years. According to Hooper agarwood is formed in young trees about 20 years old but as the infection takes time to mature, the trees of about 50 years have the highest concentration of agarwood.

Latest report on the artificial induction of agarwood is by Varma, who selected a 15-16 year old plantation, (raised in 1957-58) at Melong centre in the Mariani Range, Sibsagar Forest Division of Assam. Experiment was started in 1973. The over-wood in the plantation consisted of *Terminalia myriocarpa*. Hourck and Muell–Arg. (*Hollock*) and *Bombax ceiba* Linn. (*Semal*) planted during 1946. Effect of auxin herbicides, 2, 4-D, 4, 5-T and MCPA in concentration of 0.1 per cent and 0.25 per cent was studied on 26 healthy trees of *A. agalloocha* ranging in girth from 38 cm.–63.5 cm. Herbicides were applied in two frequencies *i.e.,* once in a year and after every 2 months. In all, there were 13 treatments including control. Herbicides were applied by making 2 holes of 1.25 cm. diam. opposite to each other, 50 cm. and 75 cm. above ground level respectively. Simultaneously, another indicative experiment was also conducted on 6 trees with grith ranging from 35-43 cm. to study effect of mechanical injury on the formation of agarwood in the tree.

Observation taken after one year did not reveal any symptoms of agarwood formation. Occular examination of the sample of wood taken from the point revealed a brownish black stain around the holes which gave faint odour of agarwood on burning. Samples taken for observation in herbicide treated experiment were distilled for 8 hours. Yield of oil varied between 0.083-0.25 per cent (yield of 0.38 per cent obtained from one treatment, according to Varma, was due to some error). The results showed that auxin herbicides treatment did not significantly induce formation of agarwood in the living tree till one year after the treatment. In the second trial involving mechanical injury external characteristics of the trees did not indicate formation of agarwood. According to Varma, due to slow formation of agarwood, more data will have to be gathered and analysed after some years.

Work done during last about 50 years by various workers while giving some indication on the probable reasons for agarwood formation have not proved conclusively the mode of agarwood formation. Unless more concerted efforts are made involving various probable aspects over a very long period, new knowledge on agarwood is not likely to be obtained.

Watt classified agar into 3 categories :

(*i*) *Charki* – (It sinks in water)

(*ii*) *Samaleh* – (It floats on water)

(*iii*) *Neem-Gharki* (Partly sinks in water) or *Samaleh–i–aala.*

These terms are not obsolete. Instead, following 4 grades are known. Yield of agarwood from the tree as also oil content is given below :

Grades	*Name*	*Colour*	*Yield*	
			Kg/ton	*Oil content*
Grade I	*Black Agar*	Black	9.45–12.6	—
Grade II	*Bantang*	Brown	—	1–1.25%
Grade III	*Phutes or Kalaguchi*	Brown interpersed with yellow coloured wood	7.9	0.78%
Grade IV	*Dhum*	Yellow with, streaks of brown and black resin	2.52–3.78	0.25–0.37%

In trade, there are further sub-grades.

Grade I is true agar and is heavier than water. Gradation of wood is done by wood collectors and is not based on any well-defined principle. One grade, therefore contains small quantities of lower and/or the higher grades.

First 3 grades of agarwood are exported to Arabia and other countries for use as incense. These are also used in India for burning as incense particularly by *Parsees.* Lower grade of agarwood are normally used for oil extraction purpose.

Some times, sapwood also becomes infested with fungus and develops some odour. This wood is called *Kolagasi.* It is used only as a cheap incense.

For the production of oil normally grade 4 is used, occasionally grade 3 is also extracted; the oil obtained is reported to be superior to that extracted from grade 4 wood. Oil content is also better in this grade. The oil has also higher specific gravity.

Distillation is done in old type of units and is supposed to be a specialised job mastered only by the few. Distillation equipment consists of the following :

1. Circular narrow necked copper still of about 75 cm. diam.
2. Still-head carrying a swan neck comprising of 2 tubes joined at an angle.
3. Cylindrical vessel made of copper (20 cm diam. and 45 cm high) which acts both as condenser and received. This is cooled in a tub (60 cm diam. & 60 cm high) containing cold water.

Details of distillation method are given by Rao *et. al.* and Gulati. Process and equipment of distillation have not changed since then.

Distillation equipment is set up. Swan neck is connected with the receiver by a hollow bamboo delivery pipe fixed in the mouth of the condenser-cum-receiver. Still is housed on a furnance while the cooling tub is embedded in the ground.

For distillation, wood is first made into chips and soaked in water for 4-5 days. Thereafter, the chips are ground into smaller pipes. This material is again soaked in water for 2-3 days and charged into stills. Ratio of material to water is about 1 : 9 to 10. The finer material better and more economical is the distillation.

Water is added till the level reaches almost the lower side of the neck. Heating is started and joints between the still and the still head and other places are closed with flour-paste. Distillation is done slowly without allowing water to boil. One charge takes 3 days involving 10-12 hours of heating per day for *dhoom* wood. Better grade take as long as 10 days for complete distillation. It is believed that valuable constituents distill over towards the end of the distillation. Most of the oil, however, distills over in first 1-2 days. Distillation is discontinued when only about 2-3 gm of oil is obtained in a day. Oil is collected from the distillate by hand. Oil and small quantity of water thus collected is kept in a bottle and allowed to stand for about 7 days in the sun. Clear oil is decanted and put in dry bottles.

On an average from 100 kg of *dhoom* wood about 700–1000 gm of oil is obtained. It is reported that odour of agarwood oil improves on keeping. Improved method suggested by Sadgopal and Varma, to the best of our knowledge, has not been adopted so far.

Solvent extraction of the agarwood did not yield a suitable product. Extraction with alcohol gave oil, resin alongwith other impurities. Treating the product with alkali and thereafter macerating with ether gave a product devoid of pleasant, agar odour. This, however according to present author needs re-examining so as to produce a commercially acceptable product.

Physico–chemicals Properties and Chemical Composition: Oil of agarwood is normally produced from Grade 4 *i.e., Dhoom wood*. Systematic work conducted at Forest Research Institute, Dehradun indicated following data from various grades of agarwood.

Quality of Agarwood Distilled	*Yield of Oil (%)*	*Specific Gravity (25°C)*	*Refractive Index (25°C)*
1. Black Agar (Heavier than water)	2.19	1.0118	1.5020
2. Superior black wood	1.91	1.0118	1.5012
3. Superior brown wood	1.11	1.0080	1.4948
4. *Dhum* wood :			
(*i*) Superior	0.84	—	1.4866
(*ii*) Normal			
(*a*) Water distilled for 30-32 hours	0.40	1.0116	1.4920
(*b*) Steam distilled for 12 hours	0.102	0.9880	1.4878
(*c*) Water distilled for further 10 hours	0.06	0.9457	1.4898
5. Brownish wood	0.23	1.0075	1.5030
6. Wood dust	0.15	0.9934	1.4841
7. Yellow wood	0.90	—	1.4940

Marc from 'c' in a laboratory test 0.22% of additional oil.

Samples of oil produced by indigenous method from various grades of wood were analysed by Bhatia and Rao.

Data are :

Sample	*Sp. Gr.*	*Ref. index (20°C)*	*Opt. Rot*	*Acid value*	*Ester value*	*E.V. after acetylation*
1. *Agar*	1.0435	1.5146	–15.8°	13.2	18.3	—
2. *Bangtang*	1.0802	1.5115	–16.2°	7.6	22.4	104.9
3. *Bhuta*	1.003	1.5113	–13.2°	9.4	25.1	110.8

(Contd...)

Sample	*Sp. Gr.*	*Ref. index (20°C)*	*Opt. Rot*	*Acid value*	*Ester value*	*E.V. after acetylation*
4. *Dhoom*						
(*a*)	0.9959	1.4910	−17.8°	8.6	27.1	115.1
(*b*)	0.9980	1.5110	−17.8°	6.8	23.6	120.6
(*c*)	0.9930	1.5095	−14.0°	8.8	24.2	108.1
(*d*)	0.9920	1.5100	−14.0°	8.7	23.9	107.8
(*e*)	0.9911	1.6090	−13.6°	9.0	24.8	111.2

It may be seen from the above that in general oil from *Dhoom* wood is lighter than water. Other physico-chemical constants do not show any pattern. Oil samples were soluble in 1.0 to 2.0 volumes of 80 per cent alcohol (20°C).

According to Rao and Bhatia, recovery of oil from agarwood by indigenous method is about 75 per cent as compared to improved method (wood distilled for 35 hours) with proper condensation of the stem oil vapours. There is light change in the physico-chemical properties of the oil samples also as given below :

	Oil sample produced by	
Physico-Chemical Characteristics	*Indigenous Method*	*Laboratory Distillation*
1. Yield of oil	0.40%	0.54%
2. Colour	Dirty brown	Light golden yellow
3. Density	0.9959	0.9811
4. Optical rotation	−17.8°	−1869
5. Refractive index	1.4910	1.5100
6. Acid value	8.6	6.6
7. Ester value	27.1	28.2
8. Ester value after acetylation	115.1	112.4

Not much work has been done on the chemical composition of oil of agarwood, Maheshwari *et. al.* found the oil to contain Agarol hydroxyagariphilone and a few related hydroxy ketones besides an agrofuran. Agarol was found to be related to costol, the characteristic constituent of costus, *Saussurea lappa* C.B. Clarke, root oil. Essential oil from the non-infected plant was found to contain iso-meric undecanes, selinane, dihydro-selinene and a few oxygenated sesquiterpenes.

Trade and Production of Agar and its Oil : Production of agarwood and its oil is rather small. However, exact statistics are not available. As per information this item has become scarce with prohibitive price of agarwood and oil. Only record of survey of its production is that of Gulati done 20 years back. According to this report quantity of agarwood available at that time was 31 tons corresponding to above 80 kg of oil (calc. at 0.25 per cent). Production of oil was confined to 3 centres *i.e.,* "Manipur, Barpathar and Dhansasi having total of 30 stills (cap. to distil about 30 tons of wood per year). Distillation was done only in Manipur and Barpathar. Recently, Bhutan has entered into this trade offering sizeable quantity of agarwood; about 6250 kg. of different grades of agarwood was offered for sale in November, 1978. Production in other regions *i.e.,* Assam and Nagaland is estimated at about 10 tons. According to an information, agarwood is also brought from Burma.

Current price of agarwood is very high; real black agar (Grade 1) can fetch about Rs. 6000 per kg. Other grades are equally expensive. Prices during 1958 were as :

Black Agar	Rs. 35 per kg.
Bantang	Rs. 15 per kg.
Bhutas (Kaluguchi)	Rs. 5 – Rs. 10 per kg.
Dhoom	Rs. 1.50 – Rs. 2.00 per kg.

Price of oil was about Rs. 1500 to Rs. 2000 per kg. Price of wood shot up to Rs. 650 per kg by 1967. The current price of wood and oil is unbelievable. It is understood from Mr. I. Sundresh that importers during their visit to India, purchase agarwood oil at fantastic prices which may range upto Rs. 10,000 or more per kg. depending on its odour.

According to one source, Burma seems to be the only regular and dependable source for agarwood.

A great deal of agarwood, however, is used for production of *attar*. While oil is produced in Nagaland, *attars* are produced in Bombay and a few other centres using sandalwood oil, or liquid paraffin as base. Data on the production of *agar attars* are also not available.

Uses : Agarwood has been reported to have stimulant cordial, tonic and carminative properties. It is also used as a cosmetic and diniment in various skin diseases. Agar power when dusted on clothes and skin acts as a preventive against flies and lice. However, its main use is in the making of fumigator and pastilles. It is highly priced and prized perfumery item.

CHAPTER 18

ESSENTIAL OILS OF CINNAMOMUM SPECIES

Genus *Cinnamomum* Blume (Fam. Lauraceae) comprises of about 270 species of shrubs and trees distributed in Asia and Australia. Most of these species are aromatic but only the following are of economic value at the present.

1. *Cinnamomum camphora* (Linn) Nees and Eberm (camphor tree).
2. *C. zeylanicum* Breyn. (Ceylon of true cinnamon).
3. *C. cassia* (Nees) Nees ex Blume (Chinese cinnamon).

About 20 species occur in India; important among these examined so far are :

1. *C. camphora* (Linn) Nees and Eberm.
2. *C. cedidodaphne* Meissn.
3. *C. glanduliferum* Meissn.
4. *C. impressinervium* Meissn.
5. *C. iners* Reinw.
6. *C. obtusifolium* Nees.
7. *C. tamala* Nees and Eberm.
8. *C. zeylanicum* Breyn.

While the above three species of economic importance are well described in literature and their oils/products have been the accepted market items, some of the species occurring in India have been worked upon only recently and may have future potential in the trade. A review of literature is summarised below :

1. *C. camphora*. commonly known in India as *Kapur, Karpur* and *Karpuram* is a large handsome evergreen tree native of Japan,

China and Taiwan. It was introduced into India and attempts were made to cultivate it on large scale in the fifties in the South. It is also found in state of nature in some parts of the country. It can, under natural conditions, attain a height of 30 m with a girth of 1.8-2.4m. *C. camphora* comprises of many forms, only a few contain camphor rich oil.

Camphor tree has been found to grow in India upto an altitude of 2100 m. However, it thrives best at an elevation of 1350-1800 m provided temperature does not fall below –9°C. It can withstand annual rainfall of 110-360 cm. The tree can, however, be cultivated in all parts of the country with annual rainfall of 100 cm. It is reported to come up well on fertile, well drained sandy loam soils. The plant can be raised from seed, layers, branch cuttings, root cuttings and root suckers. Propagation by seeds, however, is the normal practice.

Camphor is produced by distilling wood from mature trees over 50 years of age; leaves and twigs are also used for distillation. Camphor can be produced by harvesting plants of over 5 years age, but yields are low. Camphor tree stands coppicing well and can be kept in the form of a bush; 3-4 coppicing every year are possible. Leaves contain more camphor than twigs. Camphor content depends on the parent stock and environmental conditions. Other conditions being equal, trees of 2-40 years of age have same camphor content as also their green, mouldy and air dried leaves. Camphor leaves from Cochin are reported to contain more camphor than those from other parts of India. Even though all parts of plant contain camphor, maximum camphor content is found in the root.

Factors influencing yield of camphor and camphor oil have been listed by Dr. Teikichi Hiraizumi as under :

1. Naturally growing trees contain more camphor and camphor oil than planted trees.
2. Isolated trees contain more camphor and camphor oil than trees growing in dense stand.
3. Healthy trees contain more camphor and camphor oil than weak trees.
4. Sandy soils as also well drained soils enhance camphor content.
5. Old trees contain more camphor than young trees.

On an average, camphor and oil content in various parts of the camphor tree are :

Percentage of	Leaves	Branches	Trunk	Stump	Roots
1. Camphor	1.0	3.0	0.8	1.3	0.8
2. Camphor oil	0.3	0.6	1.4	1.8	2.5
3. Total yield	1.3	0.9	2.2	3.1	3.3

On an average, whole tree contains 0.8 per cent camphor and 1.5 per cent camphor oil (Oil contains 50 per cent camphor in solution). Due to adequate production of synthetic camphor in India, production of natural camphor has not been considered to be economical. However, at the current price of camphor and reported yield of about 150-190 kg of camphor and 25-32 kg of camphor oil per hectare per year, cultivation should be economical, especially when only leaves are used. Scope of production of natural camphor from this source needs re-examining.

Cinnamomum zeylanicum Breyn. (Ceylon cinnamom) commonly known as *Tamalpatra* (Sans.), *Dalchini* (Marathi, Telugu, Bengali and Hindi), *Cannal avangapattai, Ilayagnam* (Tamil), *Dalchini Lavangapatti* (Kanara), *Kurundu* (Sinhalese) is a small ever green tree usually 6.0-7.5 m high, some times 18m as in Ceylon, and gives the well known commercial cinnamon. It is reported to occur wild as also on plantation scale in the southern coastal regions of western India upto an altitude of 1800 m. It is abundant in the region 30-210 m above sea level and is fairly common upto 1050 m. Leaves have spicy odour and hot taste. Bark of tender shoots and stems is smooth and pale while, that from old and aged branches is rough and brown.

Four kinds of *C. zeylanicum* are distinguished according to taste of leaves; sweet, insipid, pungent and bitter. Leaves having pungent and bitter taste are useful for oil production. There is considerable variation from one tree to another; there are also different races.

C. zeylanicum is a hardy plant which can grow in any soil under a wide variety of tropical conditions. Quality of bark is influenced by soil and ecological factors. While sandy soil gives best quality of cinnamon, sandy soil mixed with humus is considered the best with sheltered situations upto 450m. It can withstand heavy rainfall. Plants raised in low lying areas produce bark of low oil content. Plants are normally raised from seeds though cuttings of young shoots and layering can also be used. Trees are coppiced in second or third year to induce branching.

C. zeylanicum yields 2 important products :

1. Bark and Bark oil.
2. Leaf oil.

Cinnamon bark contains 0.5-1.0 per cent oil. For distillation, chips and refuse left after collection of cinnamon quills are used. The oil is distilled mainly in Europe, and USA. Main constituents of the oil are cinnamic aldehyde, (60-75%) eugenol, benzaldehyde, methyl amylketone, –phellandrene, –pinene, –cymene, nonylaldehyde, linalool, cuminaldehyde, caryophyllene and ester of isobutyric acid, furfural, hydrocinnamic aldehyde.

Oil of cinnamon bark is one of the most valuable essential oils in the trade.

Green leaves of *C. zeylanicum* yield an oil in an yield of 0.75-1.0 percent. Oil contains 70-95 per cent eugenol for which constituent the oil is widely used.

Root bark of *C. zeylanicum* also gives an oil in 3 percent yield which is different from both bark and leaf oil. It has camphoraceous odour and camphor separates out on standing.

Physico-chemical properties of leaf as also bark oil are as under :

Characteristics	*Leaf oil*		*Bark oil*
	India	*Ceylon*	*Ceylon*
1. Sp. gravity (30°C)	1.0443 – 1.004 1.023 – 1.040		1.0620 – 1.066
2. Ref. Index	1.5223 – 1.530 1.581 – 1.591		1.5350 – 1.540
3. Optical rotation	—	1° to +3°	—
4. Acid value	14.0 – 15.7	—	—
5. Ester value	4.7 – 16.7	—	—
6. Eugenol (%)	70 – 85.5	70 – 95	4 – 10
7. Cinnamic aldehyde (%)	—	upto 4	65 – 76
8. Solubility in 70% alcohol	1 :10 vol.	—	1 :10 vol.

An oil supposed to have been produced from *C. zeylanicum* from a Bangalore specimen yielded a yellow coloured oil (0.25% on dry weight basis) having :

α^{15} 0.9225; n_D^{20} 1.5070 (α_D^{20} – 2.9°; and containing eugenol 15% γ–pinene and β–phellandrene 4%–terpineol and other terpene alcohol 3%–caryophyllena and other sesquiterpenes 62%; sesquiterpene alcohol 7%.

True cinnamon was introduced into Bhubneshwar in 1970. Plants raised from seeds (95-100% germination) gave heterogenous population; most of the plants gave eugenol rich oil (82–98%); only one plant gave an oil with no eugenol. Eight years plants attained a height of over 4 m. Flowering started after 4 years with profuse and regular fruiting. Oil content varied from 0.49-0.87% in the population. Bark contained 0.7% oil of standard quality. On the whole, cultivation trials were found encouraging.

Cinnamomum Cassia (Nees) Nes Ex Blume : *C. cassia* commonly known as Chinese cinnamon or cassia contains cinnamic aldehyde (70-90%) as its main constituent. While leaf and bark oil from *C. zeylanicum* are chemically different, leaf and bark oil from *C. cassia* are identical in chemical composition. Cassia leaf is distilled for its oil, cassia bark in ground or powdered form is a widely used spice. It is commonly known as *Cassia lignea* while dried flower is traded under the name "cassia buds"; in commerce for medicinal preparations.

For oil purpose, leaves, leaf stalks, young twigs along with various refuse products are used.

Cultivation efforts with *C. cassia* outside China have not been successful. Nearest possible oil that could be produced in India is from *C. tamala* (*Tejpat*) growing in Kumaun.

Oil yield from cassia leaf is 0.31 per cent; yields of 0.77 per cent have also been reported. Leaves have better oil content (0.54%) than the stalks (0.2%). Physico–chemical properties as reported are :

1. Specific gravity 25°/25°	1.046 – 1.059
2. Optical rotation (25°C)	–0°40′ to 0°30′
3. Refractive index (20°C)	1.6045 – 1.6135
4. Aldehyde content (Bisulphita method)	88 – 99 per cent

As mentioned earlier, out of the 3 commercially accepted oils from *Cinnamomum* species, in India oil only from leaves of

C. zeylanicum is produced from *C. camphora.* Other species of *Cinnamomum* occurring in India and on which some work has been reported are described as under :

1. *C. Cecidodaphne* Meissn : Commonly known as *Rihu* (Bengali and Lepcha), *Gundroi* (Cachar), *Malligiri* (Nepal). *Goudhori, gondserai, gunserai* (Assam), *Gam-Salu* (*Garo*) is an evergreen tree 6-7.5 m high with an average mid-girth of 1.5 m and distributed in the Eastern Sub-Himalayan tract upto an altitude of 1200 m in Assam, Manipur, Sikkim and Bhutan. It is reported to be cultivated in Dehra Dun. Seeds are non viable. Flesh of the fruits is fragrant.

Wood of the trees has strong persistent camphoraceous odour. It is light and has oily feel. Timber is highly prized for cabinet making, construction work ploughs and yokes. Unrestrained exploitation of the tree for wood has now made it a scarce item. Leaves and wood of the tree contain essential oil.

Essential oil from wood has been examined by some workers who have reported yields varying from 1-1.25 per cent to 2.95 per cent. Physico–chemical properties reported are as :

Characteristics	*Properties reported by*	
	Barua et. al.	*Finuemore*
1. Colour	Colourless to slightly yellowish	Pale yellow
2. Specific gravity	1.103 (15°C)	1.10 (23°C)
3. Refractive index	1.5312 (29°C)	—
4. Optical rotation	0°.4′	–0°.4'
5. Saponification value	0.8	—
6. Saponification value after acetylation	6.9–7.0	7.0

The oil has been reported to be very rich in, ethers, comprising of safrole (39.7%) elemicin (30.2%), myristicin (20.1%). Other constituents identified are methyl eugenol (0.3%) and 4 unidentified compounds (total 0.7%).

Work on oil from leaves is not reported.

2. *C. Glanduliferum* Meissen : It is reported to resemble *C. cecidodaphne* and is also known as *'Gondserai'* in Assam and *'Malliguri'* in Nepal. However, this is a small tree attaining a grith of about 1.35 m and growing in state of nature in central

Himalayas and Khasi Hills. Freshly cut wood has a strong smell of camphor. Wood contains decamphor. It is reported to be useful for the same purpose as *C. cecidodaphne.*

Leaves from plants growing in Nepal gave an oil having sp.gr. 0.9031–0.9058; (D) – 24°57′ to –26°12′, acid value 0.34–0.90, ester value 8.8–18.4, ester value after acetylation 46.9–55.3 Seed kernal has fixed oil. Oil produced in Urio (lake Como; 0.6% yield) has cineol 34% e–α–terpineol 10% and d–camphor. This shows clearly the difference of this species from *C. cecidedaphne.*

3. *C. impressinervium* Meissn : It is a small tree growing in Sikkim at an altitude of 1200-1800 m. Bark is used as in case of *C. tamala* (Cassia bark). It is used as an adulterant for cinnamon. No other work is reported.

4. *C. iners* Reinw : It is suspected to be a variety of *C. zeylanicum* and is a large tree occurring in evergreen forests of Western Ghats from Mysore to Coorg to the Annanallais and Kerala, Karnatak and Shevroy and Kollimalai hills. It is commonly known as *Tekhi* (Bombay), *Jangli dalchini* (Hindi), *Ranachadalchini* (Marathi), *Kattu-Kkaruvap-pattai* (Tamil), *Adavi lavangapatte* (Telegu and Kanara). No work has been reported on this except that the bark contains 0.5 per cent of an essential oil with the odour of cloves and musk and leaves contain 0.73-0.85% of oil.

5. *C. Obtusifolium* Nees : It is a large robust tree occurring in the central and outer Eastern Himalayas upto an altitude of 2100m, East Bengal, Assam and Andaman Islands and is commonly known as *Kinton, ram-tezpat, texpat* (Bengali), *Patichanda* (Assam), *Barasingoli* (Nepal) and *Phatgoli* (Kumaun). Leaves known as *Tejpat* are aromatic and put to the same use as those of *C. tamala* Bark is used in dyspepsia and liver complaints in Nepal. The bark particularly of the root resembles Ceylon cinnamon. The wood is reported to be generally used in Assam for making charcoal.

6. *C. macrocarpum* Hook F. : It is a moderate sized tree occurring in North Kanara, Western Ghats, Nilgiris and is commonly known as *Tejpatra (Sanskrit), Karuva (Tamil)* and *Lavanga (Telegu)*. No work on this has been reported except that an essential oil is obtained (part of the plant used is not mentioned) which is applied externally in rheumatic affections.

7. *C. pauciflorum* Nees : It is a small tree occurring in Sikkim Himalayas and Assam and is known as *Dingla-tterdop* (*Khasia*). The wood is reported to be aromatic. No work is reported on this species.

8. *C. tamala* Nees and Eberm : It is a moderate sized evergreen tree growing upto a height of 7.5 m and girth of 1.35 m, distributed in tropical and sub-tropical Himalayas upto an altitude of 900-2400 m, Khasi and Jaintia Hills upto an altitude of 900-1200 m, and in Eastern Bengal. It is cultivated in Jaintia and Sylhet. It is commonly known as *Tamalka tejpatra* (*Sanskrit*), *Tejpat* (*Hindi and Bengali*) *Tamalpatra* (*Gujrati*), *Talishappattiri* (*Tamil*), *Talisapatri* (*Telegu*).

This species yields the *tejpat* leaves used extensively in India as a spice as also exported in large quantities. Bark is also collected from Sikkim Himalayas and is known in trade as Indian Cassia Lignea. Export and import of cassia and *tejpat* leaves during the last four years is as under :

Export–Import of Cassia and Tejpat Leaves

Export	*Cassia*		*Tejpat*	
	Qty.	*Value*	*Qty.*	*Value*
Export				
1972-1973	652.35	33.02	23.01	0.20
1973-1974	—	—	538.54	3.54
1974-1975	1335.02	80.39	611.78	3.23
1975-1976	457.26	23.91	1142.45	7.42
Import				
1972-1973	39.30	5.60	0.89	0.008
1973-1974	23.60	4.86	—	—
1974-1975	38.50	7.01	1.70	0.007

(Qty. in tons, value in Rupees in lakhs)

From the study of literature it seems there are two chemical types of *C. tamala* occurring in India.

1. Eugenol type.
2. Cinnamicaldehyde type.

While there are references on work of eugenol type, there is only one reference on the latter.

C. Tamala–Eugenol Type : Eugenol type seems to occur more widely than the cinnamic aldehyde type. It is found in Kashmir (Kathua and Rajouri forests), Tehri Garhwal, tropical and sub-tropical Himalayas, Khasia and Jaintia Hills and Bengal, Plants are also cultivated. Seeds are sown in the nursery in March-April. Germination starts in 40-45 days. Seedlings when 4-5 years old are transplanted in rows 2m apart and about 3m between plants. Shade trees are provided in the early stage and removed after 8-9 years. Harvesting of leaves is done when trees is 10 years old and continues for about a century. Leaves are collected every year from young vigorous plants and on alternate years from old and weak trees. Collection is made during Oct.–December till March, *i.e.*, after and prior to monsoon, as rain affects the aroma of leaves. Small branches with leaves are tied in bundles, dried and sent to market. One tree yields about 9-10 kg of leaves every years. Leaves on distillation give an oil rich in eugenol. Production of oil from this source is done in Meghalaya.

Bark and leaves in Jammu gave an oil in 0.05 and 1.2 per cent yield respectively. Leaf oil had specific gravity (15°C)1.0257 and refractive index 1.5775. Recently, Sarin *et al.* reported an yield of oil of 1.2 per cent on hydro-distillation having sp.gr. (20°C), 1.025; optical rotation + 16°37′, refractive index (20°C) 1.526; eugenol 60–80 per cent and solubility in 70% alcohol as 1 : 1.2 vol. The oil resembles cinnamon leaf oil and contains d-1-phellandrene and 78 per cent eugenol. According to Sarin *et. al.* large quantities of the raw material are commercially available in various trade centres in the region.

Leaves of *tejpat* are used mainly as spice. In Kashmir these are used as a substitute for *pan* (betal leaves). Leaves are carminative and are used in colic and diarrhoea. Bark is aromatic but coarser than the bark of true cinnamon and is one of its common adulterants. Oil from the bark is pale yellow containing 70-85 per cent cinnamic aldehyde.

C. Tamala–Cinnamic Aldehyde Type : A wild growing Cinnamon in Kumaun Hills is exploited for its leaves which are sold in the market as *Tejpat.* Leaves are strongly aromatic while the bark is mucilagenous and contains only traces of an essential oil. The plant occurs in the shady forests and also cultivated in the farms in the Nainital district of Kumaun Hills. It is small evergreen tree or shrub attaining a height upto 7.6m and a girth of 1.40 m. Leaves opposite, glabrous, shortly stalked, ovate, oblong 3 nerved from the base; long pointed young leaves are coloured pink. The plant flowers in May and fruit in June-July.

Only reference on this type of oil is that of Gulati *et. al.* who studied the oil from both cultivated and wild growing plants.

Oil content in leaves from cultivated plants is more, 0.23-0.62% (0.45% average), as against that from leaves of wild growing plants, 0.13-0.43% (0.33% average). Physico-chemical characteristics of oil samples are :

Characteristics	*Oil sample from leaves of*	
	Wild plants	*Cultivated plants*
1. Specific gravity (30°C)	(i) 0.9563	0.9730
	(*ii*) 0.9843	0.9876
2. Refractive index (30°C)	(*i*) 1.5038	1.5119
	(*ii*) 1.4942	1.5419
3. Optical rotation	(*i*) –0°6′	–0°30′
	(*ii*) +0°12′	
4. Aldehyde content	(*i*) 36.54	49.50
	(*ii*) 24.56	

CHAPTER 19

EUCALYPTUS : A VERSATILE MATERIAL FOR AROMA CHEMICALS

Eucalyptus are fast growing Australian plants having about 700 species and find use in timber, pulp, paper, pharmaceutical, mineral floatation and perfumery industries. One of the most important characteristics of eucalyptus is that these can be grown with great vigour in very different ecological conditions from those in their natural range. Keeping in view their versatile nature, the eucalyptus were introduced into India. About 100 species have so far been introduced in India, particularly, to meet the demand of pulp, paper and timber industry. A few species of this genus are reported to contain essential oils like *E. citriodora, E. exserta, E. leic-how , E. maca-rthuri, E. saligna, E. sieberiana, E. radiata, E. dives* and *E. globulus*. The essential oils of some of these species could be used as a raw material for the synthesis of fine aroma and pharmaceutical chemicals.

E. citriodora

The oil obtained from the leaves has been found out to be a high grade perfumery oil. This chemotype is being cultivated on a large scale and oil produced in 3.1% yield contains up to 72.8% citronellal and 14.5% citronellol.

The existing source of citronellal is Java citronella. It is present in this oil in a yield of 35-40%. Thus *E. citriodora* scores over Java in oil yield as well as in citronellal content. Also, a large cultivation of *E. citriodora* can feed a cottage scale paper industry. Citronellal is one of the most important aldehydes. It has a flowery, honey like note but cannot be used as such because of unstable nature of the aldehyde group. However, it could be a goldmine for a number of high grade perfumery materials. Citronellal is one of the most important constituent of synthetic rose and lilly of the valley perfumes.

Citrinellol can be prepared from citronellal. This method involves reduction of *E. citriodora* oil by aluminium isopropoxide followed by efficient fractional distillation.

Rose oxide, a minor constituent of Bulgarian rose which is responsible for its typical smell is found to be a salt of geranium. It is now being used in most of the perfumes in minor quantities to impart final touches. Its true odour is green, fatty, nutty, dusty with rosaceous touch. Tetrahydro geraniol has been prepared by hydrogenation of citronellol. Various esters of citronellol have been prepared starting from citronellol, all of these possess different odours.

Hydroxy citronellal has not been isolated from any natural source so far and is entirely a new development. The composition remained a trade secret for many years. It duplicates with remarkable togetherness with the odour of lilly of the valley, when blended with other compounds, it imparts a floral note to perfumes which cannot otherwise be obtained. It is now' extensively used as an ingredient for special perfume composition. The preparation of hydroxy citronellal is not difficult, but tricky. By simple structural modification of citronellal, a series of thirty six compounds have been prepared. Citronellal was allowed to react with different alkyl and aryl magnesium halides to give higher homologues of citronellol. The alcohols were then submitted to acetylation, oxidation and hydration.

Menthol and Thymol are widely used in the pharmaceutical industry and their demand is ever increasing. Citronellal is a good starting material for the synthesis of menthol and thymol.

E. exserta and E. leichow

Both the species were imported from China and were found to be for growing suitable in Indian climate. The analysis show that these species contain 35.69% and 37.49% 1,8-cineole respectively.

E. macarthuri

This exotic species shows three forms as below :

(*i*) First type contains geraniol and geranyl acetate as the major constituents of the oil.

(*ii*) Second type contains eudesmol and geranyl acetate as the major constituents of the oil.

(*iii*) Third type contains cineole as the single major portion of the oil.

This offers a scope for selection of cineole-free strain which are liable to be superior from the perfumery point of view.

Pure genranyl acetate possesses a sweet typical rose leaf odour and is a very useful sweetening agent in floral and fancy perfume compositions.

E. sligna

E. saligna yielded 4% of yellowish coloured essential oil with characteristic camphoraceus odour. The chemical composition shows α-pinene, β-pinene, camphene (6.8%), myrcene, limolene (8.5%), 1, 8-cineole (25.2%), δ-cadinene. β-caryophyllene, α-terpineol (6.4 %), geraniol (1.0%) camphor (11.5%), terpinyl acelate (2.5%), caryophyllene oxide and few unidentified compounds.

E. globulus and E. radiata

Numerous species of eucalyptus are a rich source of cineole but these two species show favourable growth in Indian climate. Cineole can be converted to good quality α-terpinyl acetate.

E. sieberiana

This species has been introduced on an experimental basis and is thriving well. The GLC of the oil confirmed the presence of β-phellandrene (28.36%) and cineole (13.28%) in addition to α-pinene (0.71 %), β-pinene (24.-34%), δ-carene (14.47%), *p*-cymene (3.84%) and two unidentified ones.

β-pinene and carene has been exploited for the synthesis of number of perfumery chemicals while β-phellandrene could be exploited for the synthesis of 1-menthol. Selective hydrogenation of β-phellandrene is likely to afford menth-2-ene, which can be converted to menthol.

History of perfumes is advancing with the discovery of new compounds. Introducing citronellal as a starting material, new compounds were synthesized with a view to obtain new compounds with different odours. Imitation of natural fragrances is a matter of art in which both the perfumer as well as the chemist are involved.

We, as a chemist are trying to produce the natural fragrances by structural changes in a molecule. All the compounds prepared were isolated and identified with the help of UV, IR, PMR and MS.

Table 1 : Analysis of oil from *Eucalyptus citriodora* leaves

S.No.	*Name of the component*	*Mol. wt.*	*Mol. Formula*	*%*
1.	Isopropyl cyclohexane	126	C_9H_{18}	0.26
2.	Pinene	136	$C_{10}H_{16}$	0.22
3.	Pinene	136	$C_{10}H_{16}$	0.29
4.	P-cymene (Methyl-4-Isopropylbenzene)	134	$C_{10}H_{14}$	0.14
5.	Limonene	136	$C_{10}H_{16}$	traces
6.	1, 8-cineol	154	$C_{10}H_{18}O$	2.11
7.	2, 6-dimethyl-5-hepten-1-al	140	$C_9H_{16}O$	0.23
8.	1-methyl-cis-4-Isopropylcyclohexane	140	$C_{10}H_{20}$	traces
9.	γ-Terpinene	136	$C_{10}H_{16}$	0.13
10.	3, 8-Menthadiene	136	$C_{10}H_{16}$	0.21
11.	1, 4 (8)-P-Menthadiene (-Terpinolene)	136	$C_{10}H_{16}$	0.15
12.	Cis-Rose Oxide (4-Methyl-2 (2-methyl-prop-1-enyl)-Tetrahydropyran)	154	$C_{10}H_{18}O$	0.40
13.	Citronellal	154	$C_{10}H_{18}O$	72.80
14.	Menthone	154	$C_{10}H_{18}O$	traces
15.	Linalool	154	$C_{10}H_{18}O$	1.51
16.	Isopulegol	154	$C_{10}H_{18}O$	0.34
17.	α-Terpineol	154	$C_{10}H_{18}O$	0.12
18.	Nerol	154	$C_{10}H_{18}O$	traces
19.	B-Citronellol	156	$C_{10}H_{20}O$	14.51
20.	Lianlool oxide	170	$C_{10}H_{18}O_2$	0.40
21.	1, 8-Terpin hydrate	172	$C_{10}H_{20}O_2$	0.80
22.	Citronellyl acetate	198	$C_{12}H_{20}O_2$	3.10
23.	Trans-Caryophyllene	204	$C_{15}H_{24}$	0.65
24.	Humulene	204	$C_{15}H_{24}$	0.10
25.	β-Cubene	204	$C_{15}H_{24}$	0.15
26.	γ-Elemene	204	$C_{15}H_{24}$	0.14
27.	Spathuleol	220	$C_{15}H_{24}O$	0.12
28.	Geranyl citronellal	290	$C_{20}H_{24}O$	traces
29.	Methyl-cis-9-octadeceneate	296	$C_{19}H_{36}O_2$	traces

Table 2 : Analogus of Citreonellol

	Compound	Yield (%)	Formula	Odour
(*a*)	1. R = C_4H_9, R' = OH	90	$C_{14}H_{28}O$	Turmeric like
	2. R = C_4H_9, R' = OAc	98	$C_{16}H_{30}O_2$	Kerosene like
	3. R = C_4H_9, R' = OH	90	$C_{14}H_{28}O$	Minty
	4. R = C_4H_9, (Iso-butyl) R' = OAc	98	$C_{16}H_{30}O_2$	Ethereal
	5. R = C_5H_{11}, (Iso-amyl) R' = OH	90	$C_{15}H_{30}O$	Leathery
	6. R = C_5H_{11}, (Iso-amyl) R' = OAc	90	$C_{17}H_{32}O_2$	Mild garlic
(*b*)	7. R = C_4H_9	60	$C_{14}H_{26}O$	Aldehydic nutty
	8. R = C_4H_9 (Iso-butyl)	60	$C_{14}H_{26}O$	Aldehydic rosaceous
	9. R = C_5H_{11} (Iso-amyl)	60	$C_{15}H_{28}O$	Mild fatty (Isopulegol liek)
(*c*)	10. R = C_4H_9, R_8 = OH	70	$C_{14}H_{30}O_2$	Fatty nutty (less volatile)
	11. R = C_4H_9, R' = OAc	98	$C_{16}H_{32}O_3$	Mild Coconutty
	12. R = C_4H_9 (Iso-butyl) R' = OH	90	$C_{14}H_{30}O_2$	Hydroxy citronellel type
	13. R = C_4H_9 (Iso-butyl) R' = OAc	98	$C_{16}H_{32}O_2$	Fatty
	14. R = C_5H_{11} (Iso-amyl) R' = OH	70	$C_{15}H_{32}O_2$	Ester like
	15. R = C_5H_{11} (Iso-amyl) R' = OH	98	$C_{17}H_{34}O_3$	Unpleasant (Very low volatile)
	16. R = C_4H_9	60	$C_{14}H_{28}O_2$	Just fatty strong coco-nutty
	17. R = C_4H_9 (Iso-butyl)	60	$C_{14}H_{28}O_2$	Fatty liliac
	18. R = C_5H_{11} (Iso-butyl)	60	$C_{15}H_{30}O_2$	Fatty nutty

CHAPTER 20

NATURAL DYES FROM FOREST BIOMASS

Historically, natural dyes were used to colour clothing or other textiles, and by the mid-19th century chemists began producing synthetic substitutes for them. By the early part of this century only a small percentage of textile dyes were extracted from plants. Lately there has been increasing interest in natural dyes, as the public becomes aware of ecological and environmental problems related to the use of synthetic dyes.

Use of natural dyes cuts down significantly on the amount of toxic effluent resulting from the synthetic dye process. Natural dyes generally require a mordant, which are metallic salts of aluminium, iron, chromium, copper and others, for ensuring the reasonable fastness of the colour to sunlight and washing. Customers who have become accustomed to the dazzling colours and wash and light fastness of synthetic dyes are hard to convince, as only a few of the natural dyes have good all round fastness.

These dyes are thereby specifically identified as dyes of the stated colour, but it does not specify whether the dyes are derived from animals or plants. This is because it is a classification based on the dye's source and colour, and it contains no chemical information, nor does it imply that dyes with similar names but unique numbers are in any way related. It also gives no information about the mechanism by which staining occurs. This is done in order to authenticate the synthesized organic dyes along with natural dyes under the same universal classification system.

Some examples include carmine, which comes from cochineal (natural red 4), lac (natural red 25) and hematein, which comes from the logwood tree (natural black 1). The FDA in the US has taken an additional step and given colors used in foods, drugs and cosmetics their own labels (FD&C) after passing them as being fit for human consumption or use. Natural dyes are a class of colourants extracted from vegetative matter and animal residues. They can be broken down into the following categories :

Demand for Natural Dyes

In recent times the handloom industry has adopted various measures and techniques to provide high quality and eco-friendly products to the world market.

Attractive colour combinations, precious fabrics and skilled designers who are adept at mixing creativity with traditional knowledge of the crafts in unlimited variations of style are at the core of Japer's hand block printed products.

A festive, energetic pink from pomegranate skin, a vivacious yellow from turmeric roots and the sapphire cobalt from indigo are just a few examples of colours evolving out of the natural dyes, whose peculiarities lend a touch of exclusivity to these works of art.

The elaborate process of the preparations of different colours in vegetables dyes is usually done-in-house by the women. The red dye is obtained from *Keshu* flowers mixed with Harsrinagar flowers. The iron-filings, horseshoes and jaggery placed in a jar of water to a fortnight produces the black dye. When this black dye is applied on the cloth along with *Harda* (Myrobalan), it produces a yellow tint.

The maroon dye is obtained on boiling madder wood and adding alum to it. Mixing indigo and turmeric produces the green colour. Ochre is used for creating the ochre dye and, by adding together *Keshu* flowers in pink dye, the vermilion tint is created.

International buyers and connoisseurs, who value the intricacy and ethnicity of the designs, the eco-friendly natural dyes and at the same time savour the culture and heritage depicted thereon, appreciate these products worldwide. Demand for hand block printed textile in fashion garments is on the rise. Buyers from this segments demand exclusive designs that may be custom made as per the buyers' requirements.

Standardization of Ecofriendly Natural Dyes

Natural dyes are being used since very ancient time, but have lost out to synthetic dyes, as it is generally very difficult to have proper standardization and reproducibility of shades. With the recent trend of higher preference for natural dyes, fuelled by increased public awareness about environmental issues and actions like the German ban on use of some synthetic dyes, prompted Indians to have a re-look at the potential of natural dyes.

The global demand for natural dyes world over is about 10,000 tones, which is equivalent to 1% of the world synthetic dyes consumption.

Mother Nature has gifted us with more than 500 colour giving plants. In India we can use those natural material, which are available in abundance, cost effective, yield good colour and fastness properties. Some of them are as follows :

NATURAL INDIGO

Common Name : Natural Indigo, Indigo, Indian Indigo

Botanical Name : Indigofera tinctoria

Natural Dye : Blue Dyestuff for Textile.

Table : Categories of Natural Dyes

Colours	*Chemical Classifications*	*Common Names*
Yellow and Brown	Flavone Dyes	Weld, Quercitron, Fustic, Osage, Chamomile, Tesu, Dolu, Marigold, Cutch
Yellow	Iso-quinoline Dyes	Barberry
Orange-Yellow	Chromene Dyes	Kamala
Brown and Purple-Grey	Naphthoquinone Dyes	Henna, Walnut, Alkanet, Pitti
Red	Anthraquinone Dyes	Lac, Cochineal, Madder (Majithro)
Purple and Black	Benzophyrone Dyes	Logwood
Blue	Indigoid Dyes	Indigo
Neutrals	Vegetable Tannins :	Wattle, Myrobalan, Pome
	gallotannis, ellagitannins and catechol tannins	granate, Sumach, Chestnut, Eucalyptus

This is one of the most ancient natural dyes used by man in textiles. Indigo is about 2-3 feet long plant, cultivated on thousands of acres of land in India. The whole plant is used for extraction of dye and the extract form of dye is supplied in powder form. This give deep blue colour on wool, silk, cotton etc. It exhibits good fastness to washing, crocking and light. This is the only Natural Dye, which falls under Vat category of dyes. We are among the largest exporter of Natural Indigo from India.

POMEGRANATE

Common Name : Pomegranate, Anaar

Botanical Name : Punica granatum

Natural Dye : Yellow dyestuff for Textiles.

Pomegranate is one of the most largely consumed fruits in Asia. The dye is extracted from its fruit rind. The colour obtained exhibit good fastness to washing rubbing and light. At times this dye is also used as mordant and is over dyed with other Natural Dye to improve fastness of later.

MADDER

Common Name : Madder, Manjistha, Majith

Botanical Name : Rubia cardifolia (Indian Madder) Rubia tinctoria (European Madder)

Natural Dye : Red, pink and orange dyestuff for textiles.

Madder is one of the oldest Natural Dyes. In a way Indigo and Madder are the main ancient Natural Dyes used by man for dyeing textile for ages. The cultivation of Madder needs sub topical climates and prefers moist soil. It is cultivated in the foots of Himalayas in huge quantity.

KAMALA

Common Name : Kamala, Kamala Powder

Botanical Name : Mallotus phillipinses

Natural Dye : Orange yellow or golden yellow dyestuff for Textiles.

Kamala is taken from fruit of the plant Malleolus. This is an evergreen tree growing upto 25 meter high. It imparts beautiful colour on wool and silk. Kamala powder was used extensively in India and neighbouring southeastern countries and also enjoy export trade.

However, on cotton Kamala does not give as deep colour as on wool or silk.

LAC

Common Name : Lac, Lac Dye

Botanical Name : Kerria lacca

Natural Dye : Red & violet dyestuff for Textiles.

Lac is probably most ancient of all animal dyes. The dye is taken from insect Kerria Lacca. This insect is cultivated on Kusumbh and Tesu trees. This has its history of use as Natural Dye in South Asia and India since the beginning of recorded history.

The principal colouring component in this dye is identified as Laccaic Acid. The dye shade exhibits very good fastness to washing, rubbing and light.

MAYRABOLAN

Common Name : Harad, Harda

Botanical Name : Terminlia chebula

Natural Dye : Greenish yellow dyestuff and Natural Mordant for Textiles.

Mayrabolan is fruit of tree Terminalia chebula. This is used as greenish yellow dyestuff for textile. This is also used as Natural Mordant in many cases. This is also used as a substitute of Tannic acid.

CATECHU

Common Name : Cutch, Catechu, Kala Kaththa

Botanical Name : Accacia catechu

Natural Dye : Brown dyestuff for Textiles.

Catechu is purified extract of wood Accacia catechu. The plant of Accacia catechu is about 15 meter high. This is mainly composed of Catechu-tannic acid with catechin, catechu red and quercetin. Catechu is among cheap Extract form of Natural Dyes it gives light brown to good brown-with acceptable fastness rating.

HIMALAYAN RUBRUB

Common Name : Dolu, Himalayan Rubrub.

Botanical Name : Rhum emodin

Natural Dye : Yellow dyestuff for Textiles.

Dolu is about 1.5 to 3 meter high stout herb. It is grown at an altitude of about 3000 to 5000 meters high. The colour is extracted from its wood. This consists of many anthraquinone derivatives based on emodin, chrysophanol, alooe-emodin and rhein. It dyes wool in range of colour mainly in yellow and exhibit good fastness ratings.

Natural Dyes From Forest Biomass

Until the discovery of synthetic dyes in 1856, the plant and animal kingdom provided all the materials for dyeing textiles, for preparation of paints and cosmetics, and for making food-stuffs visually more attractive. Use of synthetic dyes involves release of large amount of hazardous chemicals in the environment during their production and also during their subsequent use. With the environmentally benign products becoming a top priority in recent years, dye industry has turned its attention to newer products, which cater of fashion trends as well as environmental specifications. Recent ban in India and abroad on the use of azo dyes possessing carcinogenic properties has led all major dyestuff manufacturers to search for benign alternatives. The safest may be the use of eco-friendly natural dyes.

The Technology

Research carried out at the Forest Research Institute (FRI), Dehradun has led to the development of processes for the extraction of natural dyes from some abundantly occurring plant materials of forest origin. These dyes may be used for imparting different shades on silk, wool and cotton using common mordants like alum, salts of iron, tin and chrome. Different combinations of mordants impart unique shades.

The processes employed for the extraction of dyes simple, low-cost and eco-friendly. Processes for numerous shades developed from a number of species, covered under 3 patents, are available for transfer and with the scope of natural dyes widening prominently, dye manufactures are anticipated to take them up in the near future in big way.

The Economics

It is now possible to produce marketable natural dyes, which would provide eye-catching shades, with the simple workable techniques developed at FRI. The forest biomass can be used for the production of dye of cottage scale, thus leading to employment generation for the people through value addition to the non-wood forest products along with creating an additional source of revenue. The cost of raw materials varies from Rs. 2 to Rs. 40 per kg depending upon the material used for isolation of dyes and the site of requirement. Processes have also been standardised to use the dye raw materials directly for dyeing different fabrics. The plant and machinery cost would be around 30 lakhs for a production unit of 10 kg/hour. The technology is available with National Research Development Corporation, New Delhi.

Future of Natural Dyes

The manufacturers in India are well aware that AZO free colours and dyes should be used. India has discarded the usage of banned materials in the dyeing process with safe substitutes, to ensure eco-friendliness of the products manufactured by the industry. The Indian Home Textile Industry has access to excellent infrastructure facilities and uses the latest technology and machinery.

Most of the manufacturers have set up 'state of art' modern processing houses to roll out large volumes of high quality products to meet the international market demands and expectations. They have regular marketing tie ups for the American continent, Europe, Japan as well as in CIS countries, Russia, Pakistan and Switzerland.

Azo dyes cause harmful effects during their production and subsequent use. The EEC has promulgated at "EC Control 01 Substance Hazardous to Health Act, 1989" and published a red list enumerating a number of chemicals, the presence of which in any kind of fabric has been banned. An ordinance in Germany stipulates that no garment or any other article that comes into contact with the skin shall contain any of the twenty aromatic amines named there under. Similar restrictions have been imposed in many other countries. In India, the ban on the use of azo dyes has been imposed by the Union Ministry of Environment and Forest under section 6(2) (d) of the Environment (Protection) Act, 1986 read with the Rule 13 of the Environment (Protection) Rules, 1986.

Due to environmental awareness the natural dyes obtained from plants and animals are the dyes of 21st century. Their production as well as trade is anticipated to increase manifold in the times to come.

CHAPTER 21

INDIAN LAC

India is the leading lac producing country in the world. Lac is derived from a specialised group of insects known as lac insects. These insects are cultured on specific trees known as lac hosts. The raw lac obtained after harvesting encrustations of the lac insect from the trees is refined to obtain three useful products: resin, dye and wax. Lac thus forms a natural, renewable and eco-friendly source of useful materials, which can be put to an unbelievably wide range of applications.

A good demand already exists for lac-derived materials, besides tremendous potential for much higher consumption, due to global demand for safe natural products. Moreover, India possesses a huge untapped potential for lac production. Despite all these facts, we have so far not been successful in establishing a healthy lac industry in the country.

This situation is mainly due to certain unhealthy factors obtaining in the marketing scenario of the industry. As per official figures, the Indian lac economy is still dependent on exports, as more than 80% of the country's production goes out to various countries.

This overdependence is perhaps the principal reason for the plight of the lac industry in the past as well as at present. However, there are indications of good growth of domestic demand for lac that needs to be further strengthened. Development of an efficient marketing system of lac, based on sound strategies and elimination of detrimental factors operating in the system is very important for creating a healthy industry in the country.

Production

Lac is mostly cultivated by economically weaker section of people, mostly tribal, especially in Jharkhand, Chhattisgarh and

West Bengal, on lac host trees scattered over wide areas. The growers mostly produce lac in small quantities, which makes the assembling process more difficult. Despite poor infrastructure, the lac marketing system is fairly organized. Nevertheless, interstate restrictions for the movement of lac, especially in State like Orissa, has left and unfavourable impact on lac production.

The *rangeeni* and *kusmi* insects of *Kerria lacca* contribute to most of the lac production in the country. Since these insects are bivoltine, two crops are produced by each of these insects.

Rangeeni : *Baisakhi* (harvested during April-July) and *katki* (harvested during October-November).

Kusmi : *Aghani* (harvested during June-July) and *Jethwi* (harvested during June-July).

These crops have been named after the month in which they are normally harvested. The percentage share of these reveals that *baisakhi* is the major crop contributing more than half of the total production. The arrival of raw lac in the market normally peaks with the harvesting periods of these crops. However, small arrivals keep flowing throughout the year.

Lac production base has greatly shrunk over the decades, leading not only halving of, but also great fluctuations in the lac production. Table 1 shows the lac producing areas of the country. However, much of the production appears to be concentrated only to a few pockets. Presently there appears to be a great deal of interest among farmers for lac production, especially due to remunerative price for lac in recent years.

Table 1 also reveals the wide area already available in the country where lac cultivation activities are still going on, even though the productivity may not be high in most of the areas. Thus, there is great scope for increasing the production of the country even in the existing areas of lac production.

Figs. 1 and 2 show the comparative picture of share of different States to total lac production in the past and at present, which reveals some interesting changes due to erosion of productive areas.

The *kusmi* seedlac is unique to India and fetches more price (10-15%) than the *rangeeni* seedlac, due to lighter colour and other advantages. Moreover, the productivity of *kusmi* lac on its major host *kusum* is much higher. Despite these facts, the share of *kusmi* lac to total production is only about 17%. There is a great need to boost *kusmi* lac in the country by adopting different strategies to bring about qualitative change in the natural lac produced in the country.

Table 1 : Lac producing areas (including minor) of the country

State	*District*
Jharkhand	Palamau, Latehar, Garhwa, Ranchi, Lohardaga, Gumla, Simdega, Sarai-kela, Hazaribagh, Chatra, East Singbhum, West Singbhum
West Bengal	Purulia, Bankura, Midnapur, Murshidabad
Madhya Pradesh	Balaghat, Mandla, Hoshangabad, Shahdol, Jabalpur, Indore, Chhindwara, Rewa, Seoni
Chhattisgarh	Bilaspur, Raipur, Sarguja, Bastar, Rajnandgaon, Durg, Korba, Raigarh
Maharashtra	Bhandara, Chandrapur, Gharchiroli
Orissa	Mayurbhanj, Kendujargarh, Kalahandi, Bolangir, Koraput, Sambalpur, Nabrangpur, Sundergarh.
Gujarat	Vadodara, Panchmahal, Sabarkantha
Uttar Pradesh	Mirzapur, Sonebhadra
Bihar	Gaya (Raniganj)
Assam	Karbi Anglong, North Cachar Hill, Nagaon
Andhra Pradesh	Adilabad, Nizamabad
Meghalaya	Khasi Hills, Garo Hills

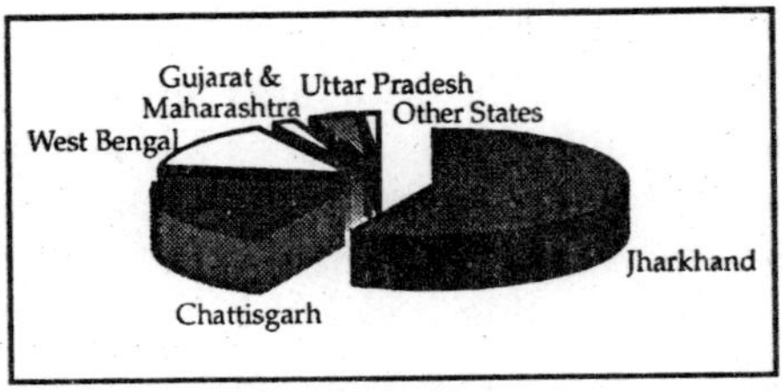

Fig. 1 : Statewise production of lac in India (1996-1999)

The production of lac in the country can be analysed in terms of long-term and short-term perspectives. Lac production in India touched 50,000 metric tonnes during 1940s. However, over the period it declined to around 20,000 metric tonnes (mt) during the first five years of 1990s. However, a major crop failure in 1998 led to a crash in lac production to around 10,300 mt.

Ever since, the lac production is slowly recovering and it is hoped that it would soon reach the 20,000 mt mark.

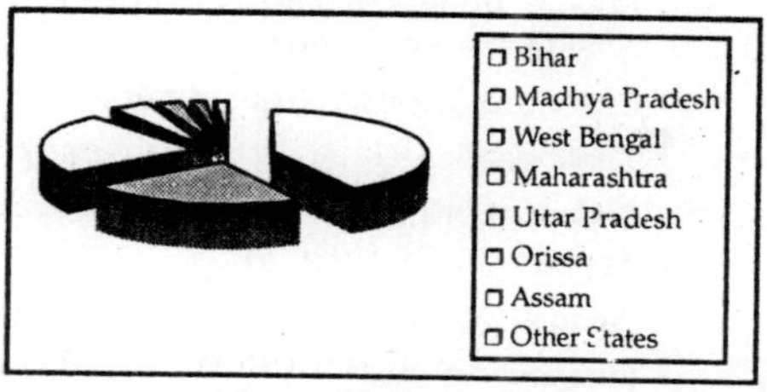

Fig. 2 : Statewise production of lac (1930-60)

As per official figures, the average production of lac during the past years had been around 15,615 metric tonnes.

Backed by years of experience in the field, we had always felt that the official figures available on the production of lac are not very reliable. It laid more emphasis on the major lac growing areas. For example, our field knowledge shows a sizeable production of lac in Orissa; but virtually no production is shown against Orissa in the official figures. Thus, the total production of the country appears to be underestimated. While reliable figures could be obtained for export of lac and lac products, only sketchy information is available on the internal consumption of lac.

Table 2 : Distribution of grower households by quantity of sticklac sold

Quantity (kg)	*Percentage of household*
Upto 10	7.6
10 – 20	53.5
20 – 40	8.8
40 – 60	7.6
60 – 100	6.9
> 100	5.6
Non-reporting	10.0
	Kabra (1983)

Table 3 : Important lac processing centres in the country

State	*Centre*
Jharkhand	Khunti, Bundu, Murhu, Daltonganj, Latehar, Garhwa, Chakradharpur, Chandil
West Bengal	Balrampur, Jhalda, Tulin, Khatra
Madhya Pradesh	Bankhedi
Chhattisgarh	Dhamtari, Khathghora, Sakti, Pendra, Rajnandgaon, Kanker
Maharashtra	Gondia
Uttar Pradesh	Mirzapur

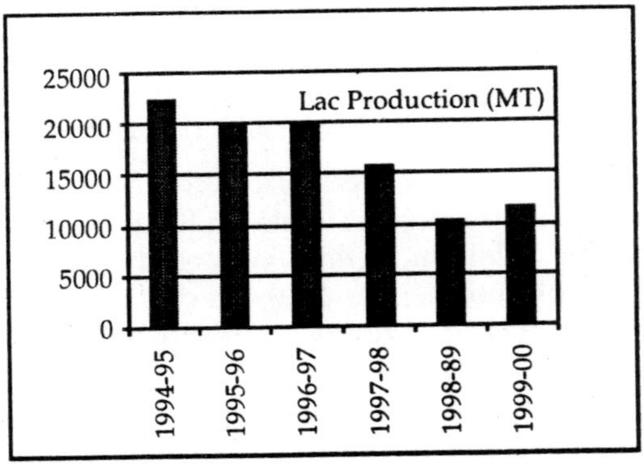

Fig. 3 : Official estimates of lac production (in terms of stick lac) in India, during the past five years.

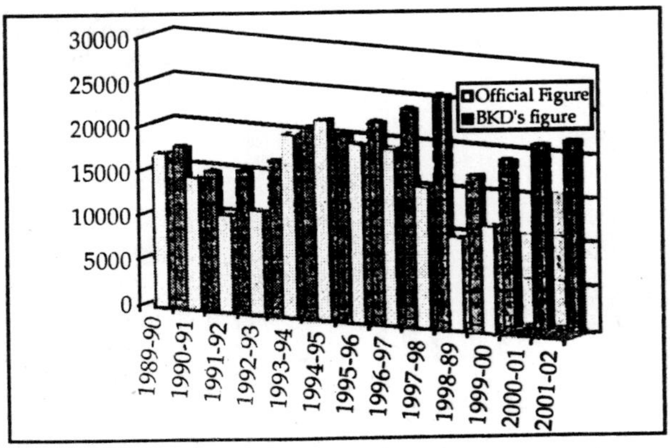

Fig. 4 : Comparative picture of official and independent estimates (Mr. B.K. Das) of country's lac production

Fig. 3 shows the official estimates of lac production in the country during 1994-95 to 1999-2000. The official figure for the subsequent period is not available. It shows a sharp fall especially during 1997-99, which is mainly attributed to major crop failure. Available information indicates that the production is slowly recovering during the subsequent period.

We have obtained independent estimates of lac production based on flow of lac through the processing units. This provides a different picture of consumption of lac in the country and exports to overseas countries. It claims an average annual production of about 21,340 mt during 1995-96 to 1999-2000 which is 37% higher than the official figures (15615 mt), for the corresponding period. Fig. 4 presents a comparative picture of official figures and independent estimates.

Unit	*Common Product*
Hand processing units	Seedlac, HM Shellac, Button lac, by-products
Small mechanized units	Seedlac, MM Shellac, Bleached lac, by-products
Big mechanized units	Seedlac, MM Shellac, Bleached lac, by-products
Specialized units	Dewaxed decolourized lac, aleuritic acid, lac dye, etc.

Marketing System

The major components in the marketing chain is :

The grower ⇨ primary middle-men ⇨ small paikar ⇨ secondary middlemen ⇨ big paikar, (*arhaliyas*) ⇨ the wholesaler ⇨ small manufacturer ⇨ big manufacturer ⇨ broker ⇨ domestic/foreign buyer.

One or more components could be skipped during the passage of material from producer to the ultimate consumer, depending on the situation. The levels involved in marketing of lac can be visualized as :

- Primary markets (village markets–*haats*)
- Secondary markets (mini assembling centres)
- Wholesale markets
- Terminal markets

The following facts, based on survey covering three representative districts of Chhotanagpur (Palamau, Ranchi and

Singhbhum), gives some picture about lac marketing scenario (K.N. Kabra, 1983). About 80% of the lac growers preferred selling their lac at *haals*. Only about 7% of lac is sold at the village itself.

More than 91% of the growers sell their produce to the *paikars* (middle men). The average production of lac per household was about 42 kg (Table 2). A large share of the production came from the top 40% of the growers. About 60% of the growers gained their knowledge of lac prices through *paikars*.

At present, we do not have reliable data based on systematic information, on where the money goes at different levels in the lac marketing chain. An old study indicates that~20% drop in the producer's share of consumer price, when lac goes through the intermediaries.

Table 4 : Major importers of lac from India (2000-01)

Country	*Quantity (Kg, metric tonnes)*	*Value (Rs. Lakhs)*
USA	402	692
UAE	108	190
Indonesia	1,423	2,299
Germany	1,090	1,796
Other countries	1,822	3,092
Total	4,728	8,069

Source : SEPC, Kolkata.

Lac Processing

The raw lac has to be refined to separate the desired portion for application in different areas.

Lac-derived raw materials can be classified into the following groups :

- Raw lac (Scraped lac/sticklac)
- Seedlac (Semi-refined lac resin)
- Shellac, button lac, dewaxed decolorised lac, bleached lac etc. (refined lac resin)
- Lac dye (by-product)
- Lac wax (by-product)

- Refuse lac *(kiri, molamma* etc.)
- Lac constituents (aleuritic acid, jalaric acid etc.)
- Lac-derived fine chemicals (isoamrettolide, exaltone, bioactive compounds etc.)

Most of the processing units are located in the following states : West Bengal, Jharkhand, Chattisgarh, Maharashtra (Table 3).

The processing units of lac can be grouped into following categories :

Kabra (1983) feels that the plight of cottage sector seedlac units is somewhat similar to that of lac growers in as much as the price spread between seedlac and shellac moves largely to the advantage of exporters. He also found a large and fluctuating price spread between sticklac prices and Kolkata shellac prices.

He observed that many factors, like the dominance of organized big exporters, affected the difference between the interrelated sticklac, seedlac and shellac prices. He found that the dominance of lac manufacturers-exporters was the cause of variations in the returns to the growers rather than the intermediaries.

The following observations were also made by Kabra (1983) on lac marketing scenario :

1. The behaviour of exporters concentrated on the value, rather than the quantity, which tended to reduce the demand for lac.
2. The sticklac prices are largely determined by manufacturer-exporter. It is done by backward working of prices.
3. Shellac manufacturer-exporter did little to plough back the profits in R&D or development of internal demand for lac.

Due to dependency of lac industry on overseas market, these exporters controlled the domestic market. Thus, the marketing scenario mostly depended on the interests of these lac manufacturer-exporters.

Quality Standards

BIS (Bureau of Indian Standards) has specified quality standards for all the common lac and lac products. ISO standards have also been laid down for certain lac grades. But the trade is not widely employing these standards in marketing. The Indian Lac Research Institute, Ranchi, provides technical support for the quality control of lac and lac products; other units engaged are : SEPC and a private firm in Kolkata.

Table 5 : Average export performance of lac and lac products (1995-96 to 2000-01)

Product	Quantity (mt)
Shellac	5,080.69
Dewaxed shelac	278.25
Seedlac	2379.50
Aleuritic acid	33.55
Bleached lac	111.03
Gasket shellac	2.10
Shellac wax	0.56

Source : SEPC, Kolkata.

Export and Imports of Lac

As reported earlier, most of the lac produced in the country is exported to various countries of the world. Table 4 gives major lac buying countries of the world.

The Shellac Export Promotion Council (SEPC), Kolkata promotes, monitors and regulates the export of lac and lac products from India.

In early twentieth century, lac used to be regularly imported from countries like Thailand, Myanmar and others. But, subsequently, due to import restrictions, this flow stopped. Recently, imports from Thailand and Indonesia have been renewed. It is hoped that this would provide additional support to the processing units in the country. But, the importers of lac have a responsible role to play. They have to ensure that these imports do not in any way adversely affect the domestic price of lac. If there is any major price reduction of domestic lac due to imports it would have negative impact on the internal lac production and would ultimately prove detrimental to the very importers, who cannot be indifferent to domestic production. Policy makers in the government will have to consider all the pros and cons of imports from other countries to develop suitable policy, which would promote a healthier industry. There had been some growth in the export value-added products of lac from the country, but much remains to be done (Fig. 5).

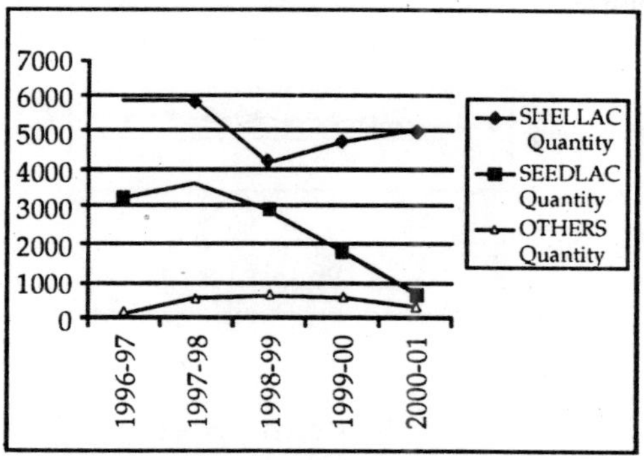

Fig. 5 : Quantities of some lac & lac products exported from India

By far, the most serious malady affecting the survival and growth of the lac industry is the highly fluctuating price. Fig. 6 indicates Kolkata price levels of two representative grades of lac over the past five years. Unless something is done about long term stabilization of the price, nothing can pull out the industry from its present plight.

As pointed out earlier, we should collect more reliable information on the internal consumption of lac. Interestingly, it is seen, from the independent estimates presented in Fig. 7 that while the exports more or less reflected the fluctuations in lac production, internal consumption more or less remained the same.

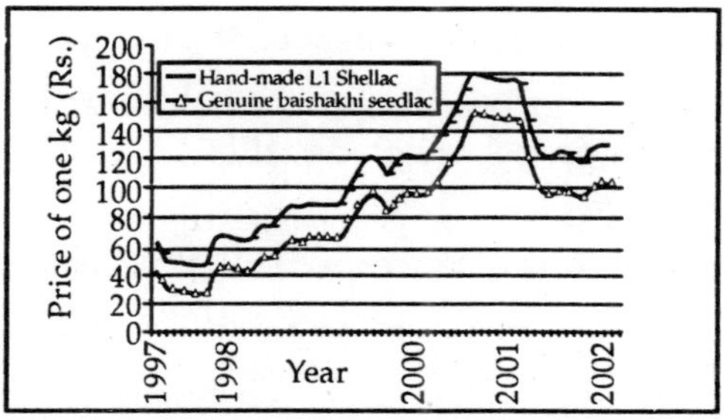

Fig. 6 : Changes in the Kolkata market price of two representative in grades of lac

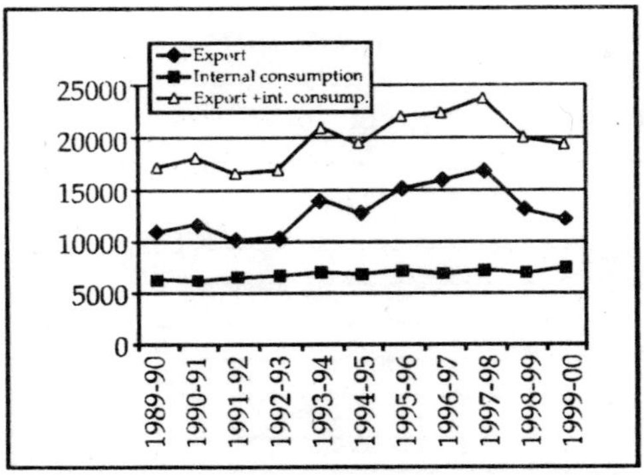

Fig. 7

SWOT Analysis of Lac Marketing Scenario

Strengths

- Large cultivation base with highly under-utilized potential
- Good manpower resource (skilled and unskilled) for production and processing
- India is the global leader in lac production
- Unique and strong R&D support to lac sector
- Favourable climate in the country for lac production

Weaknesses

Grower–Processor

- Production of lac in small quantities
- Lac hosts, used for cultivation widely scattered
- Poor infrastructure for transport of produce
- Exploitation of growers during purchase at primary level (less price, improper weighing, allowances of poor quality, distress sale etc.)
- Adulteration of raw lac, especially at secondary markets
- Dependency of processors on middle men
- Lack of effective govt. or other machinery for procurement
- Lack of effective Minimum Support Price for the growers

- Violent fluctuations in prices with correlated patterns of demand/supply
- Lack of constant impetus for lac growers to produce more

Processor Consumer

- Small and fragmented units
- Lack of modern, efficient processing
- Poor profit margin of esp. small processing units
- Less stress on value-addition
- Inadequate stress on by-product utilization
- Inadequate quality control measures
- Poor linkages of industry with R&D
- Lack of specialised and aggressive marketing machinery
- Virtually no promotional activities
- Lack of interest in domestic market potential by processors-exporters
- Unhealthy competitive trade practices
- Less stress on packaging, branding and common grade-standardization for precise product identity
- Failure to promote and encourage lac-based enterprises
- Absence of concerted efforts by the processors to create a healthy marketing system (both forward and backward)
- Mechanism for assured supply of the commodity at consistent price level

General

- Lack of adequate market information collection and dissemination system.

Opportunities

- Tremendous scope for increasing the global demand for lac
- Huge potential for boosting domestic lac consumption
- Great scope for value-addition for export market

- Scope for achieving reasonable and steady price level by adopting production and marketing strategies to strengthen demand
- A sizeable unutilized infrastructure for lac processing
- Untapped potential for greater utilization of by-products like lac dye, wax etc.
- Rising global demand for natural, ecofriendly products

Threats

- Availability of cheap synthetic resins, dyes etc. easily.
- Competition from other lac producing countries like Thailand.
- Over dependency of the industry on export market.
- Unscrupulous elements damaging the delicate marketing system.
- Imports from other lac producing countries at very cheap price.

Concluding Remarks

Lac is commodity of the future. But, concerted efforts are needed to realise the full potential of this commodity. Shellac Export Promotion Council, which addresses only lac exports, should probably also pay attention to ways and means of boosting internal lac consumption. Better linkages should be established between the industry and the R&D unit. Research efforts should focus on value-addition, technology refinement and development of products in the stronger areas of utilization of this commodity such as food and cosmetic industry, in order to boost the demand for lac.

A balanced mix of intensive and extensive lac production technologies should be promoted to ensure stable and economic lac production in the country. The cultivators should be encouraged to produce more and concomitantly a mechanism should be developed for assured purchase of lac produce at MSP. Market intervention should be made to prevent fall of growers' price below economic level during glut seasons.

All efforts should be aimed at achieving assured supply of lac to any industry at reasonable and steady price of consistent quality, with remunerative price for the lac growers. The lac manufacturers-exports who handle a huge proportion of the lac production of the country should play a more responsible role. They should lay more stress on increasing the volume and improving the quality of their products.

They should also join together with other lac manufacturers to develop a mechanism to maintain a reasonable price level and assured procurement of lac. They should restrain from indulging in market manipulation through lobbying which would prove detrimental to them and the industry, in the long run. All the agencies related to lac-producers, processors, consumers, researchers, extension workers, policy makers and financial institutions should develop a visionary protocol building up a bright future for lac.

CHAPTER 22

LAC BASED SSIs

Lac is a natural resin of insect origin produced by the colonies of a tiny insect known as *Karria lacca (Kerr).* It serves as an important source of income to more than three million tribal families as well as foreign exchange earner for the country. There are around 200 lac processing units in India and these units fall under two groups, namely non-mechanised unit and mechanised units. Non-mechanised units manufacture seedlac and hand made shellac and are scattered all over the States of West Bengal, Jharkhand, Chattisgarh, Maharashtra and Madhya Pradesh. These units employ about 5 to 50 workers. The mechanised units, very few in number, manufacture machine made shellac, bleached lac, aleuritic acid, lac dye, etc.

India is the world leader in both production as well as export of lac, contributing around 70% and 60% in production and export respectively. As a renewable source of resin, dye and wax, lac has a bright future. Because of it natural, non-toxic, renewable and eco-friendly nature and certain specialties, lac and lac-based products already developed have tremendous market potentials. Also the advanced countries have now started using lac in preference to synthetic resins. A major portion of export of lac is in the form of raw materials though the technologies of manufacturing various finished and value added products are already there. Only in the recent years, export of value added lac products has shown slow but steady rise from 1.4% of total lac exported in 1993-94 to 7.85% in 1998-99. Hence lac and its products provide ample scope for setting up of cottage and small scale industries which can tap new avenues for increasing the internal consumption as well as export of value added products of lac. It will also generate employment and investment opportunities in rural region.

Different Forms and Uses of Lac

The raw lac, which is in the form of encrustation and is obtained from the twigs by scraping, is known as sticklac. It is generally processed for its resin, wax and dye. The sticklac is refined through coarse crushing, repeated washing with water and then drying the grains. The dried lac grains are called seedlac. In this process the dye is separated from the resin and is recovered thereafter from the wash water. Seedlac is a semi refined product which still contains certain impurities. It is, therefore, further refined either by hot filtration or by extraction with rectified spirit to get the refined material known as shellac.

Lac resin, due to its adhesive, surface coating and insulating properties, has the potential of widespread applications in different types of industries. As a surface coating material it is used in

(*i*) varnish and printing ink industries for preparation of furniture polish, floor polish, sealers, etch primers, paints, etc.

(*ii*) cosmetic industries for preparation of hair spray, eye shadows, lipstick, nail polish, etc.

(*iii*) pharmaceutical industries for coating of tablets

(*iv*) food industries for coating of fruits, chocolates, coffee beans, aluminum foils, metal cans, etc.

(*v*) leather industries and

Table 1 : Production details of a seed lac plant (capacity : 750 kg/day)

Cost of production, Rs./kg	85
Capital investment for smallest viable unit, Rs.	10,00,000
Space required, m^2	225
Power requirement, kW	20
Plant and equipment, Rs.	4,00,000
Working capital per month, Rs.	7,70,000
Profit per month, Rs.	1,48,000
Employment, No.	20

(*vi*) for coating of miscellaneous items such as earthen wares, stones, jewellery, etc. Due to its insulating properties, it is used in electrical industries for coating of isolators, spark plugs, as micanites etc. In adhesive industries it is used as gasket cement, sealing wax, hot melt adhesive, polymer adhesive and making of particle board, etc.

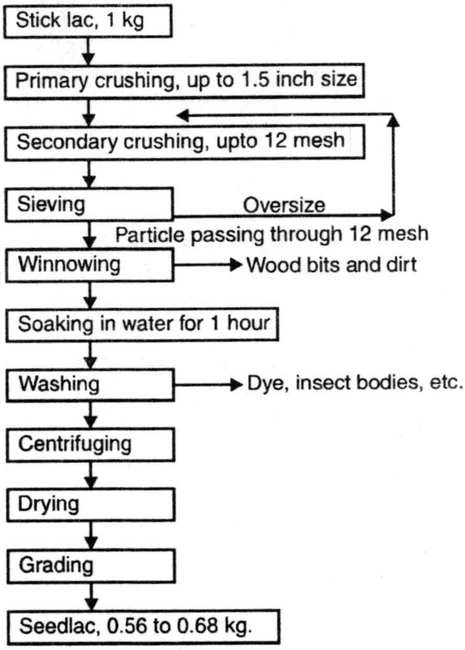

Fig. 1 : Process flow chart for making seedlac

Table 2 : Production details of a bleached lac plant (capacity : 50 kg/day)

Cost of production, Rs/kg	250
Capital investment for smallest viable unit, Rs.	10,00,000
Space required, m^2	150
Power requirement, kW	11
Plant and equipment, Rs	5,00,000
Working capital per month, Rs.	3,00,000
Profit per month, Rs	62,000
Employment, No.	10

Lac wax is used in shoe, automobile and floor polishes, fruit coating etc., where as lac dye is used in dyeing of wool, silk and also finds application in some food industries.

Seed Lac

Seed lac is the semi-refined form of lac, which is used as a starting material for preparation of various lac products. India annually exports around 3,000-4,000 tons of seedlac. The flow sheet of a seed lac plant is shown in Figure 1. Raw stick lac is crushed to suitable sizes by means of primary and secondary crushers. The crushed particles are fed to a charging tank where they are soaked in water for nearly an hour before being charged into a cylindrical washing barrel fitted with centrally rotating stirrers.

The wash water containing the water soluble dye flows out through the screen fitted on one side of the barrel. Washing is continued until almost all the dye is removed. Certain washing aid like washing soda can be added in this process. The washed seedlac is discharged from the barrel to a battery of centrifuge which remove all the free water present. From the centrifuge, the washed seedlac is transferred to a dryer for drying. Equipment such as crushers, vibrating screen sieves, winnower, barrel washer, centrifuge and dryers are required for the production of seedlac. Production cost, number of employment etc. of a 750 kg/day capacity seedlac plant are given in Table 1.

Table 3 : Production details of a lac dye plant (capacity : 75 kg/day)

Cost of production, Rs/kg	100
Capital investment for smallest viable unit, Rs.	8,00,000
Space required, m^2	2,000
Power requirement, kW	3.5
Plant and equipment, Rs	4,00,000
Working capital per month, Rs.	2,90,000
Profit per month, Rs	95,000
Employment, No.	8

Bleached Lac

Nearly half of the world's consumption of lac is in the form of bleached or white lac. The product is thus of considerable commercial importance. In western countries, it is used in the manufacture of colourless polishes and lacquers. In our country also there is great demand for bleached lac varnishes.

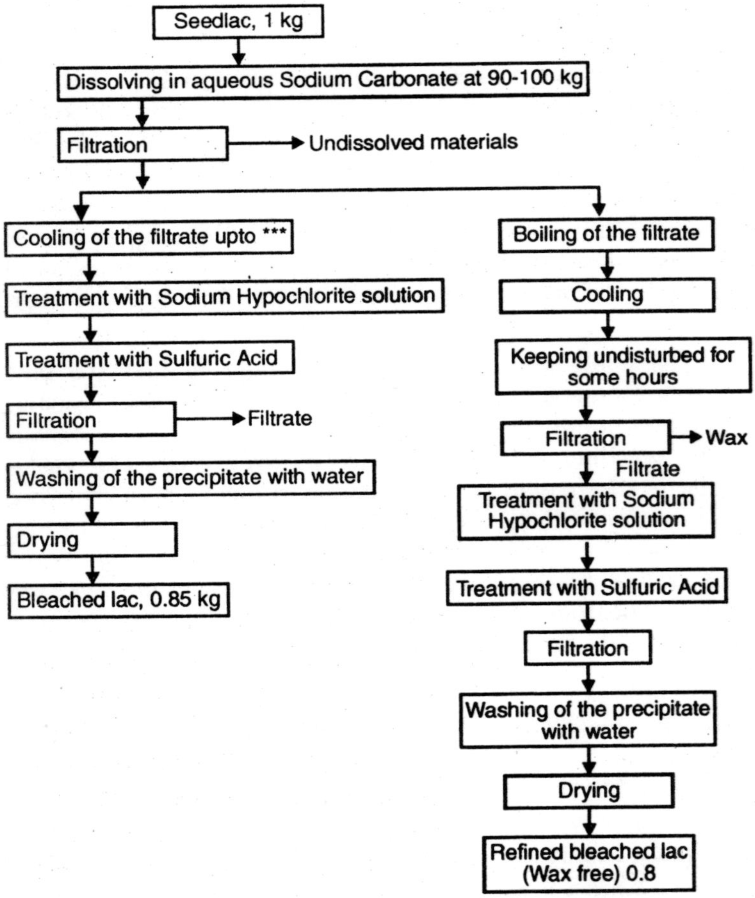

Fig. 2 : Process flow chart for making bleached lac.

Crude lac contains two colouring materials, laccaic acid and erythrolaccin. The former is more or less completely removed during the washing of sticklac into seedlac where as the latter remains, which imparts the pale yellow 10 deep orange brown colour of retained lac. A process, based on bleaching by sodium hypochlorite, has been developed and standardized for adoption at industry level. The process consists of dissolving seedlac in hot (90-100°C) aqueous sodium carbonate solution, straining of the undissolved materials and treating the filtered extract with sodium hypochlorite solutions.

The bleached lac thus formed is reclaimed from the solution by precipitation with sulfuric acid. The precipitated bleached lac is then filtered, washed free from the acid and dried. If refined (wax free) bleached lac is to be prepared, the aqueous solution of lac prepared above is heated to boiling temperature, cooled and allowed to stand undisturbed.

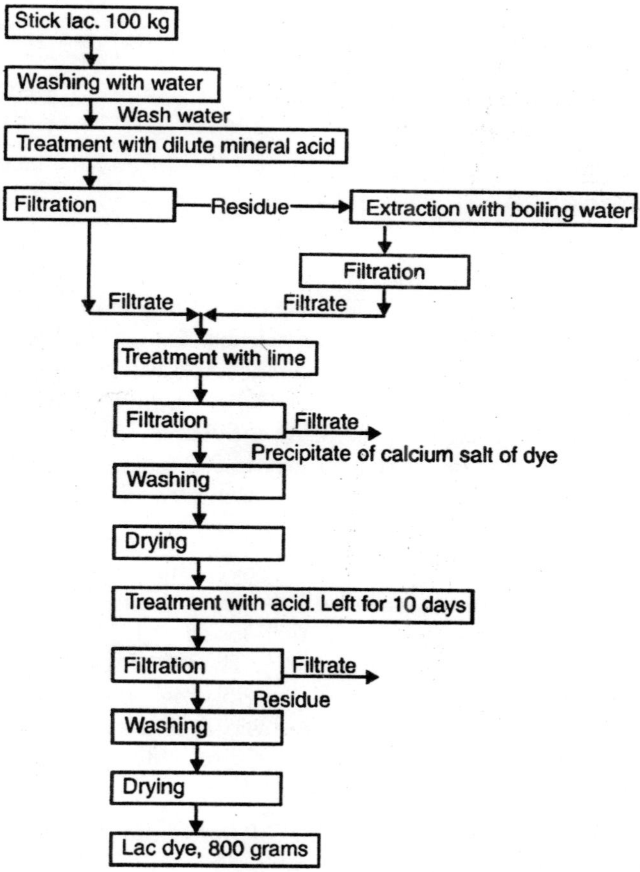

Fig. 3 : Process flow chart for making lac dye

The wax floats up and on cooling, sets to a hard crust which may then be skimmed off. The solution is then filtered and subjected to bleaching as mentioned above. Equipment such as stirrers, filters, dryers, rotary washers, dehumi difiers. etc. are required for the production of bleached lac. The process flow chart is shown in Figure 2 and production details of a 50 kg/ day capacity bleached lac plant are given in Table 2.

Table 4 : Production details of a Aleuritic acid plant (capacity : 10 kg/day)

Cost of production, Rs./kg	1350
Capital investment for smallest viable unit, Rs.	5,00,000
Space required, m^2	100
Power requirement, kW	11
Plant and equipment, Rs.	3,00,000
Working capital per month, Rs.	3,50,000
Profit per month, Rs.	28,500
Employment, No.	13

Lac Dye

Lac dye, being natural, is gaining renewed interest in international as well as domestic markets for use in textile and food industries. When sticklac is crushed and washed with water, the lac dye (present to the extent of 1%) is dissolved out and goes in the effluent of the lac processing industries. With the present level of production of sticklac around 15,000 tones per year, about 150 tones of lac dye can be produced annually, which is generally lost.

It is, therefore, high time to recover this valuable material for domestic consumption as well as export. A simple and economical process for its recovery is developed at ILRI, Ranchi. The process flow chart is shown in Fig. 3. The wash water is filtered and acidified with mineral acid to bring the pH to the desired level and then filtered again. The residue obtained is extracted with hot water and filtered. The filtrate thus obtained is combined with the main filtrate.

It is then treated with lime till optimum pH is reached when calcium salt of lac dye precipitates out which is then washed and dried. Then it is treated with hydrochloric acid and left for 10 days. The lac dye separates out from the solution and is filtered, washed and dried. It can be further purified by crystallization with water. Equipment like barrel washer, storage tanks, filters, stirrers and dryers are required for production of lac dye. Table 3 shows the production details of a 75 kg/day capacity lac dye plant.

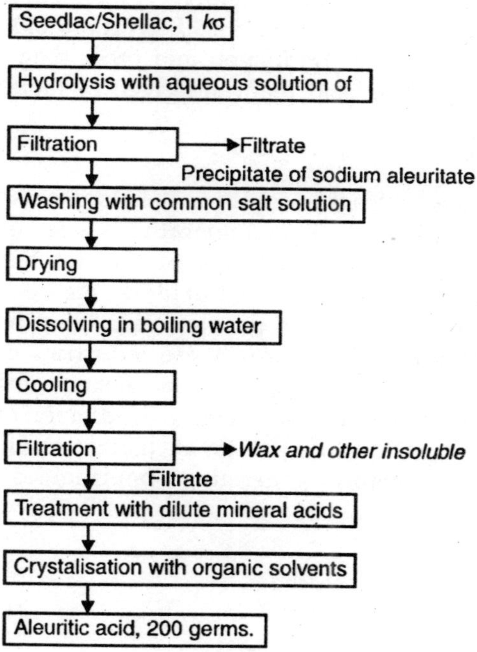

Fig. 4 : Process flow chart for making Aleuritic Acid.

Aleuritic Acid

Aleuritic acid, a colourless crystalline compound is one of the major constituent acid present to the extent of about 35% in the lac resin. This can be isolated from the resin by adopting a standard method developed at ILRI, Ranchi.

It possesses significant value for use as a starting material in the synthesis of many value added products like perfumery, pharmaceutical and bio-active compounds. It is purely an export oriented product and is in constant demand from the foreign buyers. Figure 4 shows the process flow chart for isolation of aleuritic acid from seedlac.

Seedlac/shellac is hydrolyzed with aqueous solution of caustic soda under specific conditions. The precipitate of sodium aleuritate thus formed is washed with common salt solution and then filtered and dried. The aqueous solution of sodium aleuritate is decomposed with dilute mineral acid get the crude aleuritic acid which is subsequently crystallized with organic solvents to obtain pure aleuritic acid, with a yield of approximately 20%. Details of production and other data of a 10 kg/day capacity aleuritic acid plant are given in Table 4.

Conclusions

Thus there is bright scope for development of lac based small scale industries with comparatively less investment as costly and sophisticated instruments are not required. Also technologies have been developed to utilize the by-products and wastes of these industries to develop valuable products. There is less unwanted effluents from these industries making them environmentally safe. Processing the lac in small scale industries and exporting value added products instead of the raw materials will not only lead to more foreign exchange earning but, more importantly, also help in generating opportunities for employment and to improve the economic status of the poor lac growers of our country.

CHAPTER 23

MULBERRY SILK

Low Cost High Profit

Sericulture in India is an age-old practice, and has the unique distinction on of producing all the four commercially known varieties of silk *viz.*, mulberry, tussar, eri and muga.

Madhya Pradesh has a vast dense forest area where tribals make livelihood either by collecting tussar cocoons from Shaaj and Saal trees and making silk by spinning thread produced by silk worm or by growing mulberry to feed silk worms.

The state has a number of agro-climatic zones and mulberry silk can be produced on dry and barren land in arid conditions. Consequent to these changes, the social, economic and occupational scenes in the rural areas are changing from bad to worse.

Under these circumstances it has become essential to explore viable alternatives for generating additional employment and supplementing the scanty income of the farm families getting marginalized in the fast changing socio-economic milieu. Sericulture is one among the several options that can be considered for rendering economic support to such families.

India is the second largest producer of silk next to China accounting for more than 15 per cent of the global raw silk production.

The total annual production of raw silk is around 16,000 tonnes, of which mulberry raw silk output per annum aggregates to 15.60 thousand tonnes. Mulberry sericulture is practiced mainly in the states of Karnataka (64%), Andhra Pradesh (18%), Tamil Nadu (9%), and West Bengal (8%).

"In Madhya Pradesh state, particularly in Hoshangabad, the climatic conditions are favourable for mulberry plantation and cocoon rearing, its shaaj and saal trees are main sources to tussar silk," Mr. R.K. Shrivastava, District Officer, Department of Sericulture said.

Thus, MP Silk Federation has taken up some initiatives and has encouraged silk production activities. The state is government providing almost all the infrastructures, except land area, and buy-back option for mulberry or tussar cultivation. State has a number of small silk weaving units.

Sericulture is a labour intensive industry. It provides direct or indirect employment to about one million people in India. Sericulture is a low investment and high profit activity. It provides regular income to farmers throughout the year unlike most other agricultural crops.

Sericulture plays a vital role in linking urban rich sections of the society to the rural poor, as demand for silk is largely from the higher income group. Export of silk goods generates a huge sum of foreign exchange to nation's exchequer.

Madhya Pradesh has graced up to tap the vast potential in sericulture.

"Sericulture has been a very old traditional occupation for the people of rural areas. Currently due to lack of resources and the increasing demand for fashionable clothes and silk product people in these areas are jobless but we are encouraging and reviving the activities," said Mr. Shrivastava.

With a view to restarting the silk production in various areas of the state and provide employment to the weaker sections of the society including scheduled caste and backward classes, State government has endeavoured to take up the task of mulberry silk worm rearing, reeling and spinning at various places of the state. The nearest reeling unit in a small village Malakhedi near Hoshangabad is registering constant progress.

State government is also focusing on tussar silk or Kosa and in various rural areas of Madhya Pradesh, where shaaj and saal trees are found in abundance. An 'Eri silk' project is also on the anvil. State government has taken up a number of silk-production projects and activities in various villages of Hoshangabad, Sehore, Betul, Bhopal, Narsinghpur, Guna and Balaghat districts.

"We have started a pilot project in Suktawa of Kesla block. We will be able to produce eri-silk in Madhya Pradesh by year

2005-2006. Talks are on with a Gulf based company. They have evinced interest in making an investment in the project," said Mr. Shrivastava adding, "Since eri-silk worm feeds on caster plant the area of caster-cultivation has been increased significantly."

At present, as compared to other agricultural crops, the income per acre derived from Mulberry Silk cocoon production in Hoshangabad district is as follows :

Citing the example of Shri Krishnappa, a farmer of Karnataka, engaged in silk production, Mr. Shrivastava said he makes three lakh rupees per acre annually. "We want to replicate the southern model and planning tours for our farmers to Karnataka at regular intervals," the added.

State Sericulture Department has also implemented a programme Central Development Programme (C.D.P) to encourage and motivate farmers for mulberry and castor cultivation with the help of State Government and Central Government. A number of farmers have taken rearing and cocooning as an industrial venture.

Sericulture involves backward and forward linkages. As a result it has become a vehicle to uplift the poor and economically weaker sections of the society. The activity also creates opportunities for employment, thereby, pulling up the poor people living below poverty line.

Under the programme (C.D.P.), the grants, facilities and help provided by the government, are as follows :

1. Tool kits worth Rs. 2,000.
2. Bivoltine equipment for farmers worth Rs. 7,500.
3. Rs. 25,000 for building the rearing house.
4. Rs. 13,440 for drip irrigation for the farmers of General category, and Rs. 17,280 for the Scheduled Caste and Scheduled Tribes and women.

Thus the assistance per acre for farmers of general category stands at Rs. 32,940, and Rs. 36,780 for SC/ST category. For two acres of land the government grant Rs. 62,750, and Rs. 98,000 for three acres of land.

Government also facilitates the farmers with free saplings for mulberry plantation. Once the saplings are planted, mulberry farm can feed silk worms for at least ten years. Also farmers are given disinfectant materials free of cost, and 75-85 per cent of cost incurred on cocoon rearing and free technical support.

Since silk-production involves buy-back guarantee from Sericulture Department farmers run the lowest risk in business. At present the support price of mulberry cocoons hovers around Rs. 80-130 per kilogram.

Project Outlays

S.No.	*Particulars*	*Cost*	*Subsidy as per the C.D.P. Schemes*	*Margin Money @10 pc*	*Loan Amount*
1.	Basic Infrastructure development of Mulberry Plantation	0.14	Nil	0.014	0.126
2.	Bivoltine equipments	0.15	0.075	0.15	0.06
3.	Drip irrigation	0.192	0.1344	0.0192	0.0384
4.	Rearing house	0.50	0.25	0.05	0.20
5.	Fencing	0.30	Nil	0.03	0.27
6.	Capitalised working capital	0.14	Nil	0.014	0.126
	Total	**1.422**	**0.4594**	**0.1422**	**0.8204**

Model Economics

Area under Mulberry	1 Acre ....
Rearing System	Shoot Rearing..........
Rearing Capacity	1250 D.F.Ls..............
Number of Crops in a year	4-5 ....
Batch Size (D.E.Ls. per batch)	250 ....

Rearing House Requirement

Rearing House	40 (L) X 25 (B) 1000 sq.ft.
Flour Area Requirement	
Rearing Area	2.52 sq. ft. per dfl.
Total Area	4.00 sq. ft. per dfl.
Type of Rearing House	Tiles or Cement Sheets Roof with bake Bricks.
Unit Cost of Construction	@ Rs. 50.00 Per sq.ft.

DESCRIPTION OF MULBERRY PLANTATION AND SILK WORMS

(*a*) Mulberry Plantation

Mulberry plantation needs plain land, alluvial, mixed with black soil. It is an intensive farming and needs nutrient rich soil.

If the soil has more clay then there will be less frequency of irrigation. In this condition, the drip irrigation, *i.e.*, one to two liters per plant or one inch per acre irrigation is needed.

Thus 7000 to 8000 kgs of mulberry leaves are produced per acre, under such favourable circumstances. M5, S41 and S56 are the old varieties of mulberry leaves, whereas, SI365 and BI are the new varieties of leaves, which are produced almost 25,000 kgs per acre.

At present, there is a wide scope for the silk production in Hoshangabad, Sehore, Betul, Bhopal, Narsinghpur, Guna and Balaghat.

INPUTS

(*a*) Compost–any compost.

(*b*) Biomass–twigs, residuals.

(*c*) Cow dung manure–80 cubic ft. per acre.

(*d*) Vermicompost–20 bags or 1000 kgs per acre.

(*e*) *Supplementary doses* : A mix of chemical fertilizers, namely ammonium sulphate, single super phosphate and nitrate potash (in ratio of 140 : 70 : 19) per acre.

(*f*) *Investment on manure* : Cow dung manure : Rs. 4,800 cubic ft. and supplementary doses of Rs. 1,500 make a total investment of Rs. 6,300.

Recurring Expenditure

S.No.	*Particulars*	*Years*						
		1	*2*	*3*	*4*	*5*	*6*	*7*
1.	Cost of cultivations	0.134	0.13	0.13	0.13	0.13	0.13	0.13
2.	Cost of material rearing	0.02	0.04	0.04	0.04	0.04	0.04	0.04
3.	Cost of Labour rearing	0.04	0.09	0.09	0.09	0.09	0.09	0.09
	Total recurring expenditure	0.194	0.26	0.26	0.26	0.26	0.26	0.26

(*b*) Stages of Silk Worm Rearing

The Mulberry leaves are rich in minerals and nutrients and are suitable for silk-worm feeding. At least 12 kgs of mulberry leaves are required to feed worms hatched from I D.F.L. (I D.F.L. 550 eggs). The quantity of the mulberry leaves can be less but the quality should be good.

There are five stages of rearing silk worms :

1. State Silk Federation takes care of silk worm till the first three development stages. If farmers takes it as a private venture state government provides a financial assistance of Rs. 520 for 100 DFLs. A larvae takes 12 days to reach the third stage.
2. The fourth and fifth stages of development take twelve and five days respectively. Farmers are required to take utmost care during the fourth stage regarding temperature, humidity and ventilation.

The worms are infection prone during first stage and fourth stage of 25-days development. The fifth stage of development remains risk-free.

The value addition is the highest in pre-cocoon level, and 800-900 kgs of cocoons are produced per acre approximately, and a good income can be earned by selling the cocoons itself. "A good quality cocoon fetches Rs. 1.25 lakh per 500 kgs," said Mr. Shrivastava.

(*c*) Temperature

All the twelve months of the year are favourable for cocoons production, in which 4-5 crops can be obtained. The temperature, humidity and rainfall are main factors which affects production process.

Cash Flow Statement

S.No. Particulars	*Years*						
	1	*2*	*3*	*4*	*5*	*6*	*7*
Inflow							
1. Bank loan	0.8204	–	–	–	–	–	–
2. Subsidy	0.4594	–	–	–	–	–	–

(Contd...)

S.No. Particulars	*Years*						
	1	*2*	*3*	*4*	*5*	*6*	*7*
3. Margin	0.1422	–	–	–	–	–	–
4. Income	0.3480	0.72	0.72	0.72	0.72	0.72	0.72
5. Total Inflow	1.77	0.72	0.72	0.72	0.72	0.72	0.72
Outflow							
1. Capital Cost	1.422	–	–	–	–	–	–
2. Recurring expenditure	0.194	0.26	0.26	0.26	0.26	0.26	0.26
Interest	0.08204	0.7204	0.6836	0.5469	0.04102	–	–
Principal	Nil	0.13673	0.13673	0.13673	0.13673	0.13673	0.13673
Total Outflowed	1.504	0.46877	0.46509	0.45142	0.43775	0.42407	0.41042
Surplus	0.266	0.2504	0.25491	0.26858	0.28225	0.29593	0.30958
O/S Liability	0.8204	0.68367	0.54694	0.41021	0.27348	0.13675	0.0000

Production of Silk when the Cocoons are Ready

(*a*) *Boiling*

After 29 days when the cocoons are ready, they are taken to the raw silk production unit, where these cocoons are boiled. Boiling is also done in four stages in different temperatures, *i.e.*

(*b*) *Reeling*

After boiling the cocoons arc transferred to reeling machines, where the size of the thread is set in deniers.

(*c*) *Re-reeling*

The raw silk thread reeled out from the first machine is again transferred to other machine for winding and spinning. The weaving of cloth takes thereafter.

Silk Weavers Plight in West Madhya Pradesh

(*a*) Lodhikheda

Lodhikheda is located 77 kms away from Chhindwara on the Chhindwara-Nagpur Road, under Sausar Tehsil, Chhindwara

District Nagar Panchayat. Earlier weavers of this area made cotton saris of 20-24 counts and cotton-silk saris of 40 counts.

There are only 1,800 families with the population of almost 9,700 in Lodhikheda, including people of different communities, *i.e.* Koshtha, Garhewal, Dha, Momin, Harijan, Dhangarh, Dheemar, Matang and Backward Class. Of them sixty per cent are weavers. The rest are dependent upon agriculture and labour.

Lodhikhdea area is known as cotton-belt barely 25-30 per cent are landholders. Strangely in absence of government or social support, the number of weavers' family has dwindled from almost 1100 families to a negligible number and now struggling to survive with whatever left as 90-100 looms.

(*b*) Paratsingha

Paratsingha is located 75 kms away from Chhindwara in Sausar Tehsil. The small village has left with only four weavers who are engaged with 'Khadi Gramodyog Board' in a blanket weaving center and a showroom. These weavers weave a traditional raincoat *'Ghoughada' (woolen* blanket) at present, on local demand.

Paratsingha, with the population of 2000 people of 400 families is losing its distinction of being a weavers' village. Most of the weavers have now become farmers or labourers. The silk weaving activities that once supplemented income of the villagers where cultivation was possible only once a year has now vanished and most of the people are jobless.

Villagers of Kurmi, Dhaogad and Adivasi community have migrated in search of job in the nearby town Nagpur. Kurmis are traditionally dependent upon agriculture, Adivasis on forest produce and labour activities. Only Dhangad community engaged in weaving, but most of them have become farmers or labourers. The traditional Ghoghda weaving activity is now dwindled to an insignificant level.

(*c*) Sausar

Sausar is situated on Chhindwara-Nagpur Road, 60 kms away from Chhindwara. Sausar is a tehsil headquarters of Nagar Panchayat Chhindwara district. Sausar, a village of 11000 people, boasts of a rich history and culture of Koshtha and Halwi community of weavers. During the past six decades, Sausar is almost losing its identity as weavers village of western Madhya Pradesh.

People of Sausar have migrated to towns in Vidarbha region to find work at various power-looms in the Nagpur district.

'Benarasi Saris' of 10-20 counts and blouses of other designs were made in Sausar till 1950. The weavers here used to make 'Puari Saris' for the people of 'Puar' community (farmers) that dominated the area.

The design of Puari Sari resembled that of 'Saindri Sari' of Balaghat. The Puari Sari was popular for its 'Gomkatari' and 'Arrow' pattern. Apart from this, cotton, silk and Tapshila saris were also made here, which were based on the 'Popati Choukada' and 'Goonji Choukada' pattern in 'red and black' and 'yellow and black' colours respectively.

With the introduction of 1951 policy of controlling yarn, the weavers of Sausar switched over to mines, but they found themselves back to weaving art when they started weaving 'Nagpuri saris' of 40 counts.

They kept the tradition alive by 1960, by weaving saris of 60-80 counts. But misfortune chased them and their skill when 'Janata Sari' scheme launched in 1970 when weavers tried to find jobs in powerloom.

The weavers and their handloom made saris lost markets everywhere and after a number of unskilled weaver for power-loom sector become jobless. Of 500 looms only 125 looms are operational among 700 weavers family. The plight of the weavers is pathetic in absence of regular work orders.

Smoothening the Way to Silk Revolution

It is well said that 'action speaks louder than words'. Similarly, the word caterpillar denotes a small creature (silk-worm), but when it acts, it gives us 'SILK', the charm of which entices the whole world. On this Earth when God had send the most intelligent and powerful living thing 'Man', he has also send the smallest of the creatures 'Catterpillar', armed with the weapon (SILK) which dominates this modem era.

"In this world of fashion and Science and Technology, this tiny creature (caterpillar) gives the satin touch to the world of glamour and blitz. But due to the lethargic attitude of some entrepreneurs these days. These tiny creatures are endangered," laments Mr. R.K. Shrivastava, the District Officer, Sericulture, Hoshangabad.

Worried over the fast disappearing traditional silk workers and weavers of the state of Madhya Pradesh, the officer has decided to bring a revolution in the field of sericulture.

Concerned over the lack of good entrepreneurs in the state, Shrivastava informs", "in Sausar, Lodhikheda and Paratsingha, a mix of people of different communities could be seen, who practiced the traditional silk weaving some few years back. But due to lack of funds and resources, these tribals slowly had to stop this work and are today rendered unemployed, which led them to hunger and starvation."

In state of Madhya Pradesh there is a wide scope of cultivation of mulberry for the production of silk. Potentials of commercial as well as food crops are very well recognised in Hoshangabad district and other areas of MP for silk production.

Madhya Pradesh is famous for its rich tribal culture, beautiful hills, hillocks, dense forests, caves, perennial rivers, springs and valleys and rich source of precious minerals and ores.

The main source of income of tribes for their livelihood is forest and agriculture. Apart from timber forest products also include wild silk, etc.

"Earlier the tribals of MP specially those of the remote areas like Paratsingha, Lodhikheda and Sausar used to rear silk worms with the help of natural resources and make silk products with the traditional methods. But today, with the increasing demand of silk, infiltration of multinational companies in the country, lack of awareness about the government schemes for their good, tribals are rendered unemployed due to their helplessness and inability to compete with the other companies."

Mr. R.K. Shrivastava informs this with great remorse, and has enthusiastically started a project of mulberry silk, wild silk (tussar), etc.; for the upliftment and rehabilitation of these tribals.

"It is painful to see the helpless tribals and people of other communities in Lodhikheda, Paratsingha and Sausar. These people used to weave silk on the earthen pots with their traditional methods, and then market them at very low prices," adds the District Officer.

Earlier, these tribals used to export raw silk from this region due to lack of knowledge about its technologies, as well as its market. Looking to the rich biodiversity in the region there is a need of systematic survey of available plants suitable for rearing silk worms, including identification, uses, valuation, etc.; considering the agro-climatic conditions and the interests of the tribals, involving the farmers and traditional silk weaver.

Emphasis should be laid on the indexing of plants useful for silkworm rearing, cultivation, collection centers, preparation of DFL (Disease Free Lays), marketing and export of cocoons or production of raw silk, dyeing, etc.

Looking at the increasing demand for silk exclusives in the country as well as abroad, Mr. R.K. Shrivastava, is determined to re-start the traditional method of silk weaving by providing every facility to the tribals.

Eri Silk

In Madhya Pradesh silk worm rearing is done on mulberry leaves and 'tasar silk' is produced with the help of the leaves of 'Saaja', 'Saal' and 'Arjun' trees.

The work of silk worm rearing provides employment to almost 13,500 families of the state every year. There is no 'Eri Silk' production done in the state. The 'Eri Silk' worm rearing is mainly done on the 'castor' leaves. Commercial Eri silk worm rearing is done in almost one month, and 5-6 crops of silk worms can be produced in a year. Along with the Eri Silk worm rearing, castor seeds and pupae can also be a good source of income.

Those districts of the state, where the climate is unfavourable, and water level is low for the mulberry plantation, the Eri Silk production can be the main source of income for the people of such areas. The castor seeds can be sown along (parallel) with the 'Kharif crops' in the state. Apart from the seeds production, the leaves of castor can also give an income of almost Rs. 6,250 per hectare under this scheme.

Since the Eri Silk thread is short, it is spun as that of cotton. Thus, this work of spinning the thread provides a lot of employment in the rural areas. In the present scenario this can also be used as an alternative for wool.

According to the agricultural statistics of the Government of India, castor seeds are sown in around 3000 hectares of land at present. This yields an income of almost Rs. 3,600 annually per hectare of land. At present in the state the Eri Silkworm rearing is not done, but the District Officer, Mr. Shrivastava, has a project ready for the year 2005-2006, with a view to raise the employment opportunities for the people.

Fact Sheet

Area, Production and Yield of Oil Seeds in District

Area under cultivation	293700 (44%)
Average Yield per hectare	1707 kg
Gross Average Income @Rs. 7 per kg	Rs. 11949 per hect.
Area under oil seed (Soyabean)	211385 hect.
Average productivity per hect.	1022 kg
Gross Income per hect.	Rs. 8176.00

Vision

If 25% area of below average productivity is proposed under castor cultivation and Eri culture, projections would be as following.

Proposed Area, 211385 X 25%–52846 Hect.

Estimated Production

1. Castor Seed	264230 MT
2. Eri Shell	1321.150 MT
3. Eri Pupae	6605.750 MT

Value of the Production

1. Castor Seed	Rs. 3699.2 Lack
2. Eri Shell	Rs. 5284.6 lakh
3. Eri Pupae	Rs. 660.6 lakh
Grant Total	Rs. 9644.4 lakh

Per Hectare Gross Profit upto pre cocoon level Rs. 18250

Value Addition

Post Cocoon Sector

1. Cost of Eri Shell	Rs. 5284.6 lakh
2. Cost of Eri Yarn (1057.2 Kgs)	Rs. 8457.6 lakh
3. Employment Generation (2-1)	Rs. 3773.0 lakh

Post Yarn Sector

1. Cost of Fabric (148 lakh mtr) @Rs. 150 per/Mtr.	Rs. 10572.0 lakh
2. Cost of Eri Yarn (105.96 lakh kgs)	Rs. 8457.6 lakh
3. Employment Generation (1-2)	Rs. 2114.4 lakh

Total Value Addition

1. Pre Cocoon level	Rs. 9466.4 lakh	64.6%
2. Post cocoon level	Rs. 3173.0 lakh	21.2%
3. Post Yarn level	Rs. 2114.4 lakh	14.2%
Grand Total	Rs. 14931.8 lakh	100%

Average Value addition per hect. Rs. 28255 through Intercropping and Eri culture

Note - Source : HDR of M.P. 2002 and district handbook of statistics.

Conclusion

According to the agricultural statistics of the Government of India, castor seeds are sown in around 3,000 hectares of land at present. This yields an income of almost Rs. 3,600 annually per hectare of land. At present in the state the Eri Silk-worm rearing is not done, but the District Officer, Mr. Shrivastava, has a project ready for the year 2005-2006, with a view to raise the employment opportunities for the people.

Emphasis should be laid on the indexing of plants useful for silkworm rearing, cultivation, collection centers, preparation of DFL (Disease Free Lays), marketing and export of cocoons or production of raw silk, dyeing, etc.

Looking at the increasing demand for silk exclusives in the country as well as abroad, Mr. R.K. Shrivastava, is determined to re-start the traditional method of silk weaving by providing every facility to the tribals.

CHAPTER 24

CANDELILLA WAX

Euphorbia antisyphilitica, a shrub belonging to family Euphorbiaceae, is indigenous to USA and Mexico. In its natural habitat, it grows best on rocky and sandy soils with good drainage and requires only 100-500 mm of annual rainfall.

The leafless stems of the plant yield a wax known in commerce as candelilla wax. It is used as a hardener for soft waxes in candles, polishes, chewing gum, leather goods, varnishes and lacquers, sealing waxes, paper sizing, dental mold and electrical insulating materials. It is imported at present by cordite its industrial potential, a technique has been developed for the isolation and refining of wax, meeting cordite factory specification from the leafless stem of the plant, introduced at Central Arid Zone Research Institute, Jodhpur, in 1963, from Mexico. As 92 per cent survival of *E. antisyphilitica* raised in 1977 was recorded, simultaneous raising of this species in different habitates in 1978 was undertaken to study their performance and adaptation for further large scale plantations.

The present study discusses the physico-chemical properties of candelilla wax obtained from plants grown at Jodhpur and also their performance and adaptation in different habitats.

Climate and Soil : Jodhpur, Kailana, Pali and Bikaner were the four different habitats chosen for the experimental plantation of *E.antisyphilitica*. Table 1 gives the agro-climate of the region.

The soil is alkaline to saline-alkaline in nature. The pH of the soil varies from 7.5 to 9.0. The soils are poor in humus which reportedly favours good growth of *E. antisyphilitica*.

Transplanting and Harvesting : The species were raised on a large scale in Polythene bags from single stem/branch cuttings in February, 1978 and transplanted in the field in August, 1978. Table 2 in indicates the performance and adaptation of *E. antisyphilitica* in different habitats.

Table 1 : Physical Settings and Environmental Characters of Different Regions

Regions	*Soil type*	*Average rainfall mm*	*Average max. temp. °C*	*Average min. temp. °C*
Jodhpur	Sandy	366	33.3	19.3
Pali	Shallow sandy loam	490	33.5	18.3
Bikaner	Deep sandy	305	33.6	18.2
Kailana	Rocky	366	33.3	19.3

Table 2 : Performance of *E. antisyphilitica* in Different Habitats

Regions	*No. planted*	*Survival (%)*	*Mean height (cm)*
Jodhpur	520	92.0	44.3
Jodhpur	1376	95.0	32.8
Pali	300	97.6	35.2
Bikaner	108	61.0	29.3
Kailana	138	54.0	21.0

The leafless stems of the species were harvested from 14 years and one year old plants respectively, raised at Jodhpur under rainfed conditions.

Yield of Stems and Wax : The average dry yields of stems from 14 years and one year old plants were 6 kg and 450 gm per plant respectively.

The dried plants were crushed and extracted with petroleum ether in a soxhlet for 36 hrs. On concentration crude wax separated out which after suitable refining gave candelilla wax representing 2.5 per cent and 2.8 per cent of dry stems of 14 years and one year old plants respectively.

Physico-Chemical Properties : The wax samples were examined for their physico-chemical constants and these were compared with those reported in literature for commercial refined Mexican candelilla wax.

Wax samples, obtained from 14 year old plants were sent to cordite factory, Nilgiris, for their evaluation. As per their report (Table 4), the samples were found suitable for and use in the manufacture of explosives. Except for two specifications which have yet to be carried out, wax samples from one year old plants also conform to cardite factory specifications.

Table 3 : Comparison of Physico-chemical constants of Indian and Mexican candelilla wax

Particulars	*Standard Maxican wax*	*Wax samples obtained at CAZRI from E.antisyphilitica*	
		14 years	*1 year*
Colour	Grey-brown	Yellowish-brown	Yellowish-brown
Odour	Aromatic	Aromatic	Aromatic
Melting point °C	67-68	72	70
Dielectric constant	2.50-2.63	—	—
Refractive index at 71°C	1.4555	1.4554	1.4554
Specific gravity at 15°C	0.982-0.993	0.98	0.98
Effective conductivity	19	—	—
Moisture per cent	Small	0.61	0.402
Volume resistivity	120	—	—
Acid Number	11-19	15.51	15.4
Saponification Number	46-66	48.71	52.2
Iodine Number	15-36	14.2	—
Ester Number	—	33.20	36.8
Unsaponifiable Matter per cent	65-67	—	—
Ash per cent	—	0.11	0.17

Table 4 : Comparison of Physico-Chemical constants of Candelilla Wax with Cordite Factory Specifications

Particulars	*Result on samples*		*Cordite factory*
	14 years	*1 year*	*Specification*
Description	Passes	Passes	Clear and free from impurities
Volatile matter per cent	0.601	0.402	1.5 max.
Total acidity mg KOH per g. of wax	15.51	15.4	22 max
Inorganic acidity mg KOH per g. of wax	Nil	Nil	Nil
Matter-insoluble in carbon tetrachloride	0.21%	0.94%	1.0 max
Saponification value mg KOH per g. of wax	48.71	52.2	45 to 65
Setting point °C	63	66	63 to 68

(Contd...)

Particulars	Result on samples		Cordite factory
	14 years	1 year	Specification
Ash-Total percent	0.11	0.17	0.5 max
Ash retained on 63 micrometre 16 sieve percent	×	×	0.01 max
Ash retained on 250 micrometre 16 sieve percent	×	×	Nil

15-36 × to be determined.

Results and Discussion

The recovery of wax by the extraction procedure, used in Mexico, is only about 2 per cent of the plant material. Using solvent extraction method and suitable refining technique, it has, however, been possible to obtain candelilla wax representing 2.5 per cent of dry leafless stem of 14 years old plants. Dry stems of one year plants give still better yield (2.8 per cent) of candelilla wax. The yield of wax is likely to go up further as according to a report the plants require 2-5 years to produce wax in quantities sufficient for commercial use.

Thus the yield and quality of candelilla wax obtained from plants grown at Jodhpur is very good. It also conforms to the specifications of both commercial grade Mexican wax (Standard) and cordite factory. Encouraged by the results, plantation of *E. antisyphilitica* to study its performance and adaptation in different habitats has been taken up. The results obtained so far are quite encouraging. Hopefully we can look for developing an indigenous source of candelilla wax not only for cordite factories but also for polishing, coating and chewing gum industries.

CHAPTER 25

BEE VENOM CAN ALSO BE ELIXIR

Bee venom is a clear liquid with a sharp smell (reminiscent of honey), bitter and burning taste and aromatic odour. It is stored in the poison sac attached to the sting of worker bee. Newly emerged worker bees have very little venom but its amount gradually increases with age to about 0.3 mg in 12-15 day old worker bee *(Apis mellifera)*. Secretion of venom ceases after 20 days as no additional venom is produced thereafter and the weight of venom in the poison sac remains unaltered.

Composition of bee venom is complex and it is an aqueous mixture of non volatile proteinous material and volatile organic compounds. The non volatile components are carbohydrates, lipids, amino acids, peptides, protein and enzymes whereas ethanol, formic acid, *n*-butylacetate, n-amylacetate and isoamylacetate constitute volatile fraction. The enzymes present are mainly phospholipases and hyaluronidases but acid and alkaline phosphatases and B-esterases are also present. Venom of Indian honey bee *(Apis cerana)* has been reported to be two times toxic than that of *A. mellifera*.

Commercial method of obtaining bee venom consists of stimulating large number of bees by electric shock to sting through different media. One way is, by stretching over the collection board, a sheet of nylon parchment taffeta above two sheets of thin polyethylene (0.025 mm thick). The venom from each sting is deposited as a droplet on the under side of the lower sheet from which it is easily removed by extracting the water. Each colony is "milked" for 5 minutes; and 20 hives produce about 1g venom. Bees which release the venom survive but the colony gets greatly irritated. Another method is to pass electric current through the sheet of gel or silicone and bees standing on this react by stinging in to it where it is deposited. Presently various modifications for extracting venom have been done in the existing systems by the introduction of cooling systems or by varying the currents.

Bee venom has been found to be very useful for curing diseases and disorders. Rheumatism can not be cured by any other system of medicine but majority of its patients can be cured by bee venom. When used correctly, it acts as a therapeutic and prophylactic remedy. It has been found that bee venom dilates capillaries, decreases viscosity and coagulability of blood, has a stimulating effect on heart muscles, influences metabolism and decreases cholesterol level. It also possesses antibiotic properties.

It has been used to cure polyarthritis, infectious-polyarthritis, spondlylosis deformans, rediculiti, neuritis, trophic ulcers, neuralgia, endo arthriti, rheumatism, certain eye diseases like iritis (inflammation of the iris) and irridocyclitis (inflammation of the ciliary body and iris), skin diseases (tuberculosis of skin), St. Vitu's dance, inflammation of the sciatic, facial and other nerves, hypertension and even malaria.

Bee venom has been found most effective when injected intracutaneously by means of natural bee sting in and around the affected regions. The area of skin to be treated should first be washed with warm water and soap. The bee is then placed on the skin by means of special tweezers for injecting the skin (generally outer aspect of the shoulders and hips are used for bee stings). The contractions of the stinging apparatus force venom out of the poison sac which is instantly taken up by blood and on reaching the blood stream, its effect is felt throughout the organism.

Two courses of treatment are generally followed. During the first course, venom from 55 bee stings is given (first day one, second day two, third day three stings and so on for ten days). Then after a break of three of four days, treatment can be continued with three stings daily for six weeks. During the second course the patient receives the venom of approximately 140-150 bees. Treatment should then be stopped if the patient is not cured or there is no improvement in the patient's condition. Intracutaneous injection of apitoxin, with due regard to patient's condition, has also been used.

For bee venom to be toxic, its dose must be 10 times greater than curative one and to be fatal, it must be hundred times greater. There are many reports of deaths of human beings and animals from multiple wild bees *(Apis dorsata L)* stings.

There have been several instances of deaths of animals like elephants and buffaloes by these bees. According to a report from Ceylon, five *A. dorsata* stings are regarded as equivalent to the bite of cobra there.

Some people show allergic reactions to be venom and for them one sting is sufficient to cause general indisposition, severe headache, vomitting and diarrhoea but with the passage of time, majority of them become immune. Generally, stings cause a local reaction only (reddening of skin, swelling, burning sensation, etc.,) but 200-300 simultaneous stings may cause disturbance in cardiovascular and nervous system and 500 or more stings cause even death. The remedy advisable at this juncture is the immediate extraction of sting together with its poison sac so that poison does not pass in to blood stream. Diphenyl hydramine hydrochloride is recommended for treating severe effects of bee stings. The drug propranolol also counteracts most of the changes induced by whole venom or its components.

CHAPTER 26

MAKE MONEY WITH HONEY

Honey holds high place in the esteem of mankind from time immemorial. Honey is a sacred substance. It is used in all religious ceremonies, from birth till the death, as a food and as a medicine. This signifies the importance of honey in human life. Daily use of honey as food ensures long and good health. Just as good manure is essential for a good crop, quality honey is essential for good health and good living. Honey is sweets refreshing, soothing and sedative relieves and stimulates digestion of those with weak stomachs and loose bowels. It removes poisonous effects from the body.

Honey as food has been mentioned in the mythical saying in Bible and Quran. Honey has certain nutritive and energetic advantages like :

- better energizing action
- mild appetizing effect
- easier assimilation and digestion of other foods
- desirable influence on the assimilation of calcium in children
- An action of preferential retention of magnesium which explains why children nourished with honey grow well.

Honey consists essentially of different sugars, predominantly glucose and fructose. It also contains protein, amino acids, enzymes, organic acids, minerals, pollen and other substances and may include sucrose, maltose, melezitose and other oligosaccharides (including dextrin) as well as traces of fungi, algae, yeast and other solid particles resulting from the process of obtaining honey.

The colour of honey varies from nearly colourless (transparent) to dark brown. The consistency may be fluid, viscous or partly to entirely crystallized. The flavour and aroma vary, usually derived from its plant origin. According to estimated the average consistency of honey contains water (17%), sucrose (1.30%), fructose (38%), glucose (31 %), other sugars (9%), total acids (0.57%), ash (0.17), nitrogen (0.04%), and undetermined (2.92%).

Honey is standardised by Agmark authorities of Government of India. This standard honey is called "Agmark" honey, and is sold in sealed packed bottle with Agmark label. The three types of grades as specified by the Agmark authorities are Special, Grade A and Standard. In all cases, the presence of methyl furfuryl should be negative by aniline chloride test.

However, before giving the details of properties of Standardised honey it may be mentioned that, Agmark specifies a specific period for consumption or there is specification of expiry date. This raises certain eyebrows since it is generally believed in India that the older the honey, the better for health. There is however, no specific base behind this belief.

Ayurvedacharya Shibkali Bhattacharyaa in his famous book *Chiranjeeb Banoushadhi* mentioned that older honey normally warms up the body quickly due to the presence of alcohol. Therefore, fixing date for honey while packing obviously has reasoning. The chief characteristics of three grades of honey prevalent in India are depicted in Table 1. However, the characteristics required for table honey in some advanced countries are shown in Table 2.

A survey of the consumer's attitude in the important beekeeping areas of Himachal Pradesh was conducted. The results based on a small sample of 25 honey consumers are shown in Table 3.

The socio-economic and personal profile of the target group of consumers is an important factor in this regard. It can be observed from the table that larger chunk (44%) of honey consumer fell in the middle age group of 26-40 years. The education variable showed that honey consumption increased with the level of education. Majority (52%) of the honey consumers possessed degree level education. People with no formal education also consumed honey but their proportion was low (4%). The average size of family was estimated at 4.60 with sex ratio of just 756 females per thousand males.

Table 1 : Characteristics of Agmark Honey

S.No. Characteristics	Grade of Honey		
	Special	*Grade 'A'*	*Standard*
1. General characteristics	Well ripened natural product, extracted with machine/ extractor, free from objectionable flavour/aroma, free from foreign matter	Natural ripened product, extracted with machine/ extractor, free from objectionable aroma	Natural well ripened product, extracted with machine/extractor, free from objectionable flavour/aroma
2. Colour	Light to dark brown	Light to dark browk	Light of dark brown
3. Specific gravity at 27°C (Min)	1.40	1.40	1.35
4. Sucrose percent (Max)	5.00	5.00	5.00
5. Ash percent (Max)	0.50	0.50	0.50
6. Moisture percent (Max)	20.00	22.00	25.00
7. Total reducing sugars percent (Max)	65.00	65.00	65.00
8. Fructose/Glucose ratio (Max)	1.0	0.95	0.95
9. Acidity (calculated as formic acid) percent (Max)	0.20	0.20	0.20
10. Fiche's test	Negative	Negative	Negative
11. Aniline chloride	Negative	Negative	Negative
12. Hydroxyl-methyl furfuryl ppm (Max)	30	40	40

On occupational front the study observed that majority of honey consumers were agriculturists (48%) followed by service class (40%) and businessmen (12%).

The frequency and amount of honey consumed depends on several factors such as climate, health of family members, age, educational status, nutritional status, economic status and availability of honey in particular area. As far as honey consumption is concerned, three types of honey consumers were found. The table shows that majority (56 %) of the respondents consumed honey once a week followed by 28% daily consumers.

The proportion of honey consumers was 16 who consumed it once in a month. The per capita per annum honey consumption on all types of households was estimated at 517 gm. The quantity consumed at a time or in one go, however, ranged between 5-10 gm in each and every selected district.

Table 2 : Characteristics required for table honey in some countries*

S.No.	*Component of test*	*Value acceptable*
1.	Moisture content	18% or less
2.	Sucrose content	3% or less
3.	Reducing sugar as invert sugar	7% or more
4.	Dextrins	5% or less
5.	Acidity or meg/kg	5% or less
6.	Ash	Between 0.1 and 0.25%
7.	Water insoluble in the solids	0.1% or less
8.	Diastege, Gothe test	Between 8 and 10
9.	Fiche reaction	Negative
10.	HMF	40 ppm or less
11.	Lund reaction precipitate	Between 0.6 and 3%
12.	Lugol reaction	Negative
13.	Trace metals	Below certain limits
14.	Polarimetry	Laevorotatory between-21' and 2'.

* Values acceptable in Argentine, Austrailia, Brazil, Canada, Latin America, New Zealand, Switzerland, Austria, Germany, Spain, France, U.S.A., Italy, Mexico, Sweden and United Kingdom.

Table 3 : Socio-economic and personal characteristics of respondents

S. No.	*Variable*	*Category*	*No. of respondents*	*Percentage*
1.	Age	(*i*) Young (upto 25 years)	6	24.00
		(*ii*) Middle (26-40 years)	11	44.00
		(*iii*) Semi old (41-60 years)	5	20.00
		(*iv*) Old (Above 60)	3	12.00
2.	Education	(*i*) Illiterate	1	4.00
		(*ii*) Primary school	3	12.00
		(*iii*) Middle standard	3	12.00
		(*iv*) High school	5	20.00
		(*v*) Degree	13	52.00

(Contd...)

S. No.	Variable		Category	No. of respondents	Percentage
3.	Family structure	(*i*)	Male/family	2.62	–
		(*ii*)	Female/family	1.98	–
		(*iii*)	Average family size	4.60	–
		(*iv*)	Sex ratio (Females/ 1000 male)	756	–
4.	Occupation	(*a*)	*Main*		
		(*i*)	Agriculture	12	48.00
		(ii)	Service	10	40.00
		(iii)	Business	3	12.00
		(*b*)	*Subsidiary*		
		(*i*)	Agriculture	6	24.00
		(*ii*)	Service	2	8.00
		(*iii*)	Business	5	20.00
5.	Honey consumption	(*i*)	Daily	7	28.00
		(*ii*)	Weekly	14	56.00
		(*iii*)	Monthly	4	10.00
		(*iv*)	Per capita honey consumption (gms/annum)	517	
6.	Preferential trait for honey consumption	(*a*)	*Form of honey*		
		(*i*)	Granulated honey	12	48.00
		(*ii*)	Liquid honey	13	52.00
		(*b*)	*Colour of honey*		
		(*i*)	Transparent	5	20.00
		(*ii*)	Red	2	8.00
		(*iii*)	Brown	11	44.60
		(*iv*)	Yellow	7	28.00
		(*c*)	*Type of honey*		
		(*i*)	Mono-tlora	6	24.00
		(*ii*)	Multiflora	19	76.00
7.	Factors governing honey's price-consumer's view	(*i*)	Cost of production	5	20.00
		(*ii*)	Demand and supply	9	36.00
		(*iii*)	Cost of production plus markup of profit margin	11	44.00
8.	Consumer's view points for raising the demand for honey	(*i*)	Advertisement	2	8.00
		(*ii*)	Creating awareness about its uses	22	88.00
		(*iii*)	Provision of subsidy	3	8.00
		(*iv*)	Quality standard/ Agmarking)	3	12.00
		(*v*)	Paking in suitable packs	3	12.00
9.	Use of honey	(*a*)	*Health*		
		(*i*)	Curing cough	12	48.00
		(*ii*)	Curing asthma	1	4.00
		(*iii*)	Other ailments	12	52.00
		(*b*)	*Religious purposes*	4.00	

The next important question was to find out the preferential traits of respondents for honey consumption. The data in respect of preference for form, colour and type of flora (base of honey) were collected and analysed. It is evident from the Table 3 that 52% of consumers preferred liquid honey.

The proportion stood at 48% for granulated honey. Most liking for liquid honey was due to the reason that it can be easily taken out of the containers in comparison to freeze or granulated honey which is not palatable to many consumers.

The colour of honey is another important trait contributing towards its preference. The honey was normally available with four colours : transparent, red colour, brown colour and yellowish. The analysis reveals that on an average brown colour was largely (44%) preferred. Regarding the choice of type of honey, multiflora based honey was the preferred most by 76% of honey consumers. Mono flora based honey was hardly available with the beekeepers as this is to be obtained mostly under controlled conditions, yet 24% of the honey consumers expressed their choice for mono flora honey.

The price of honey is an important issue in the policy decisions. The price of honey depends on various factors of which cost of production, demand and supply situations and cost of production plus mark up margin are important. The information gathered on these aspects reveals that 44% of consumers held their view in favour of fixing price which may take care of cost of honey production and mark up margin. Strong opinion with regard to this factor was given by consumers of Kangra district (H.P.) from where most of the beekeepers formed part of detailed study conducted on economic aspects of 200 sample migratory and stationary beekeepers.

Demand for a commodity depends on various factors which include beside prices the income price of related goods, advertisement, subsidy, quality standard and packing in convenient/suitable packs etc. The information on such variables was collected and views of honey consumers were analysed. It can be seen from Table 3 that majority (88%) of the consumers favoured creating awareness among the people about its uses through personal motivations.

This means that creating awareness should be accorded the top most priority to raise the demand for honey which may lead to sale of the unsold stocks of honey, if any quickly, thus being a stimuli to beekeepers. Supply of honey at subsidised rates to consumers was also addressed by 8% consumers. Advertisement, quality standard (Agmark honey), and packing in suitable packs of 100 gms, 500 gms and 1 kg, were the other suggestions put forth by honey consumers in ultimately raising demand for honey locally and nationally. Since in India the honey consumption is too low (2.3 gm/capita/annum) in comparison to about 1 kg in certain developed nations like Germany the demand has to be raised in India.

It is interesting to note that very less proportion (4%) of the sample respondents reported its use in religious activity. This observation was in contradiction with the general feeling that as in every religious activity honey was used. This response came in the context of quality of honey being used for religious purposes : the respondents explained that since honey is used on rare occasions in small quantity they did not count its use as a significant one. On the other hand the use of honey whatever was reported mainly meant for maintaining good health by way of protecting themselves against certain diseases.

CHAPTER 27

WHITE GOLD MUSHROOM CULTIVATION

The cultivation of edible Mushrooms is generally assumed to have begun in France around 1650. Gradually the cultivation spread first in Europe and in 1823, it was introduced in Netherlands.

Mushroom is a group of macroscopic fungi and lack of chlorophyll depends upon prepared nutrients. The terms Mushroom is often used loosely, however the common view is, the umbrella-shaped fungi belonging to ***Basidiomycotina,*** which are edible be called mushroom. In nature, mushroom grows wild in every country from snowy mountains to sandy desert, waste land, forest etc., and appear in all season but mainly in rainy season because organic matter decompose very easily.

Mushroom has to derive its nutrients from the organic matter on which it grows in common with many other. Mushroom is capable of using such indigestible substances as cellulose and lignin. Besides carbohydrates mushroom needs nitrogen but they can utilise nitrogen only in the form of protein and lignin. There are 1000 species of edible mushroom but use those that cultivated by own hands or species you are familiar with and avoid all others. Earlier Mushroom was not particularly easy to cultivate it was because their growing conditions were not properly known and it was practically impossible to influence those condition. It was only after the second World War that mushroom cultivation really began to develop in Netherlands as an Agro Industry. Around 1950, the first mushroom house equipped with ventilation and heating system was built.

Advantages of Mushroom Cultivation

The artificial cultivation ensures that mushroom sold is truly edible. They are now regarded as useful food in modern diets, complementing the staple diet. Dieticians, to counter obesity and other syndromes associated with present-day eating habits, often prescribe them.

Plant residues such as straw, leaves, and also wastes from agriculture, forest and industry mostly remain unused. These are disposed off by burning or by some other wasteful and sometimes harmful methods.

However, by suitable treatment, these residues can be converted into substrates for the cultivation of mushrooms and the spent substrate can be utilized as manure.

Another advantage in mushroom growing is that they are grown in rooms, for which the wasteland may be utilized. Being grown in vertical stacks, for mushroom come into production very rapidly; makes mushroom growing a profitable venture.

There are various types of edible cultivated mushrooms prominent among them are :

Species	*General Name*
Agaricus bisporus	Button Mushroom
Entinus edobes	Shiitake
Volvariella volvacea	Paddy Straw mushroom
Pleurotus Sajor-Caju	Oyster Mushroom
Auricular Species	Wood Ear Mushroom
Flamblina Velu	Winter Mushroom
Tremelia Fuctormis	Silver Ear Mushroom
Pholiota Namelx	Hameko/Vascid Mushroom

The most commonly favoured variations in India are button, oyster and paddy straw mushroom.

Mushrooms contain higher amount of nucleic acid as compared to conventional foods. The comparison is presented as follows :

Food Item	*% Nucleic Acids (on dry weight basis)*
Mushrooms	2.1%
Cereals	1.1%
Meat and Fish	2.2% to 5.7%

Mushroom protein is comparable to the high quality animal protein. It is superior to vegetables and cereals proteins. Hence it can substitute the non-veg diet and can provide a valuable source of high quality proteins to vegetarians. Mushrooms are regarded as one of the world's greatest untapped resources of nutritious and palatable food for future.

Comparison of Nutritive Value of Mushrooms with Other Foods

Food Acid (FAO)	*Essential Amino Acid Index (FAO)*	*Amino Score*
Mushroom	98.6	90.0
Beef	100	98.0
Chicken	100	98.0
Milk	99.0	91.0
Cabbage	72.0	63.0
Turnip	69.0	38.0
Tomatoes	44.0	

Research data shows that mushrooms are highly nutritive food. Among all varieties of mushrooms, oyster mushrooms are virtually calorie free with high protein content. An analysis of some common edible mushrooms showed that on fresh weight basis the moisture content was 89-91%; ash 0.97-1.26%; protein 2.78-3.94%; fat 0.25-0.65%; crude fibre 0.09-1.67%; carbohydrates 5.3-6.28% and energy value 24.4-34.4K cal.

Agaricus bisporus contain 0.95% of manifold; 0.28% of glucose; 0.04% of pentoses; 0.59% of glycogen; and 0.91 % of crude cellulose. Carbohydrates of mushrooms do not contain starch and cane sugar (sucrose). Glucose is present in very small quantity and like any other vegetable, mushrooms contain fat in small amounts. Therefore, mushrooms can be recommended as a 'Slimming diet'.

Many vegetables and cereals are deficient in some of the essential amino acids. Mushroom protein contain adequate quantities of most of the essential amino acids compared to those of egg protein, which is supposed to be a standard complete protein. A few essential amino acids, which are lacking in cereals are present in high concentrations in mushrooms. Therefore, mushrooms when supplemented with cereals in the diet will help in overcoming amino acid deficiency, particularly lysine. The mushroom protein is easily digestible and its overall quality is intermediate between vegetable and meat protein.

Most mushrooms are very poor in vitamin A and have high riboflavin and nicotinic acid, particularly thiamine, biotin and pentothenic acid. Folic acid, a B-vitamin known as a blood-building vitamin, is also available in mushroom. Thus, mushroom can go a long way to ward-off malnutrition.

Mushrooms also have high mineral content, particularly phosphorus and potassium. Iron and calcium are in low concentrations, slightly less than a third of the total iron in mushroom is in the available form and can be readily utilized by the human body.

Mushrooms are a good source of several vitamins such as:

- Thianine
- Riboflavine
- Niancin
- Ascorhic acid.

Protein is the most critical component of food. Protein deficiency in the world is most serious human nutritional problem. The quality of dietary protein is determined largely by essential amino acid present in these proteins.

Cereal and vegetarian diets which provide the bulk of proteins to the maximum population of the India are deficient in some of the amino acid particularly Lysine and Triptohen, Dietary proteins are split during digestion into their constituent units amino acids.

There are at least 20 amino acids, commonly found in food. Eight of them are classified as essential and they must be present in the diet. These eight essential amino acids can be listed as follows :

1. Isolucineine
2. Leucine
3. Lycine
4. Methionine
5. Phenylalanine
6. Thronine
7. Tryptophen
8. Valine

In addition to them, Cystine and Tyrosine are usually included in essential amino acids as they have a sparing action on Methonine and Phenylalanine respectively.

Salient Features of Mushroom Cultivation

1. **100% risk free project** : Heavy plant and machinery is not required, hence the capital investment is low. Similarly running cost is also very low due to absence of any recurring maintenance expenses.
2. **Low working capital** : The raw material used for cultivation of mushroom is mainly agro waste like paddy and/or wheat straw. Hence the requirement of working capital is very low.
3. **Latest technical know how** : The latest cultivation techniques, know how etc.; for quality mushroom would be supplied by various agricultural universities, IARI constituents, CSIR labs etc.
4. **Very high returns** : Agro waste like paddy straw, wheat straw, sugar cane waste etc.; are converted into compost, which is used as bed for cultivation of mushroom. Thus agro-waste is converted into an income generation activity.
5. **Low execution period** : The project can be executed from concept to commissioning within four months.
6. **Easy financing** : Mushroom cultivation is considered to be an agro project which is thrust area for development. Hence finance is available from Banks on top priority.
7. **Long term business** : International as well as domestic market is growing at a faster speed. Presently there is huge gap in demand and supply. By a conservative estimate the domestic demand for mushrooms would be about 3,000,000 tonnes per annum by the year 1998. Presently only 5000 tonne is being supplied in domestic market.
8. **Pollution free environment friendly project** : The mushroom cultivation project is free from any type of pollution, rather it is helping the nature by maintaining the bio-cycle.

Importance of Mushroom Cultivation for India

(*a*) Mushroom cultivation is a highly labour intensive activity.

India has abundance of rural labour, which is suited for such a project.

(*b*) Mushroom harvesting cannot be effectively automated : Manual operations are best suited for Indian industry as it ensures creation of more employment.

(*c*) Mushroom helps in maintaining the cycle of nature by decomposing agro residues. This agro waste is available in abundance in our country.

(*d*) Mushroom is a good source of high quality proteins and rich in vitamins and minerals. It is an ideal food for vegetarian population in India.

(*e*) Mushroom cultivation provides excellent opportunity to educated rural youth to become entrepreneur and provide jobs to others.

(*f*) Mushroom cultivation provides opportunity to use wastelands except water logged land, which is abundantly available in India.

(*g*) Mushroom cultivation provides excellent opportunity in rural areas near by the metropolitan cities to become economically sound by selling mushrooms in urban areas because of great demands of mushrooms in metropolitan cities.

(*h*) The rural women whether they are educated or uneducated can easily handle such type or projects very easily along with their house jobs and become economically sound.

Types Suitable for Cultivation

At present three mushrooms are being cultivated in India. These are : white button mushroom (Agaricus bisporus), paddy-straw mushroom (Volvariella volvacea) and oyster mushroom (Pleurotus sajor-caju). Of these, A. bisporus is the most popular and economically sound to grow and is extensively cultivated throughout the world.

However, due to its low temperature requirement, its cultivation is restricted to the cool climatic areas and during the winter in the plains of Northern India. In summer, the tropical paddy-straw mushroom is suitable for growing in most suitable parts of India.

Oyster mushroom can grow at moderate temperature ranging from 22° to 28°C. Therefore, it is suitable for most of the places in India.

In North India, the climatic conditions prevailing during different seasons can be exploited for growing mushrooms throughout the year. To this a year-wise production schedule is suggested :

❒ Mid-November to mid-March	Agaricus bisporus
❒ February to mid-April	Pleurotus sajor-caju
❒ Mid-June to mid-September	Volvariella volvacea
❒ September to November	Pleurotus sajor-caju

Technology for Cultivation of Button Mushrooms

The first step in the production of button mushrooms is the formation of compost. The process of showing and then actual growing is as follows :

Manual Compost Making

It is prepared on well-cleaned pucca floor, which is raised from the ground level to allow the excess wider to drain out. It can be done in open, but it should be protected from rain by covering it with polythene sheet.

Formula for Making Synthetic Compost

Wheat straw	1000 kg
Wheat/Rice bran	100 kg
Ammonium sulphate/can urea	12 kg
SSP/SOP/MOP	25 kg
Molasses	16 kg
Gypsum	100 kg

Composting Method

1. **Wetting the straw :** The first step in the composting process is to wet the straw. Fresh dry straw resists water absorption, and unless it is persuaded to absorb water, it will not soften; and unless it softens it will take more water later. In practice, the straw is spread thinly over the entire floor of the composting yard.

It is then gradually wetted by sprinkling water gently, till the straw takes no more water. The straw is then turned for even wetting. Again water is sprinkled till it can absorb no more. At this stage water content is 75% and for most compost this point is reached when the compost is just saturated and before any run off occurs. One tonne of dry straw will require almost 5,000 litres of water to bring in saturation.

2. **Mixing and heaping :** After the straw is wetted, the supplements excluding gypsum are uniformly scattered over the straw and mixed. Some growers prefer to mix half the supplements at the beginning of composting and the remaining half after the first turn.

It is not known whether this practice is in any way beneficial. After mixing, the mixture is finally stacked in a heap. A heap one metre high, one metre wide and of indefinite length has been found to be suitable for Delhi during September-October. The straw can be stacked manually or with a stack mould. The straw should be firmly but not compactly compressed into the mould.

The dimensions of the heap can be adjusted according to the size of straw and air temperature. The principle is that longer the straw, bigger the heap. If composting is done in the cooler months when the temperature ranges between 10° and 18°C, a small heap would be unable to retain heat and moisture and the composting would be unsatisfactory.

During the hot weather generally and in particular in tropical and sub-tropical regions, the temperature difference between inside of the compost and the surrounding air is too small to produce chimney-effect necessary for compost ventilation. Core ventilation does not take place. As a rule undesirable acid zones occur inside the compost. In such cases, relatively narrow heaps would be more suitable.

Turning Schedule

It is important to ensure that the heap attains sufficiently high temperatures (70°-75°) to bring about the correct composting, otherwise the compost will lack the necessary nutritive value so essential for a good crop. Care must also be taken to see that over composting does not take place. Open the heap and remake it a number of times. For this purpose, the time schedule suggested is :

Day zero	wet, mix and stack the heap
4th day	first turning
8th day	second turning
12th day	third turning
16th day	forth turning
20th day	final turning and filling of the trays

Nitrogenous supplements and carbohydrates are mixed on day zero. Gypsum is usually mixed at the third and forth turning in equal quantities. During the final turning, 40 ml Malathion diluted in 20 litres of water is sprinkled.

Preparing the Beds

However, the compost should not be too dry or too wet at the time of filling the beds. The moisture should be sufficient as determined by the 'palm test'. If relatively dry, the water should be sprinkled and, if too wet, the excess water may be allowed to evaporate from the compost. It is preferable to use the compost as soon as it is ready, because it gradually deteriorates on being allowed to remain in the heap for too long.

The mushroom houses are constructed with built-in shelves on which the beds are laid. The shelf structure consists of wooden or steel framework and wooden shelf boards. The heavy compost should not be filled to more than 14 cm thick and the bed made from light compost should be 20-25 cm thick. In this system spawning and cropping take place in the same room.

Growing Process

This constitutes the spawn running, the first stage of case running the second stage case running and the growing period. The total process takes about 70 days. The mushroom farm is designed to give 4 to 5 crops per annum.

The temperature and humidity for different rooms is generally as under :

Room Type	*Temp*	*R.H.*	*Resh Air*
Spawn Running	21-24	90-95%	NIL
First stage case Running	-Do-	-Do-	NIL
Second stage case Running	-Do-	-Do-	NIL
Growing Process	16-18C	90%	Variable

Duration of Each Cycle

Spawn Running	12-15 Days
Case Running	15-20 Days
Growing (Cropping)	35-45 Days

On completion of each cycle the rooms are thoroughly cleaned, disinfected and sterilised. Likewise the air handling equipments and metal cost are also cleaned before the start of next cycle.

Market Position of Mushroom

World Production	*1986*	*1990*
World production of cultivated mushroom	21,80,000 MT	40,00,000 MT

Total cultivation of mushroom increases 72.5% in four years.

96.0% for Agarious (Botton Mushroom)

4% for Pleurotus spp. (Oyster Mushroom)

Estimated demand increase by experts @24.5% per annum

Major cultivated production species	*% Share in total production 1990*
Agaricus (Cotton Mushroom)	37.7
Pleupotus (System Mushroom)	24.1
Stakes	10.6
Aurioularia	10.6
Volvanila Volvacea	5.5

Cost of Project

Fixed Capital

1.	Building cost of structure 5 No. (25' × 25') including Cost of Cultivation Shelves	3,10,000.00

Plant, Machinery and Equipments

(*a*)	Air, handling unit, humidifier, water pump, Spray etc.	70,000.00
(*b*)	Water tank, equipment, basket, pipe-lines, power installation etc.	20,000.00
(*c*)	Misc. Including furniture etc.	15,000.00
		1,05,000.00
	Sub Total (*a*)	**4,15,000.00**

Working Capital (for six months)

Cost of Raw Material

(*a*)	Cost of compost	1,00,00 kg.	Rs. 2.25 kg.	2,25,000.00
(*b*)	Cost of spawn	000 kg.	Rs. 45.00 kg.	27,000.00
(*c*)	Casing soil			8,000.00
				2,60,000.00

Utility and Power

(*a*)	Power	25,000.00
(*b*)	Maintenance and Repair	9,000.00
		34,000.00

Staff and Labour

		Qty.	*Rate per month*	*Duration in Month*	
(*a*)	Skilled Worker	1	3000	Six	18,000.00
(*b*)	Unskilled Worker	4	2500	Six	60,000.00
					78,000.00
	Production overheads				40,000.00
				Sub total (*b*) =	**4,12,000.00**

Total Capital Investment

(*a*)	Fixed Capital	4,15,000.00
(*b*)	Working Capital	4,12,000.00
		8,27,000.00

Contribution by Owner	
(Approx. 25% of Total Capital Investment)	2,05,000.00
Loan Required	
(Approx. 75% of Total Capital Investment)	6,22,000.00
Sales	
(Expected Production Rate @Rs. 38/kg	25,000 Kg.
Total Sales proceeds (Fox Six Months)	25,000 × 38
	9,50,000.00
Gross Profit	
(*a*) Sales	9,50,000.00
(*b*) Less Production cost	– 4,12,000.00
	5,38,000.00
Administrative and Other Expenses	
(*a*) Interest on Loan @15%	93,000.00
(*b*) Depreciation @15% on fixed capital	62,000.00
(*c*) Admn. Expenses	25,000.00
	1,80,000.00
Net Profit	
(*a*) Gross Profit	5,38,000.00
(*b*) Less Administrative expenses	1,80,000.00
Other Expenses	**3,58,000.00**

Net Profit Ratio

$$\frac{\text{Net Profit} \times 100}{\text{Sales Turover}} \quad \frac{(3,58,000 \times 100)}{(9,50,000.00)} = 37.8\,\%$$

Return of Investment

$$\frac{\text{Net Profit} \times 100}{\text{Total Investment}} = \frac{(3,58,000 \times 100)}{(8,27,000)}$$

Fixed Cost	
(*a*) Depreciation @15% on fixed Capital	62,000.00
(*b*) Interest on Loan @15%	93,000.00
(*c*) Staff and Labour	78,000.00
(*d*) Overheads (For Six Months)	37,000.00
	2,70,000.00

Break Even Point

$$\frac{\text{Fixed Cost} \times 100}{\text{Fixed Cost} + \text{Profit}} = \frac{2,70,000.00 \times 100}{2,70,000.00 + 3,58,000.00}$$

$$\frac{2,70,000.00 \times 100}{6,28,000.00} = 43\,\%.$$

CHAPTER 28

BANANA CANNING INDUSTRY

India produces 45.49 million tonnes of fruits annually which accounts for about 10 per cent of world's fruits production. However, for various reasons, this abundance of production is not fully or satisfactorily utilized as it should be. About 25 to 30 per cent of it is wasted due to improper handling, lack of transportation and storage facilities and by spoilage. To avoid the wastage, the surplus could be processed and preserved properly.

Food processing plays a very important role in increasing the shelf life of the products besides considering the nutritional aspects.

In 1996, world production of most important fruits was around 400 million tonnes. Bananas compete with grapes for second place behind citrus accounting for 13-14% (*i.e.*, 52-56 M tonnes) of total world fresh fruit production. Banana production has been increasing by around 3% per year over the last decade. India's contribution to world production of banana is about 15.02 per cent (*i.e.*, 7.9 M tonnes).

Banana has a very special place in diet, low in fats, cholestrol and high in calories. About 100 g of banana provides about 100 calories which is 50% more energy released by other fruit like apples, citrus fruits and others.

Keeping banana as a whole fruit at room temperature for long time is not feasible due to its poor shelf life quality. The processed product from banana are also very limited at present and there is a good market potential for the banana product. Hence a study on canning of banana was taken up with a view to commercialize the production by selling up an agro processing industry.

Materials and Methods

Canning is a process of heating the foods in hermetically

sealed containers for sufficient time at a high enough temperature to eliminate the pathogenic microorganisms that endanger public health and those microbes and enzymes that deteriorates the food during storage.

Major Machineries Required

- Fruit washer
- Fruit peeler/slicer
- Can reformer
- Can flanger
- Can seamer
- Exhausting and processing retort with time and temperature control.
- Thermocouple with selector switch
- Other accessories *viz.*, weigh balance, thermometer, refractometer, vacuum oven etc.

The levels of variables for canning of banana

	Factors	*Level*
1.	Variety	(3) Poovan, Rasthali and Red banana
2.	Syrup type	(2) Sugar and Jaggery
3.	Syrup concentration	(3) 200B, 250B , 300B
4.	Months of storage	(4) 10, 1.5,3, and 4.5.
5.	Replications	2

Canning Procedure

The thermal process time determined using Ball's formula method for the particular experimental set-up was found to be 25 minutes. The fruits were peeled and sliced into uniform thickness of 10 mm and put in 301 × 205 OTS (Open Top Sanitary) cans procured from industry. Sugar and jaggery syrup each of 20, 25 and 30°B were prepared and poured into the cans at a fruit syrup ratio of 1: 1 by weight.

The banana slices were dipped in 0.1 per cent potassium metabisulphite solutions before canning. Citric acid of 0.2 per cent concentration were added to the covering syrup as preservative. The filled cans were exhausted in the hot water bath till the can centre reached 79°C. The exhausting time taken by the cans were 8 minutes.

The exhausted·cans were immediately seamed using can seamer. The cans were processed in a pressure cooker whose sterilizing temperature was 120°C for 25 minutes.

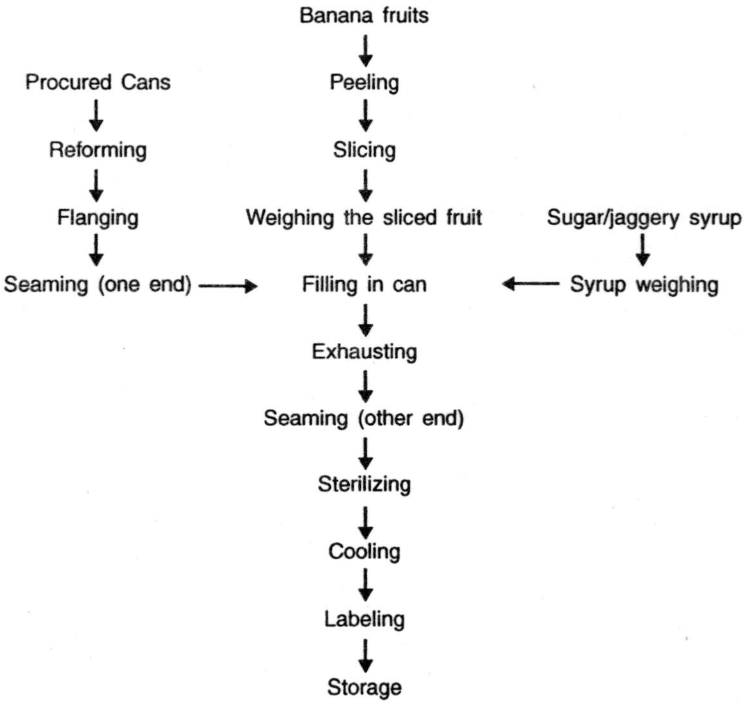

The processed cans were cooled and stored in clean and dry place at ambient conditions for storage studies. The chemical, textural and microbial analysis were made for canned banana for regular interval of one and a half months of storage.

The drained weight and moisture content decreases with increase in storage period. The cutout examination of cans revealed that there was an reduction in syrup TSS and increase in fruit TSS for all the varieties due to water loss and solid gain by the fruit slices during the period of storage. There was a reduction in syrup pH and increase in fruit pH due to the leaching out of fruit acid from fruit to syrup.

The total and reducing sugar content of fruit slices were increased and the ascorbic acid content decreased. This may be attributed to oxidation loss of ascorbic acid during processing. The textural analysis revealed that the fruits stored in lower sugar concentration (20°B) were harder than those stored in higher sugar concentration.

Also as the period of storage increases, the slices become more softer and delicate and its shape gets deteriorated. The microbial counts were within the permissible limits and hence the canned banana was microbiologically safe upto 4.5 months of storage.

The sensory evaluation showed that Red banana stored in sugar syrup was found to be the best in flavour, taste and overall acceptability.

Rasthali stored in sugar syrup was best in colour and Poovan in sugar syrup was best in texture.

The recipe (fruit pudding) made from Poovan stored in 25°B jaggery syrup was acceptable. The overall acceptability was found to be very good. The cost of canning per can was calculated to be Rs. 12.35 which represents the value addition to banana.

CHAPTER 29

IMPROVED FOOD FROM BANANA

The formative years of childhood are of crucial importance with regards to health and quality of life and therefore adequate care should be taken of this section of our vast vulnerable population. Weaning food plays a vital role in the all round growth, development and mental health of children in the formulative years.

Several types of weaning foods are being marketed in many countries including India. Most of these foods being nutritious blends of cereals, legumes and milk are excellent supplements to child's milk intake and are convenient to feed.

Most of the proprietary weaning foods are prepared from wheat, rice, corn, pea, lentil, chickpea, soyabean, peanut, gram, milk and tapioca, and prepared by roller drying, extrausion cooking and dry blending processes but the cost of the products are high and beyond the reach of lower income group and weaker sections of the society.

These also contain high fibre making it difficult to digest, and hence satisfy hunger but fail to meet the nutrition.

Keeping in mind the above points and understanding the potential of Bhimkol banana which is being used in North-Eastern region and West Bengal as weaning food (pulp) from time immemorial, this research work was undertaken to develop a banana (Bhimkol) based weaning food with tapioca starch and green gram using low cost simple technology.

In brief, it is a dry-blending process, wheat flours of banana, tapioca and green gram were mixed properly after unit operations and the required treatments. Fortification with recommended vitamins and minerals would enhance the Protein Efficiency Ratio (PER) and Food Efficiency Ratio (FER) parameters further.

Table 1 : Nutritional composition of the weaning food

Components	*Composition level*
Acidity (%)	1.47 ± 0.15
Moisture (%)	5.42 ± 0.03
Total Sugar (%)	26.36 ± 0.56
Reducing Sugar (%)	12.42 ± 0.25
Non-reducing Sugar (%)	13.94 ± 0.16
Carbohydrate (%)	76.30 ± 0.43
Starch (%)	34.13 ± 1.76
Ascorbic acid (mg/100g)	3.18 ± 0.13
Protein (%)	14.14 ± 0.14
Fat (%)	1.04 ± 0.06
Fibre (%)	1.35 ± 0.32
Total ash (%)	1.74 ± 0.07
Energy (K. cal/100g)	371.12 ± 1.24
Phosphorus (mg/100g)	188.00 ± 8.50
Potassium (mg/100g)	260.00 ± 11.14
Calcium (mg/100g)	68.42 ± 7.07
Iron (mg/100g)	10.70 ± 0.18
Sodium (mg/100g)	25.14 ± 1.24
Magnesium (mg/100g)	82.73 ± 2.87

The Process

Fully ripe Bhimkol (Banana) fruits were washed, peeled and peeled fruits were blanched in water (3 min) with potassium metabisulphite (0.1 per cent). After blanching, pulp was prepared by separating the seeds using Screw-Expeller Press (Mather & Platt).

Banana pulp so prepared was mixed with tapioca starch (Wet process production); at 20% level, the mixture was spread uniformly on trays at about 0.5 cm thickness and dehydrated in a Cabinet Dehydrator (Mather & Platt) at 60°C for 7 hours.

The dried mixture was grounded and sieved through a 50 mesh sieve. The mixture was immediately packed in transparent polyethylene bags (150 gauge) and stored for further analysis and final preparation of the weaning food was done the next day by mixing it with 40 per cent green gram powder (cleaned green gram seeds were dehulled, dried at 70°C for 2 hours, grounded to fine form of powder (50 mesh).

Table 2 : Amino acid composition of the weaning food

Amino acids	*Composition level (mg/100g)*
Aspiratic acid	230
Theonine	674
Serine	326
Glutamic Acid	865
Proline	728
Glycine	804
Alanine	412
Valine	655
Methionine	337
Leucine	1230
Tyrosine	514
Phenylanine	368
Histidine	260
Lysine	808
Arginine	288

This weaning food formulation can be prepared by sun drying and tapioca starch can be replaced by rice-flour.

This can then be used by the rural population using the Bhimkol they grow in their home-stead garden because almost every household in villages and semi-urban areas still grow Bhimkol banana; and in places like Goalpara, semi-hill areas and remote villages glut condition is often created leading to spoilage of the fruits.

The nutritional composition of a preferred Laboratory prepared Weaning Food is given in Table 1.

Table 3 : Viscosity measurement of the formulated weaning food at various slurry concentration

Food	*Viscosity (CP) + SEM % Slurry concentration*		
	5.0	*10.0*	*15.0*
Bhimkol based weaning food	7.0 + 0.46	34.2 + 2.3	420.4 + 6.02

In our further studies, it has been discovered that atleast 15 amino acids beneficial for body growth are present in the prepared Weaning Food. Table 2 shows the details of amino acid composition of the banana-based (ripe) Weaning Food.

Since weaning food is used in the form of slurry composition, it was decided to study the viscosity details of the above formulated Weaning Food at various slurry concentrations.

Since the Weaning Food is intended for use by infants, it was also decided to evaluate various quality details.

Evaluation of Protein Quality of the Weaning Food

In addition to the extensive evaluation, it was decided to study the details by giving the Weaning Food young children.

Acceptability trial of the Weaning Food on young children:

The acceptability of the Weaning Food was evaluated by developing water soaked foods (hot water) and testing them on pre-school children (1-3 year group).

The mean intake of the food samples by the children is found to be 85.25 ± 2.89 gram.

Table 4 : Weight gain, food intake and food efficiency ratio of albino rats fed on the weaning food

Diets	Weight gain (g)	Food intake (g)	*FER
A. Casein	45.40 + 1.13	147.60 + 5.60	0.31 + 0.006
B. Bhimkol based weaning food	43.45 + 1.19	143.73 + 4.30	0.30 + 0.006

* FER = Food Efficiency Ratio.

In order to complete all investigations the shelf life and other studies were carried out which are given here.

Shelf Life Studies

The weaning food was stored for 120 days in plastic bin, metal contained polyethylene bag and aluminium foil pouches. The effect of the storage time and packaging materials on the nutritional parameters of the Wearning Food were studied.

Aluminium foil pouch was found to be the best packaging material which gives optimum protection against moisture absorption and also showed maximum retention of nutrients up to 120 days of storage.

Table 5 : Weight gain, Protein intake and Protein Efficiency ratio of Albino Rats fed on the weaning food

Diets	Weight gain (g)	Food intake (g)	*FER
A. Casein	45.40 + 1.13	14.76 + 0.56	3.09 + 0.06
B. Bhimkol based weaning food	43.45 + 1.19	14.37 + 0.430	3.03 + 0.06

* PER = Protein Efficiency Ratio

Table 6 : Organileptic evaluation for colour, flavour, taste and general acceptability of the weaning food

Maximum Score Value	Colour	Flavour	Taste	General Acceptability
10	6.85 + 0.36	7.15 + 0.43	7.35 +0.48	7.65 + 0.38

Table 7 : Vitamine Composition of the Weaning Food

Vitamin	Amount
Vitamin A	2922 (IU/100g)
Thiamine	0.018 (mg/100g)
Riboflavin	0.006 (mg/100g)
Niacin	0.058 (mg/100g)
Vitamin C	3.18 (mg/100g)

Studies on the Plate Count for Bacterial Contamination in the Weaning Food

The bacteriological plate count was carried out at 0 day and after 120 days. The bacteriological plate count at the initial phase was 670 colonies per gram of food. At the end of 120 days the colonies was found to be 2,330 colonies per gram of food which is an acceptable level.

Recommended Fortification Level

- The Weaning Food is to be fortified with the following Vitamins per kg.

- Vitamin B_1 (Thiamine chloride hydrochloride) = 10 mg
- Vitamin, B_2 (Riboflavin) = 10 mg
- Niacin (Nicotinic acid) = 54 mg
- Vitamin C (L-ascorbic acid) = 468.2 mg
- Vitamin D = 4000 IU.

The cost of production is about one third of that of the commercial Weaning Food now available in the market. The Banana skin peeled out in the preparation of the Banana pulp can be, if desired, processed to give animal food supplement and be used for single-cell-protein (SCP) production or preparation of Glucose syrup.

CHAPTER 30

GROW MONEY ON COCONUT TREE

The Coconut palm is one of the most useful plants in the world. It is grown in more than 70 countries of the world. Indonesia and the Philippines are the first and the second largest coconut producing country in the world. India being the third largest coconut producing country having an area of about 1.80 million hectares under the crop.

Annual production is about 7600 million nuts with an average of 5300 nuts per hectare. South East Asia is regarded as the origin of Coconut. It is mainly cultivated in Philippines, Indonesia, India. Sri Lanka, Thailand, Malaysia, Fizi, Samoa, Solomon Islands etc.

The major coconut growing states in India are Kerala, Tamil Nadu, Karnataka, A.P., Maharashtra, Orissa, West Bengal, Assam, Goa, Daman and Diu, Lakshwadeep, Gujarat, Kerala tops in production accounting for 55 per cent of total production in the country.

Coconut trees are grown mainly for the high oil content of the endosperm, which is widely used in both food and non-food industries (*e.g.* margarine and soaps).

Total world productivity has increased, substantially from 35 million tonnes around 1980 to almost 50 million tonnes now. Yield varies from region to region (3,500 to 6,000 nuts/ha/year), which is due to a number of factors. One tree may yield on an average 70-100 nuts to a maximum of 150 nuts per year.

The kernel (copra, coco-water and shell) comprises 65 per cent of total weight, while the husk contributes 35 per cent. Besides the valuable contents of the nuts, the palm yields husks, shells, leaves and the stem which are used domestically as raw materials for many products from fuel to building materials.

Coconut Production and Productivity in India

Kerala is the largest producer of coconuts in India producing 5500 Million nuts followed by Tamil-Nadu, Karnataka and Andhra Pradesh as they have taken 2nd, 3rd and 4th position respectively. Central Kerala town Kochi is a major trading center for coconut oil and copra. Giving focus on productivity, Maharashtra comes in 1st position (914548 Nuts/ha.), West Bengal in 2nd position (13490 Nuts/ha.) and Andhra Pradesh in 3rd position (10660 Nuts/ha.), Kerala being a major coconut producing state its productivity is not so high (5870 Nuts/ha).

Unlimited Products from Coconut

It is noticeable that almost all the parts of freshly grown coconut are eatable. Coconut water or milk is an excellent natural soft drink for all. It is useful for diabetic and heart patients. It is very useful to people suffering from diarrhoea and vomiting. It helps in increasing blood circulation in the kidneys. Oil is extracted from dry copra. Copra contains about 65 to 75 per cent oil. Copra is also used in the preparation and decoration of cakes, sweets such as *Ladoo, Barfi* etc. are prepared from it.

The unopened spathe is tapped for toddy. This toddy can be converted into jaggery, vinegar and sugar. Kernel (wet meat) is mainly used in making curries: chutney, toffee, sweet and for other cooking purposes. Convenience products from Coconut Kernel Desiccated Coconut (DC), Coconut Cream, Coconut Milk and Spray Dried Coconut Milk Powder are manufactured in the country. DC is used as a substitute to grated raw coconut in various food preparations. Defatted desiccated coconut is also available in the country.

Processed Coconut cream/coconut milk are used in various food preparations as a substitute to milk extracted from raw kernel in the traditional method. Spray drying is the best method for the preservation of coconut milk. The product has advantages such as less storage space, bulk packaging possible at low cost and long shelf life. Spray dried coconut powder is manufactured by one unit in the country.

Coir, the fibrous husk of the coconut, is used in a surprisingly large number of ways. Ropes and yarns, aquarium filters, car seat, covers, flower pots, soundproofing, mulch for plant growing, heat insulation, brushes, bristles, mattresses, door mats and matting, rugs, carpets... the list goes on and on. Leaves are used for making thatch, jhaps and for other purposes while leaflets are collected and composted. Coir pith (or) coir dust is used as a soil conditioner (manure). Midribs and leaf-petioles are used for making brooms. Inflorescence bunk stalk, stipules and dried spathe are used for warming water or for cooking purpose.

Coconut shell is used for making fancy items, household utensils etc. Decoction obtained from roots is used as mouthwash and gargle. Coconut oil is one of the most important edible oils for domestic use.

Coconut Shell Powder

The shell charcoal is the raw material required for the manufacture of activated carbon. Shell charcoal is manufactured by burning shells of fully matured nuts in limited supply of air sufficient only for carbonisation, but not for complete destruction. The output of charcoal in the traditional pith method is just below 30 per cent of the weight of the original shells. In India the average output in the traditional method has been found to be 35 kg of charcoal from 1000 whole shells or about 30,000 whole shells yield 1 tonne of charcoal.

Sometimes, especially when the processing is defective, the output is still lower and nearly 50,000 shells are required to produce one tonne of charcoal. To obtain good quality charcoal, fully dried, clean, mature shells should be used. Now several modern methods are in vogue for the production of charcoal. In the modern waste heat recovery unit the heat generated by the burning of coconut shells is used for drying copra and shell charcoal is obtained as by-product.

The manufacture of coconut shell powder is not an organised industry in India. The product finds extensive use in plywood and laminated board industry as a phenolic extruder and as a filler in synthetic resin glues, mosquito coils and agarbathis.

Coconut shell powder is preferred to other alternate materials available in the market such as bark , powder, furfural and peanut shell powder because of its uniformity in quality and chemical composition, better properties in respect of water absorption and resistance to fungal attack.

The product is manufactured in sizes ranging from 80-200 mesh. Keeping in view of the vast industrial uses, the demand for coconut shell powder appears to be promising.

Process

Coconut shells free from contamination of coir pith, etc., are broken into small pieces and fed into a pulveriser. The powder from the pulveriser is fed into a cyclone and the parallel product is collected in bag filters.

The shell powder is then fed into a vibrating sieving machine and packed according to mesh size requirements for various end uses.

The rejects from the sieving machine can be recycled in the pulverizer for size reduction. The main requirements for consistent good quality of coconut shell powder are proper selection of shell of proper stage of maturity and efficient machinery.

Machinery

- Hammer mill
- Storage bin
- Impact pulveriser
- Cyclone
- Bag filter
- Air blower
- Sieving machine

Manufacturing Process Through Drum Method

A M.S. Drum kiln is used for carbonization of shells. The drum consists of three sets of six 1" dia holes provided at its bottom, middle and upper layers and a lid. A detachable chimney is provided which is installed on the lid after closing the drum. The manufacture of charcoal requires optimum carbonization of raw shells in a limited supply of air so that there is neither unburnt shell nor ash due to complete combustion.

The steel drum is filled with raw shells after placing temporarily a four inch diameter wooden pole in the centre of the drum. The wooden pole is then removed, leaving a hollow space in the centre, which allows the flow of smoke during carbonization. To start carbonization, a piece of burning rag is dropped to the bottom of the drum through the hollow space in the centre. When the fire is well underway, the lid with the chimney is placed into position and the upper and the middle sets of holes are closed. Carbonization which starts at the bottom progresses as it goes up as well as radially from the hollow space. When carbonization is complete in a particular zone, a persistent glow can be seen in all the six holes of a set

When the bottom most set of holes indicate this situation, the middle set of holes is opened and the bottom set closed. The stoppage of air flowing into the bottom region avoids over-carbonization in that region. The progressive carbonization results in reduction in volume of contents and therefore more shells are added from the top.

When the middle region is carbonized well, the top set of holes is opened while the middle set is now closed. A further addition of raw shells is done to fill the volume reduction to maximise capacity for burning. When the top region is well carbonized, the top set of holes is also closed, resulting in complete stoppage of air inflow to the drum. The drum is then cooled for about eight hours after which the product is ready for discharge.

Product Specification

- Fixed carbon 72% (minimum)
- Volatile matter 15% (maximum)
- Ash 2% (maximum)
- Moisture 10% (maximum)
- Size not more than 5% shall pass through a 0.63 cm mesh
- Colour uniformly black

Sources of Machinery/Equipment

- A.L. Jacob & Sons, A.L. Jacob Road, Ernakulam, Cochin-28.
- Vivega Engineers, 143-C, Nava India Road, Peelamedu, Coimbatore - 641004.
- Ananth & Co. Bhayana, Shakthi Nagar, lrinjalakuda, 680125.

- Plant India, Vytilla Road, Palarivattom, Kochi-682025.
- Kaps Engineers, 831, G.LD.C. Makarpur, Vadodara - 390030.
- Pilots India, Kalletumkara - 680683, Irinjalakuda, Kerala.
- Universal Enterprises, 99/77/1B Sri Rampura, H.O. Court Road, Mysore - 570008.

Tender Coconut Water

The water of tender coconut, technically the liquid endosperm, is the most nutritious wholesome beverage that nature has provided for the people of the tropics to fight the sultry heat.

It has caloric value of 17.4 per 100gm. "It is unctuous, sweet, increasing semen, promoting digestion and clearing the urinary path," says Ayurveda on Tender Coconut Water (TWC). Numerous medicinal properties of tender coconut water reported are:

1. Good for feeding infants suffering from intestinal disturbances.
2. Oral rehydration medium.
3. Contains organic compounds possessing growth promoting properties.
4. Keeps the body cool.
5. Application on the body prevents prickly heat and summer boils and subsides the rashes caused by small pox, chicken pox, measles, etc.
6. Kills intestinal worms.
7. Presence of saline and albumen makes it a good drink in cholera cases.
8. Checks urinary infections.
9. Excellent tonic for the old and sick.
10. Cures malnourishment.
11. Diuretic.
12. Effective in the treatment of kidney and urethral stones.
13. Can be injected intravenously in emergency case.

14. Found as blood plasma substitute because it is sterile, does not produce heat does not destroy red blood cells and is readily accepted by the body.
15. Aids the quick absorption of the drugs and makes their peak concentration in the blood easier by its electrolytic effect.
16. Urinary antiseptic and eliminates poisons in case of mineral poisoning.

"It's a natural isotonic beverage with the same level of electrolytic balance as we have in our blood. It's the fluid of life. The major chemical constituents of coconut water are sugars and minerals and minor ones are fat and nitrogenous substances.

Coir Fibre

Coir fibres are extracted from the husks surrounding the coconut. In most areas coir is a by-product of copra production, and the husks are left on the fields as a mulch or used as fertilizer because of high potash content. India is the main country where coir is extracted by traditional methods for the commercial production of a variety of products, including brushes and brooms, ropes and yarns for net and bags and mats, and padding for mattresses. However, world wide only a small part of the fibres available are currently used for these purposes.

The average fibre yield is dependent on geographical area and the variety of the coconut tree. Husks are composed of 70 per cent pith and 30 per cent fibre on a dry weight basis. The ratio of yield of long, medium and short fibre, respectively, is on average 60:30:0. Based on these data and combined with the production data in the maximum total world production of coir fibre (included short fibres) can be estimated to range between 5 and 6 million tonnes per year. Only a small part (less than 10 per cent) of this potential enters commercial trade.

Coir Markets

Coir despite to comparatively low trade value, the fibres provide significant economic support to populations in specific areas of the producing states of Kerala, Karnataka, Tamil Nadu and Andhra Pradesh, and in western India. Women in these areas are particularly dependent on coir production for their livelihood.

Traditionally, coir has been processed into a range of products such as yarns for the production of floor coverings, mats and malting, cordage and nets, bristle fibres for brooms and brushes, and for use with domestic mattress and upholstery industries.

These markets have been dwindling in recent years due to strong competition from synthetic products. However, there is a firm trend in the industrial countries towards the production and use of more environmentally benign products and systems, which may help to mitigate the adverse ecological affects of current production methods.

The effects of chemical industries, atmospheric degradation, global warming, fast-declining natural resource base, deforestation, waste production, pollution and similar global issues have increased the demand for environmentally benign products.

Renewable raw materials such as plant fibres and products, therefore, may have good market perspectives if they can be produced at an economically competitive price and on a scale where quality and supply can be guaranteed. Currently, expanding export markets for coir can be seen in the demand for erosion control mats and other geotextile applications for civil engineering, or in the demand from the automotive and mattress industries for rubberised coir pads.

Coir Market Diversification

Cost effective and environmentally safe technologies have been investigated for drying, bleaching, softening and dyeing of fibre and yarn, and for fast printing of coir products to encourage increased demand for coir products (fibre, yarn, door mats, matting, runners and carpets).

To upgrade coir products to meet the standards required of export markets, efforts must start at village production levels. Marketing of ecologically safe coir products should imply and include, however, the whole production chain from fibre extraction to end product and disposal or re-use when no longer required. It follows that production methods should be safe for the health of the workers involved in the coir industry without negative effects on production performance. Demand for geo-textile products increasing and a large number of alternative end uses for coir based products may become feasible, for example, for the production of fibre reinforced composite materials, fibre board and similar building materials.

Coir Pith

Coir pith a waste product obtained during the extraction of coir fibre from husk is very light, highly compressible and highly hygroscopic. As a by-product of coir fibre extraction large quantities of pith are obtained, which have been accumulating at production sites over the years. The extraction of 1 kg of fibre generates more than 2 kg of coir pith. Recently, however, the product has gained commercial interest as a substitute for peat moss in horticultural substrate cultivation.

Low susceptibility to biodegradation and a highly porous structure enables coir pith to absorb large volumes of water (more than 50 per cent by weight), which makes it highly suitable in a potting mixture. For horticultural use, the product has to meet specific chemical and biological standards of pH electrical conductivity and elemental composition. Repression of sodium and potassium from the cation complex of the coir may be desirable for many sensitive horticultural products. Technical information to describe microbial contamination and product safety is another concern for users.

Coconut Vinegar

Coconut water is being wasted in copra and DC industry in the country. The CFTRI, Mysore has developed a technology for the production of coconut vinegar from matured coconut water using vinegar generators. The vinegar generator assembly comprises a feed vat, an acetifier and a receiving vat for collection of vinegar. Vinegar has extensive use as a preservative in the pickle industry and flavouring agent in food processing sector. Now natural vinegar enjoys export market in place of synthetic vinegar.

Product Specification

	Appearance	*White smooth creamy consistency*
•	Flavour	Coconut
•	Fat	23%
•	Protein	4%
•	Sugars	5%
•	Others	1%
•	Water	67%
•	Machinery/Equipment	Hammer mill

Process

The matured coconut water consisting of about 3 per cent sugar content is concentrated to 10 per cent level by fortifying with sugar.

The fortified coconut water is then fermented by inoculating the solution with yeast, Sacharomyces cerviseae. After alcoholic fermentation for about 4 to 5 days, the clear liquid is siphoned off and inoculated with mother vinegar containing acetobacter bacteria. The alcoholic ferment obtained is then fed into a vinegar generator where the feed is uniformly sprayed over the surface of the porous packing medium (corn cobs). Here the alcoholic ferment is oxidised to acetic acid.

The product is run out from the packing medium by gravity flow into the receiving vat from where it is recycled into the vinegar generator and the process of acetification is repeated until a strength of 4 per cent is attained.

This acetified vinegar is then aged before bottling.

Product Specification

- Acidity as acetic (g/100 ml) 5.26 - 5.76
- Total solids (%) 1.83 - 1.92
- Total ash (%) 0.38 - 0.42
- Sp. gravity 1.012 - 1.008
- Alcohol Nil
- Equipment/Machinery

- Feed trough
- Vinegar acetifier
- Receiving trough
- Wooden storage drums

Spray Dried Coconut Milk Powder

Spray dried coconut milk powder if dissolved in water will result in coconut milk which can be used in place of fresh coconut milk for food preparations/beverages in households and food industries by dissolving it in water. Central Food Technological Research Institute, Mysore has developed the technology for spray dried coconut milk powder which is available to entrepreneurs at a total know-how fee of Rs. 5 lakhs.

The total capital investment for plant and machinery along with working capital margin comes to around Rs. 2.36 crores. The internal rate of return works out to be about 16 per cent with a pay back period of about 3 years. On an average one thousand coconuts would yield about 99.6 kg of spray dried coconut milk powder. Spray dried coconut milk powder has a tremendous market potential in India and in international market. Direct employment potential is around 40 personnel.

Coconut Cream

Coconut cream is the processed milk extracted from fresh matured coconuts. This is an instant product, which can either be used directly or diluted with water to make various preparations such a curries, sweets, desserts, puddings, etc. It can also be used in the manufacture of bakery products and for flavouring food stuff. Processed and packed coconut cream has a, shelf life of six months and once opened it should be stored in refrigerator for subsequent use. The technology has been developed by the RRL, Thiruvananthapuram under a sponsored project of the Board.

Process

The first step is breaking the dehusked nuts into halves. The split nuts are deshelled to separate the kernel. These two operations usually are done manually. Kernel is washed and then blanched by immersing in hot water at 80°C for 10 minutes. The next step is communication of kernel into small gratings using a hammer mill. The gratings are subjected to pressing using continuous screw press to extract the milk. The coconut milk thus obtained, filtered by passing through a vibratory screen.

Food additives such as emulsifiers and stabilizers, are to be added to the milk to obtain a stable consistency and texture. For this purpose, permitted emulsifiers and stabilizers are mixed with hot water separately and mixed thoroughly. This is added to the co onut milk and then subjected to emulsification using a mechanical impeller emulsifier.

The emulsified milk assumes a creamy consistency. The coconut cream is then pasteurized at 95°C for 10 minutes in a plate heat exchanger. The pasteurized coconut cream is hot filled in cans using a mechanical volumetric filling machine followed by steam exhausting. The cans are seamed using an automatic can seamer.

The seamed cans are sterilized in a rotary retort at 15 psi for 20 minutes. The cans are then cooled in running water. The residue obtained after the extraction of milk is dried in a hot air dryer to 3 per cent moisture level.

The residue forms a base for making coconut *burfi*, coconut cookies, curry and *chutney* powders. Other by-products like coconut water and coconut shell could be processed into value added products.

Vinegar and Nata-de-coco can be manufactured from coconut water. Shell charcoal and shell powder can be manufactured from coconut shells. The utilization of by-products would improve the economic feasibility of the process.

Elevator, screw press, coconut milk storage tanks, vibrating sieving machine, coconut residue mixer, Additive mixing tank, Emulsifier, Homogenizer, Pasteurizer, Volumetric filling machine, Exhaust box, Can seaming machine, Horizontal rotary retort, Hot air drier, Agro waste vertical boiler, Sterilization tank, Coconut residue storage bins.

Coconut Cake

Coconut cake is the residue left after the extraction of oil from copra which is mainly used as a cattle feed. Coconut cake contains 4-5 per cent oil which is extracted by solvent extraction process.

This oil is generally used for industrial purpose and deoiled cake is used to make mixed cattle feed. There are a few such units in the country especially in Kerala. Coconut Toddy tapping is an organised industry in traditional coconut growing tracts in the country. Coconut jaggery is made from sweet coconut toddy.

Toddy on fermentation becomes an alcoholic drink. Arrack and vinegar are also manufactured from coconut toddy.

Coconut Wood Based Products

The coconut wood because of its distinct grain characteristics is ideal for making wall panels, furniture, doors and windows, show pieces, etc. There are several small scale units manufacturing a variety of articles from coconut wood. Coconut leaves are plaited and used for thatching houses and sheds in rural areas. It is also used for thatching 'honeymoon huts' and such huts in town and cities.

Technology is available with Regional Research Laboratory, Thiruvananthapuram, Kerala for extending the life of leaf thatch from one year to four years. Plaited coconut leaves are also used for making baskets, headgears and for erection of temporary fences. Plaiting of coconut leaves is a cottage industry in traditional coconut growing States. Midribs of leaves are used to make brooms of different types which are used for cleaning rough grounds and floors. Brooms of midribs of coconut leaves are manufactured on a commercial scale in Tamil Nadu and Karnataka.

Value Addition in Coconut Industry

Currently more than 70 products are commercialized and nearly 50 products have entered into the international market. There is growing demand for coconut products in dietary foods, functional foods, nutriceutical, pharmaceutical, cosmetics, biofuel, bio-lubricants, as virgin coconut oil against viral, fungal, bacterial diseases. By products like geo textile coir composites, coir pith have wider uses in soil erosion control, soil moisture conservation. Coconut wood as forest timber substitute can be used. All possible value additions can be explored as niche market products. Increasing the productivity, reducing the cost of production, integrated farming, farm level processing, proper value addition, product diversification and by-product utilisation coupled with effective marketing strategies and market promotional activities can definitely make the coconut industry more competitive and sustainable.

While coconut cultivations is done by the individual farmers at their farms they joined together as a society can take up farm level processing. The coconut farmer by selling coconuts to the society gets his first income and then the society can take up processing of coconuts into husk, shell, coconut water and coconut meal. The husk can further be separated into fibre and coir pith. The coir pith will be composted using vermiculture or fungus and can be sold to another industry which can go for coconut water concentrate making. The unused coconut water can be used for biogas plant for generating energy at low cost.

Similarly, the coconut shell can be used for shell gasifier for producing energy, or for making handicrafts or can be sold for making charcoal, activate charcoal or shell powder units. Charcoal making can also be taken up by the society. Coconut meat can be used either for making copra and coconut oil, or can be used for extracting coconut milk and selling milk to the hotels and marriage places or making virgin coconut oil. Coconut meat can also be used for making chips which with proper packaging and labeling is an exported item as well as having domestic market.

Export Potential in Coconut Industry

Coconut is exported in several forms. The young fruits (nuts) are exported for consumption as a food. The old nuts are exported both as food and planting material. Germinated nuts and young seedlings are exported for planting as well as ornamental purposes. Coconut products exported range from desiccated coconut, coconut cream, milk (santan) and oil. The export conditions vary depending on their form.

The export of coconut products indicate that India has done well in coil products, while in the export of other coconut based products, its share is very negligible compared to other coconut growing countries of Asia and Pacific. However, the export of coconut products over a decade showed that there is considerable increase from 1990 to 2002 but as stated earlier, coil and coil products are the major products. The export of other products needs to be augmented.

Future of Indian Coconut Industry

In India, the coconut area over a period of five decades had increased from 0.62 million ha (1950-1951) to 1.900 million ha (2002). Similarly, the production has increased from 3,282 million nuts to 12,822 million nuts and productivity of coconuts from 5238 to 6776 nuts/ha/year. The global demand for coconuts and coconut products has increased considerably. It has been assessed that 9 per cent increase per annum for desiccated coconuts, 5 per cent increase/annum for fresh coconuts, 45 per cent increase for coconut milk, 45 per cent for activated carbon, eight fold increase of coir products and 100 per cent increase in coir dust will be there for next 6 to 8 years.

As the population is growing world over, the consumption also will increase in both edible and industrial uses of coconuts. Uses of coconuts and its products as functional foods, nutraceutical, pharmaceutical and cosmetics as well as bio-fuel have greater market in the years to come. The use of medium chain fatty acids and the lauric acid in producing antiviral, antifungal, antiprotozoal, and anti microbial medicines, particularly use of lauricidin in the cure of HIV/AIDS virus will go a long way in creating market potential for coconut oil and virgin coconut oil.

The tender nut water as energy, nutritional, health and sports drink has unlimited market potential. The organic products which are emerging new markets have huge market potential in western and American countries. It is projected to grow, by 20-25 per cent in the most important markets of Europe and North America. The functional foods are catching up with an anticipated growth rate of 7-8 per cent through 2 per cent. Coir and coil products have made good progress and major export is from such products.

The geotextile and coil composite for erosion control, and soil moisture conservation and coir pith for growing horticultural plants have unlimited market in Europe and U.S. which can be exported. There is still scope for high tech coir products for sound proof system in buildings, making life safety belts, etc.; need to be explored. In all these areas, coconut can compete provided the productivity is increased and production cost is reduced, and quality standards are maintained at international standards. Similarly, the price trend of coconut oil is always fluctuating but the prices of various coconut products showed that the value-added products like coconut milk, powder, cream, desiccated coconut, etc.; had reasonably good price to coconut oil. The other-value added by-products like coconut water, coconut water concentrate vinegar, charcoal, activated charcoal, coal coil, coir pith, coir based products also have higher price and good market demand.

The Need

There is a need for strengthening coconut industry by promoting product diversification and by-product utilization, to ensure adequate, appropriate, timely and concurrent attention to all the links in the production, post harvest and consumption chain, to maximise economic, ecological and social benefits from the existing infrastructure created for coconut development, to disseminate technologies using participatory approach through demonstration and promotion.

The government should converge and synergize all the efforts through vertical and horizontal integration of existing technologies and address the problems and bridge the gaps through appropriate interventions to ensure adequate, appropriate, timely and concurrent action. This would help develop a mechanism which makes coconut based industries competitive and ensures reasonable returns.

CHAPTER 31

APRICOT KERNELS

Prunus armeniaca (Rosaceae) which yields the apricot fruit includes a number of botanical varieties occurring both in wild and cultivated forms in the Indian states of Jammu & Kashmir, Uttar Pradesh, Himachal Pradesh, West Bengal and to a small extent in Punjab. There are numerous wild varieties without any specific name which are used as parent stock for grafting of superior varieties. The total apricot production of the Ladakh region of the J&K states is estimated to be about 1450 quintals per year and among the horticultural produce plays a vital role in the socio-economic life of the people of the region. Almost all parts of the fruit are of economic value. The pulp used in food products is a source of vitamins, sugars and minerals; the seed cake serves as an important article of commerce; the seed cake serves as an animal food and manure whereas the pit shells are used for making activated carbon.

The seed cake is reported to contain pangamic acid (Vit. B_{15}), a necessary food factor which forms the topic of present discussion. Various reviews have appeared on the status of research on pangamic acid and its chemical and therapeutic efficacy established.

Pangamic acid (Vitamin B_{15}) has been reported to be safe and effective in the treatment of patients sufferings from ischemic heart disease in doses of 150 mg/day for 20–30 days with improved heart contraction and coronary and systematic circulation. In the treatment of severe cases of aetherosclerosis, it brings down the blood cholesterol and lipo–protein level. A vitamin preparation containing vitamins A, E and B_6 has been used as enteric coated pills in severe cases of aetherosclerosis. Its usefulness in doses of 100–300 mg has been established in infective hepatitis in children upto the age of 15. It has been claimed to foster a greater adaptation of muscular activity if administered orally to highly trained athletes for 3 days prior to moderate or intense exercise at a dose ranging between 100–300 mg/kg. It has also been used in treatment of patients suffering from various dermatoses *e.g.* scleroderma, psoriasis, echzema, urticaria, etc.

Keeping in view the clinical importance and the nutritional status of pangamic acid, it has been considered to explore the possibility of the utilization of apricot kernels as a source of this vitamin.

Pangamic acid can be recovered from the apricot kernals process. The process includes extraction of the kernels and isolation by fractional precipitation followed elution. The eluate is purified by dialysis and converted to the sodium salt by addition of metallic sodium.

Besides apricot kernels, pangamic acid has been reported present in rice bran, brewer's yeast, ox blood and horse liver. Telegdykevats *et. al.*, have reported it to be present in wheat bran 31, barley grit 12, corn grit 150, oat grits 106, wheat germ 70, wheat flour 8–12, rice bran 200 mg/100 g as confirmed by *i.r.* spectrophotometry. Eva *et. al.* have developed a method for determine small amounts of vitamin B_{15} in both free and combined state in natural substances by reacting it with mel to form quanternary salt oxidizing the reaction product and determining the amount of vitamin B_{15} liberated spectrophotometrically. Telegdykovats *et. al.* have given a method of its separation from vitamin B_1 and B_2 by and methods of isolation from carcal products containing 80-200 mg and 100 g. They have also reported its identification on the basis of multiple TLC and *i.r.*, spectra of the eluates. Skupin *et al.* have also dealt with the aspects of identification and determination of pangamic acid on the basis of TLC.

Pangamic acid is a white crystalline compound having a molecular formula $C_{10}H_{19}O_8N$ (mol. wt. 281), soluble in water but insoluble in non-polar organic solvents. It was chemically identified to be *d*–gluconodimethyl aminoacetic acid by Mischeau *et al.* by correlating spectral properties through model compounds. It almost completely decomposes after 3-5 minutes in alkaline solution at higher temperatures and is stable in acid solution. Bukin *et al.* have also obtained a patent for the preparation of calcium pangamate prepared by esterifying gluconic acid or its lactone with dimethyl glycine and subsequent neutralisation with calcium carbonate.

CHAPTER 32

TAMARIND : THE ECONOMICAL TREE

Tamarind tree (Tamarindus indica linn.) is an arboreal and edible fruit-bearing tree of Legumoisea family (fabacae). The tree is indigenous to India and tropical Africa and it grows wild as well as cultivated. The tree grows all over Indian sub-continent and in adjoining countries (Sri-Lanka, Thailand, Pakistan etc.) and some parts of Africa. In certain tropical South American countries, Costa Rica, Mexico and West Indies some tamarind is grown.

In India, it particularly grows well in the eastern region and south central Indian peninsula *i.e.*, the tribal region of the Indian states of Chhatisgarh, M.P., Andhra, Karnataka, Tamil Nadu, Bihar, Jharkhand, Orissa and eastern U.P. India exports tamarind pulp and seed products to Pakistan, Arab countries, Europe and North America.

Tamarind has been recognized as an underutilized crop by the International Centre for Underutilized Crops, Southampton, UK, with a suggestion for its better utilization to boost rural economy.

Why grow Tamarind?

Evergreen tamarind tree grows to a height of 20–25 meters and has a life span of about 80–100 years. The tree provides a perennial cover, protecting soil and adds to storage and recycling of nutrients. The tree has pinnately compound (feather-formed) leaves; about 2 cm long. The yellow flowers, about 2.5 cm across, are borne in small clusters. The tamarind fruit is a plump pod 8-20 cm long, enclosed in a thin, brown, brittle shell. The pod does not split open on ripening, as certain other legume pod do. The pods, ripened on the tree, are harvested merely by shaking the branches and collected from the ground. A tree can continue to bear fruits for 60–80 years.

The pod, weighing 15-30 g, contains 1 to 12 large, brown flat seeds embedded in a soft, brownish pulp. A less common red pulp bearing tamarind is also known, which being more sweet, is preferred for making preserves and fetch higher price. The sweetness of thesis red variety of tamarind is due to lower free acid (part of tartaric acid is present as salts) content. Being highly cross-pollinated crop, wide variability is found in tamarind. Attempts should be made to select and develop improved quality and high yielding variety of tamarind tree.

Though all the parts of this tree are useful, the pulp portion of the fruits is most valued and widely used in the Orient in foods, beverages, and medicines. Extensive chemistry, technology and uses of tamarind products have been reviewed in this article, which can be of use to the industry based on this tree and the tribal population of the region which collect tamarind-fruit and sell it for processing or directly to the end user.

The other commercial product of tamarind, *e.g.*, the tamarind seed kernel powder is described in another article in this issue.

Economics of Industry

A tamarind tree begins to bear fruits at the age of 8-12 years. Trees, which are grown from grafts and budding can, bear fruit, only in 5-6 years. A full-grown tree produces in a year about 200-400 kg pods, which gives about 110-250 kg of pulp. The pulp can fetch about Rs. 20/kg to the whole-seller and Rs. 35-40/kg for a retailer. It is not unusual for a tribal house to have one or two tamarind trees in courtyard or several of these on a farmland. Each tree can give an income of Rs. 1500-2000 per year. Processor of tamarind pulp and tamarind kernel powder (TKP) get much larger, additional income. Thus tamarind tree is a sustainable resource.

It has been estimated that the potential availability of tamarind pod in India is about 300000 Tonnes, which on processing yields 1,50,000 Tonnes of pulp and 1,20,000 Tonnes seed/year. On processing the seeds can produce about 1,00,000 Tonnes of TKP powder. There is not much variation in the annual production of pulp or TKP, because it is a product of a perennial tree and its annual production is only slightly subjected to variation on climate and annual rainfall in the growing regions.

During the past decade tamarind trade has increased and is likely to continue to expand. The export of tamarind based products from India has been Rs. 50-60 crore per annum.

Tribal and Forest Related Industries

Tamarind products based industries in India, are classified as forest related industries and according to the government policy, these should aim at benefiting the tribals. It may be mentioned that the practice of procuring forest products, using cheap human labour is not confined to India only, but is prevalent in other developing countries as well.

It has been amply proved that without participation of private industries or individuals, the full and successful utilization of forest or any other raw material is not possible. Non-tribal owns most of the tamarind related industries and only a few of them are in tribal region. Under these circumstances it is the importance, that the industrialists do not unduly exploit the forest-wealth, which is supposed to be belonging to the tribals of a particular region.

Recently there was a policy notification from the M.P. Government, which aims at establishing cooperatives to procure and sell tamarind, collected by the tribals. This shall give better returns of forest products *e.g.*, tamarind to the tribals.

Tamarind Pulp

On an average the pod is composed of 15-25% outer shell, 45-55% pulp, 30-35% seed and 10-15% fiber. Pulp in ripened pod contains 60-70% moisture, which after deshelling and drying, comes down to 15-35%. This is the marketable from of tamarind pulp.

The composition of this economically valued product *i.e.*, the dried pulp is given in the Table 1. Minor constituents of tamarind pulp have been identified and these include phenolics pigments (flavonoids, anthocynins), which have antioxidant properties, vitamins (B–complex thiamin and niacin and C), minerals (Potassium, Calcium, Iron and Phosphorous) and flavouring constituents.

Colour of tamarind pulp is due to the presence of anthocynin pigments. The aroma in the pulp is a mixture of citrus note and spice like flavor and some of the flavor constituents have been identified by gas-liquid chromatography followed by mass spectra.

Table 1 : Composition of Ripend, Sun Dried Edible Pulp

Constituent	*% Composition*
Moisture	15–30
Total sugars	55–65
Sucrose	2–4
Reducing Sugars (Glucose 70%, Fructose 30%)	25–45
Pentoses	4–6
Proteins	2–8
Lipids	0.25–1.0
Crude Fiber (Cellulosic)	2–18
Total Acidity (As tartaric acid)	17–20
Tartaric Acid	9–18
Ash (minerals)	2–3
pH	3.2
Pectin	2–4
Starch	5–6
Minor constituents - tannins, vitamins – (B-group and C), carotenes.	

Tamarind pulp has a unique sweet-sour taste due to the natural occurrence of sugars and plant acids together. Additionally the presence of viscosifiers (pectin and starch), produce a unique after-taste in tamarind pulp-paste, which no other fruit or a blend can produce.

Tamarind pulp should preferably be used within a year, because, besides insect damage, its brown-red colour darkens to black due to the setting up of Maillard reaction of the free amino acids and sugars present in it. As a result of pectin degradation, the pulp becomes soft and sticky in humid weather.

Products Processed from Tamarind Pulp

Dried pulp is extensively used in Indian cuisine and home-kitchen in preparations *e.g., sambhar, rasam, chat* and *chutneys.* This involves pre-soaking and water extraction of the deshelled pod, from which seeds have been removed.

Removal of shell and seeds from ripened tamarind pod in India is generally carried manually, though mechanical methods have also been developed. Water solubles from deshelled tamarind pods can be directly extracted in hot water, keeping water : tamarind ratio as 2:1, in which case the seeds are also separated.

On filtration through cloth, followed by concentration to 80% solid level, pulp concentrate, with a good fruit flavor is obtained. Industries have now come up with bottled and ready to use tamarind juice extract and pulp powder, which eliminates these operations and increases the shelf life of the products.

Tamarind juice and pulp concentrates are made by hot water extraction of 60–70% soluble solids (%~tartaric acid 13, sugar 50, protein 2, pectin 2, and moisture 30) from the pulp, followed by evaporation of excess water, till it has a consistency of thick non-flowing paste.

Tamarind concentrate is a hygienic and convenience product, which can be dispersed and reconstituted in hot water. Spray drying of the extract gives free-flowing, water dispersible tamarind pulp powder, which contain only 3–8% moisture and added starch (20–40%). Technologies for making tamarind juice concentrate and powder have been developed and standardized by the Central Food Technological Research Institute (CFTRI) at Mysore, Indian Standard (IS : 5955 : 1993) for tamarind concentrate has been prepared.

Tamarind pulp has been used to polish, copper brass and silver ware.

Other Products from Tamarind Pulp

Two major constituents of tamarind pulp, namely tartaric acid and pectin are commercial products, and these can be separated from pulp, while the pulp sugars can be fermented into alcohol.

Technologies for preparation of tartaric acid and its salts, pectin and ethanol from tamarind pulp have been developed by the CFTRI and the National Chemical Laboratories, Pune and patented. Commercial application of these technologies has not been made to any significant scale.

At present, a by-product (cream of tartar) of wine industry provides the raw material for making major portion of world's natural tartaric acid (dextrorotatory from or, L(+) tartaric acid). Since India does not have a large wine industry, most of the natural tartaric acid is imported into India. Some synthetic tartaric acid cracemic and meso-forms) is also produced by hydroxylation of maleic acid.

Tartaric acid and its salts have extensive applications in food and beverage, pharmaceuticals and other industries. Though tartaric acid is poorly absorbed in the intestines and being non-metabolized, it is excreted in the urine, yet it is a good souring agent for food and beverages.

Compared to grapes, used in making wine, the amount of tartaric acid in tamarind is high. It has been suggested that tartaric is produced in tamarind leaves by oxidative breakdown of sugars formed by hydrolysis of starch.

From leaves it is trans-located into the pod, where it is stored, partly as its salt. Potassium bitartrate (~ 2%), being least soluble (solubility 0.84g/100 water), crystallizes out from the boiling the water extract of tamarind pulp and it can be recovered by filtration.

The filtrate on concentration and alcohol precipitation can yield 2-3% pectin, which is similar to apple pectin. Additional tartaric acid (~ 12%) is recovered by lime precipitation as calcium tartrate from which free tartaric acid is regenerated on acidification with sulfuric acid and recrystallized into pure acid. Upto 12% alcohol can be produced by the fermentation of pulp sugars, present in the filtrate after recovery of tartaric acid. Thus it is possible to setup an integrated, economical unit for production of these three products from tamarind. Such an industry can provide pectin and tartaric acid from an indigenous source, which are imported.

Tamarind Based Beverages, Jams and Sauces

Tamarind pulp is an excellent source the making soft drinks. Pectin in the hot water extract of tamarind juice can be recovered as already described or alternatively fungal pectinae enzymes can decompose it. This results in loss of viscosity and formation of a clear juice. Sugar and tartaric acid concentrations can be adjusted to any desired level to give tamarind syrup and soft drinks or it can be fermented into wine like drinks. In Latin American countries, tamarind pulp based beverages, 'Jugo' or 'Fersco' de tamarindo are very popular.

Pulp from ripened tamarind fruit has been used for making mixed fruit jam, where all three major constituents are utilized. The pulp has also been used in confectioneries and candies. Tender green tamarind fruits have been used for making pickles.

Tamarind pulp has been used in Ayurvedic System of medicine and it is reported to have refrigerant, carminative and laxative effect. Diluted and with added salt, pulp juice has been used to prevent dehydration and heat exhaustion. Tamarind leaves are a good source of vitamin–C and Beta-carotene.

Conclusion

Harvesting of tamarind fruit from the tree is a laborious and time consuming work, but carrying it in an organized way can help rural, particularly economy of tribal region of India. There is a need to develop and transfer tamarind product processing technology to the enterpreneur and rural mass for better and economical utilization of this sustainable natural resource.

CHAPTER 33

INDUSTRIAL UTILIZATION OF TAMARIND SEED

In a separate article the industrial uses of the pulp from tamarind pod was described. It was mentioned that the use of the pulp from tamarind pod in Indian cuisine has been an old practice. With advancement in food technology, the applications of tamarind pulp can now be made in the form of packaged and ready to use tamarind pulp paste or powder. Industrial products *e.g.*, tartaric acid, pectin, sugars and alcohol can also be produced from the pulp.

For quite sometime, *i.e.*, till 1943 to be exact, tamarind seed did not find any major food or non-food applications. Only during the famine periods, the tribal in India mixed the roasted and dehusked tamarind seed powder with the flour of other food grains and used it as a famine food and also as a feed for animals. Before the emergence of scientific wisdom, many traditional experiences, gained by the tribal *e.g.*, the use of tamarind seed, as an emergency food has been a correct choice. Thus, it is now known, that in composition of tamarind kernel powder (TKP) is much like a cereal seed and it contains ~20% proteins, having recommended level of essential mino acid, and ~8% lipids of edible quality. However, the reserve polysaccharide in tamarind seed is not starch, but cellulose like linear beta-glucan (amyloid), which is solubilized due to the presence of numerous short side chins. It is not metabolized in the human system and hence does not contribute energy. TKP contains some anti-nutrients *e.g.*, tripsin-inhibitor, which of course can be deactivated by proper processing.

Historical

During the Second World War, to meet the scarcity of starch (an edible product) for the Indian textile industry, research was carried out at the Forest Research Institute (FRI), at Dehra Dun (India), to find a substitute for starch for sizing cotton yarn and cloth. As a result, TKP was found to be a reasonable substitute for starch, as a sizing material for cotton and jute.

It may be recalled that among the exigencies of the war years, was a strident industrial demand for gums which could serve as thickeners and stabilizers for food and also serve industrial purpose of acting as a sizing agents for paper and textiles. Besides tamarind seed, another agricultural product, which received attention as a source of industrial hydrocolloid at that time, was also of Indian origin, namely the galactomannan polysaccharide gum from guar seed, which successfully replaced locust bean gum in paper and other industries.

The difference between the two discoveries, was that guar gum was developed as an industrial commodity in the USA by the research carried out at the Institute of Paper Chemistry, Appleton, Wisconsin, and subsequently by Professor, R.L. Whistler's Group, at the Department of Agricultural Chemistry, Purdue University, Agricultural Experiment Station, Lafayette, Indiana. In contrast to this, early developments, related to tamarind seed gum, were carried out, by a group of scientists at FRI and subsequently by the research group of Dr. H.C. Srivastava, at the laboratories of Ahmedabad Textile Industry's Research Association (ATIRA), in India.

Tamarind Seed Availability for Processing

It has been estimated that the potential availability of tamarind seed in India alone is about 1,40,000 Mt./year, which can produce about 1,00,000 Mt. of TKP. There is not much variation in the annual production of TKP, because it is a product of a perennial tree and its annual production is not subject to variation such as that of an annual crop *e.g.* guar. The annual production of guar gum in India, has been estimated to be about ~ 6,00,000 Mt. This makes TKP as the second largest, hydrocolloid gum of Indian origin.

Tribal and Forest Related Industries

This has been discussed at length in another article. Besides tamarind pulp, the industry based on utilization of tamarind seed can help in development of tribal region in the tamarind growing states. Besides harvesting of tamarind pod, the separation of seed from the pod is carried manually and provides employment to local population.

As long as labor cost remains low in under-developed and developing countries, where gums are produced, the price of raw gums will remain low. It is also evident that developing countries are now entering into a phase of economic and cultural growth, resulting in rise of living standard and increase in the wages is bound to take place. This shall ultimately result in increased labor cost and the cost of gum products. Although the increase in the cost of gum products is one aspect, there exist another possibility *i.e.*, the labor force can get diverted towards more lucrative and betterpaid jobs. In that case collection of the forest products, which involves working under harsh conditions of jungle can decrease. In that case, the commercial availability of a forest products *e.g.*, TKP and gum arabic can be reduced, in spite of the fact that the commodity is there.

To overcome such eventualities, the developed countries are working to find substitute of such gums, the availability of which is likely to reduce or there is an unusual increase in its price, due to one reason or the other. This should be taken as a warning, because, world-over such changes have taken place in the past also, *e.g.*, the use of locust bean gum in paper industry, was completely given up in favor of guar gum, which again got threatened to be ousted in favour of cationic starch. It may also be noted that the choice of a gum or a hydrocolloid is mainly determined by its physical behaviour *e.g.*, rheology, particularly when being used in a non-food application. It is often possible to chemically modify cheap and abundant polysaccharides into a derivative, which can substitute a costly and scarcely available gum.

Economy of Tamarind Seed Products

The seed forms about 30% of the ripened tamarind pod. Earlier, the retailers sold some of the tamarind pods, without removing the seeds and the seeds of thus sold tamarind pods were not collected for extraction of TKP. With the emergence of a well-organized tamarind pulp industry, in the southern and central states in India and with increasing domestic and export demand for TKP, the collection and processing of tamarind seed has increased during past two decades. Because of the difference in the fields of applications of tamarind pulp and TKP, it is a normal practice to process the pulp extraction in a different and exclusive industrial unit, which sells its by products, the seed, to a TKP extraction unit.

Tamarind pulp and TKP processing industries are mainly located in Karnataka, Andhra Pradesh, MP, Chhattisgarh, and Tamil Nadu. About 30-40 kg seeds, produced from 100 kg of the pods are sold at the rate of Rs. 4-6/kg to a TKP processor. The processing cost can be Rs. 3-4/kg and 20-25 kg of TKP produced is sold at the rate of Rs. 10-15/kg. Deoiled TKP is sold at higher price, while there is a value addition of –10% in terms of –6% oil recovered.

About 18-20 kg testa (husk) produced in the process is burnt in kilns for making bricks. Currently the testa is not processed to recover the water/alcohol or ethyl acetate solubles, which are –20% tannins and 8-10% low molecular weight phenolics. The solvent extracted portion of testa consisting of phenolic compounds, have very good antioxidant properties and it can be used to prevent rancidity of edible oils and deep-fried foods.

Constituents of Tamarind Kernel and Seed Coat

Tamarind seeds have a dark brown outer seed coat called testa, which encloses the creamy white kernel, composed of two halves of cotyledon. Compositions, of tamarind seed kernel and seed coat are shown in Table 1. It will be observed that, unlike many other legume seeds, the seed kernel or the cotyledon of tamarind contains higher amounts of lipids and proteins. The polysaccharide of tamarind seed is contained in granules, called albumen, which has protein and lipid intricately mixed with polysaccharide.

Typical gum bearing seeds, *e.g.*, guar or locust bean have most of the protein and lipids confined, only in the germ portion of the seed, which is distinct and easily separated from the endosperm. In case of tamarind seed, the protein and lipids are distributed throughout the cotyledon and this makes the isolation of pure tamarind seed polysaccharide difficult. Fortunately, in textile sizing and printing, removal of both these constituents (lipid and proteins) is not necessary.

Though tamarind seed lipids are easily removed by solvent extraction of cracked cotyledon (~1-2 mm size) removal of proteins is rather difficult. Deolied TKP is now available in Indian Market and it is also being exported. Oil bearing TKP is readily fermented, due to air oxidation of unsaturated fatty acids present in it. This causes rancidity and development of bad smell. Deoiled TKP is more stable, whiter in color and it is non-lumping. The oil-extracted from TKP compensates cost of deoiling process.

The recovered oil resembles groundnut oil in composition, and it is used for edible as well as non-edible purposes. The tannins extracted from the seed coat can be used for leather tanning.

Besides tannins, which are polymerized phenolics, the testa also contains solvent extractable bioflavonoids, which are very good antioxidants for food additives. These have been shown to be very effective in preventing spoilage of food and rancidity of fat.

Manufacture of Tamarind Kernel Powder

For the manufacture of TKP, the seeds are washed in water to remove adhering pulp and hollowed seeds, which float in water, are removed. The seed kernel is tenaciously held by the seed coat and for its removal; the seeds are perched in an oven or with hot sand at 150°C for 10-25 min., when the seed coat becomes loose and brittle.

Care is taken that the kernel is not charred during heating. When partially charred, the TKP produced is darker in color and has lower viscosity. The testae are then removed by light pounding over winnowing. This operation is generally carried out manually, although mechanical disintegrators have also been devised.

A process for simultaneous decortications and pulverization of the kernel has also been devised, where the lighter, and powdered testae are removed by blowing it off from the kernel powder with air. The kernel, which easily breaks in two cotyledon halves, is further broken down into 2-3 mm pieces. At this stage, the oil can be removed by solvent extraction (hexane or chlorinated-hydrocarbon) of coarsely powdered kernel. Presence of ~8% oil reduce the shelf life of TKP, but, deoiled TKP can be stored for over one year.

Broken kernel pieces are sometimes bleached using sulfur dioxide or sodium bisulfite, to improve the whiteness of the final powder. This treatment is also a safeguard against bacterial infection during processing. Finally the broken kernel is powdered using impact-cum-attrition type pulverizer or hammer mill.

Most of the TKP is processed in small or medium scale industries and the product coming in the market is not of uniform or reproducible quality. Since the product is basically meant for non-edible purpose, microbial control is not carried. Generally the TKP is of such fineness, that > 90% of it pass through 250 standard US-mesh *i.e.,* of 60 microns fineness.

The powder is generally mixed with appropriate preservatives *i.e.,* chlorinated phenols, benzoate and bisulfite. The product is packaged into 25-50 kg. bags, and it is recommended for storage in dry place. According to one estimate 60-80 thousand tons of TKP is processed in India. The export earning from TKP is estimated to be Rs. 50 crores.

Properties of TKP

Good quality of deolied TKP is a colorless or off-white, free flowing powder, having a faint smell. It is dispersible in cold water, but hydrates only on heating to > 60° for 25-30 min. A 1% solution of TKP, when completely hydrated by heating at 60-75°C, will produce a viscosity of 50–80 cps, when measured by Brookfield Viscometer. The increase in viscosity with concentration is logarithmic and the viscosity of 8% solution can be 30,000-45,000 cps. When measured by a Brabender Amylograph, the viscosity change pattern is somewhat similar to that of starch. A 5% solution of TKP show a viscosity of –1600 Brabender Units. For use as a size, this viscosity is very high compared to that of 5% cornstarch, which is about 300 only. Hence depolymerized TKP is preferred for sizing. Solutions of TKP have a non-Newtonian (Pseudo-plastic or shear-thinning) rheology and the viscosity is reduced with rise in temperature. The structural organization of tamarind seed granule appears to be similar to that of starch in being subjected to disruption by heat and alkali.

Compared to many plant polysaccharides, TKP pastes are more stable to heat and shearing. In preparation of TKP printing pastes for polyester, with dispersed dyes, cooking of the paste for a rather long period and often over night is done TKP pastes are also more resistant to acid and alkaline depolymerization, compared to other gums.

Chemical Structure of Tamarind Polysaccharide

Tamarind polysaccharide has a cellulosic, β, 1 → 4-linked glucan backbone. Nearly three-fourth of the glucose units in the backbone are substituted by a-linked xylopyranose group at C–6 of glucose in the main chain, and nearly one out of every three-xylose units, is linked to a (3–galactose unit at C–2 of xylose (Fig. 1). The ratio, glu: xyl: gal is 4:3: 1–1.5 which is somewhat like a cell-wall polysaccharide, though it is a reserve polysaccharide. Thus tamarind polysaccharide is regarded as a galactoxyloglucan. High degree of substitution of glucan chain, produce a stiff, extended conformation for tamarind polysaccharide molecule, with large volume occupancy in solution.

```
{-4-β-D-glc (1→4) -β-D-glc (1→4) -β- D-glc (1→4)-β-D-glc (1→}n
        6                 6                 6
        ↑                 ↑                 ↑
     α-D-xyl           α-D-xyl           α-D-xyl
                          2
                          ↑
                       β-D-gal
```

Fig. 1 : Idealized repeat unit of tamarind polysaccharide.

Stability of tamarind polysaccharide to heat and mild acids and bases, can be attributed to the high degree of substitution of the backbone, which, to some extent, shields it from hydrolyzing agents. High substitution of glucan backbone produces a ribbon like conformation for the polysaccharide. Cellulase enzymes (from aspergillus oryzaeniger) hydrolyze the glucan backbone and reduce the viscosity, ultimately producing mainly 7-9 oligomers. Hydrolyzed tamarind oligo-saccharides, along with non-sugar sweeteners *e.g.*, aspartame, have been recommended as non-calorie sugar mimics, which can replace major part of sugar in food products, without a significant change in the original texture of a food.

Modified and Derivatised Tamarind Polysaccharides

TKP is the powdered kernel of tamarind seed form which purified polysaccharide, free from protein, fiber and oil, can be isolated. Both wet and dry methods, have been used for isolating commercially purified (> 95%) polysaccharide, in 50-60% yield. The protein has been separated by precipitation by heat coagulation or decomposed by concentrated alkali or enzymes. Polysaccharide is then obtained by alcohol or salt precipitation, followed by spray or drum drying. Physical methods are based on difference in size and density of protein and polysaccharide particles in TKP and these appear to be more attractive. Currently only Japan (Dainippon Pharmaceutical Co.) is making purified tamarind polysaccharide (trade name Glyloid), which is claimed to be a substitute for pectin, for making jams and jellies. This is an important field of research for India, looking to the under-developed pectin industry and availability of raw material. As a gelling agent, tamarind polysaccharide can function even better compared to pectin, as is evident from the gel strength data (Table 2).

TKP and tamarind polysaccharide have been derivatised into anionic (carboxymethyl) and nonionic (hydroxyethyl) derivatives, which show improved solubility and washability in textile printing. A cold water-soluble TKP is also known.

Major Applications of Tamarind Polysaccharide and TKP

Textile Applications : Being one of the cheapest gums, TKP has been used as a substitute for other hydrocolloids, wherever it can sever reasonably well and is cost effective. Currently its largest application is in textile, where it is being used as a sizing material for cotton and jute. Lowest grade TKP goes in textile sizing. It is also used, sometimes blended with other polysaccharides, in printing of polyester, with disperse dyes. The results are reasonably good and it works well. However, it is unsuitable in printing cotton with vat dyes, due to the formation of insoluble complex with dyes. It is also unsuitable in cotton printing with fiber reactive dyes. The presence of large amounts of proteins in TKP based thickener, which also reacts (via amino and hydroxyl groups) with reactive dyes in preference to cotton cloth, reduce the color yield. Purified tamarind polysaccharide and its anionic derivatives, are likely to act as suitable thickener for reactive printing, where these may prove comparable to sodium alginate. No work has so far been carried out on these lines. For making TKP print paste, heating is required, but cold water hydrating TKP has also been produced. Anionic derivatives have been shown to offer some advantage over regular TKP in print paste thickener.

Table 1 : Composition (%) of TKP and Seed Coat

Constituent (Cotyledons)	*Seed Kernel Testa*	*Seed Coat*
Moisture	11-15	9-11
Protein	15-21	–
Lipid	4-8	–
Fiber	3-8	20–22
Polysaccharides	65-72	–
Ash	2.5-4	7-8
Tannins	Nil	20-22

Food Applications : Purified tamarind polysaccharide can be used as a general, low viscosity thickener and a non-calorie bulking agent in food. Tamarind polysaccharide based thickners produce better freeze-thavy stable products and emulsion stabilization in better. At low water activity *i.e.,* at sugar concentration >60%, it forms gel just like HM-Pectin, and its use in making jams and jells is done. It has been used in confectioneries, and as a binder in pharmaceutical tablets. Being non-metabolized, it acts as a soluble food fiber. Enzymatically depolymerized tamarind oligosaccharides have been used to replace sugar in low calorie food products. Overall there can be good value addition in making purified, food grade tamarind polysaccharide as has been done in Japan. The starting material is cheap and in substituting high priced pectin it can give good returns to a processor. Additionally, if the protein rich fraction is recovered, it can be used in animal feed.

Table 2 : Strength of some fruit pectin's and tamarind gels

Fruit	*Gel strength, g*
Lemon	340
Orange	361
Apple	253
Tamarind polysaccharide	630

Adhesive Paints and Binder : TKP can substitute starch in many adhesive applications, and as a binder in particle board, corrugated board and books. It has been used in briquetting of coal powder and palletizing of minerals. It has also been used for stabilizing soil. In paints, TKP can act as a rheology control additive and it can be used in water-based distempers, rubber and synthetic emulsions.

Other Applications : Because of its better thermal stability, TKP is being evaluated as thickener in drilling fluids for petroleum oil, including by O.N.G.C. in India. Use of purified tamarind polysaccharide can be done in cosmetics. Scientists at the C.D.R.I, has developed absorbable dressing for wounds based on purified and sterilized tamarind polysaccharide. It has been used as a creaming agent in latex.

Conclusion

TKP, cannot be considered to be a commercially pure polysaccharide, but it is a cheap sources of making industrially pure tamarind polysaccharide, a hydrocolloid of certain desirable and valued functional properties. Recent increase in export of TKP is an indication that it is being used for making value added products and being employed as a cost effective substitute for other hydrocolloids. Tastae of tamarind seed should be better utilized for extraction of tannin and anti-oxidant biofkivonoids. Unfortunately in India, the place of its origin, it is still being used mainly as a cheap sizing agent in textile and in printing. There is an urgent need to have a dedicated research centre to develop value added products not only of tamarind but other polysaccharide, hydrocolloids of India *e.g.*, guar. This will not only help the industry, but also boost the export earning from agro products. Organization such as APEDA, should seriously consider this.

CHAPTER 34

SOME IMPORTANT NTFP OF INDIA

(Medicinal Plants)

India is a country with vast reserves of non-timber forest produces. In almost all the forest types of India NTFPs of one or other forms are available providing livelihood to hundreds and thousands of forest dwellers as well as people far from the forests. The parts of various species used vary from underground parts like tubers, roots to leaves, stems, fruits, flowers, barks and what not. When some NTFPs are valuable for one community others undermine them. There is a famous *Santali* saying, *"Chinle ran–Bankhan bir sahan'* which means, "If you are ignorant of medicinal plants they are as good as fuel-wood". The indigenous knowledge (IK) of different communities in using NTFP resources is an important factor, which they have developed for thousands of years of their close association with the forests of their surroundings.

In the following descriptions of plants (species wise) with their names in major vernacular languages in India, habit, habitat, their phytography (description of a plant), phenology (the study of periodicity of flowering and fruiting), distribution with particular reference to India, present status and uses are provided. Utmost care has been taken to describe the plants in common language avoiding technical terms unless it is essential. Flowering and fruiting *i.e.,* phenological duration has been taken for many of the species from herbarium sheets deposited at Central National Herbarium (CNH) at Kolkata.

- *Abelmoschus moschatus* Medic. (Malvaceae)

  Syn. : *Hibiscus abelmoschus* L.

  Sans. : Lata Kustari,

  H. & B. : Musk dana, E.-Ambrette plant or Musk mallow.

  Habit : Shrub

Habitat : Terrestrial

Phytography : Herb with leaves–alternate, simple, lobed; flowers in racemes, yellowish; fruit–capsule.

Fl. & Fr. : May–November

Distribution : Plains of India.

Status : Rare in wild state, partially cultivated.

Uses : Seeds known as *ambrette* seed. Used as flavouring agent, toxic stimulant, and carminative. Tender leaves and shoots are eaten.

- *Abroma augusta* L. (Sterculiaceae) **(Plate-1)**

H. & B. : Ulatkambal.

Habit : Shrub

Habitat : Terrestrial

Phytography : Shrub with leaves–alternate with long petiole, simple lobed; flowers–in racemes, purple; fruit–capsule.

Fl. & Fr : July–November

Distribution : Throughout India.

Status : Wild but presently cultivated.

Uses : Roots used as a uterine tonic, and emmenagoge. Leaves used in uterine disorder, diabetes, and rheumatism. Seed lowers cholesterol in blood.

- *Acacia caesia* W. & A. (Mimosaceae)

Syn. : A. *intsia* Willd.

H.–Aila. *B.*–Dextari

Habit : Small tree

Habitat : Terrestrial

Phytography : Small tree, spines paired; leaves–pinnately compound; flowers–in spikes, small; fruit–pod.

Fl. & Fr : May-August

Distribution : Throughout India especially in sub-Himalayan tract.

Status : Wild, infrequent

Uses : Stem bark used as substitute for soap as hair wash.

- *Acacia catechu* Willd. (Mimosaceae)

  S.–Khadirà

  H. & B. : Khair, *E.*–Catechu tree

  Habit : Medium size tree

  Habitat : Terrestrial

  Phytography : Middle size tree, spines paired; leaves-pinnately compound; flowers in spikes; fruit-pod.

  Fl. & Fr : June-September

  Distribution : In dry deciduous forest areas.

  Status : Wild, frequent

  Uses : Gum is the source of catechu.

- *Acacia sinuta* (Lour) Merrill. (Mimosaceae)

  Syn. Acacia concinna DC.

  H.-Ritha, *B.*-Banritha

  Habit : Medium size tree.

  Habitat : Terrestrial

  Phytography : Stunted tree; leaves–pinnately compound; flowers in heads; fruit–pod.

  Fl. & Fr : January–September

  Distribution : Found in tropical forest especially in Deccan.

  Status : Wild, infrequent.

  Uses : Pods are the source of *shikakai* used as detergent. Tender acidic leaves used in chutneys.

- *Acacia auriculiformis* A. Cunn. ex Benth. (Mimosaceae)

  H. & B. : Sonajuri

  Habit : Tree

  Habitat : Terrestrial

  Phytography : Tree; leaves–phyllode; flowers in spike, yellow; fruit–pod.

  Fl. & Fr : February–August

Distribution : Found as plantation stands created for afforestation.

Status : Planted, quick growing plant.

Uses : Fire wood, poll etc.

- *Acacia farnesiana* Willd. (Mimosaceae)

  H.–Kikar, *B.*–Gaya babul. *E.*–Cassie flower

  Habit : Small tree

  Habitat : Terrestrial

  Phytography : Stunted tree; branches armed with paired spices; leaves-pinnately compound; flowers in heads, brightly yellow, fragrant; fruit-pod, cylindrical, pulpy.

  Fl. & Fr : January–September

  Distribution : Eastern Ghats, Deccan and Other parts of South India.

  Status : Wild, abundant.

  Uses : Flowers are the source of *lassie* perfumes; pods astringent contains 30% tannin.

- *Acacia nilotica* (L.) Willd. ex Delile Sub sp. *indica* (Benth.) Brenan (Mimosaceae)

  Syn. : A. *arabica* Willd.

  H.-Kikar, *B.*-Babul.

  Habit : Tree

  Habitat : Terrestrial

  Phytography : Medium sizes tree with stipular spines; leaves–pinnately compound; flowers in axillary heads, yellow; fruit–pod, deeply constricted.

  There is three recognized types of the species : *Cupressiformis* Steward, *Vedians* Cooke and *Telia*.

  Fl. & Fr : August–April

  Distribution : All over India.

  Status : Common, wild

  Uses : Gum is the source of 'gum Arabic' and source of tannin. Leaves contain 32% tannin. Gum is diuretic and bark is astringent.

- *Acacia senegal* Willd. (Mimosaceae)

  S.-Sneta

  H. & B. : Swet Khadir

  Habit : Tree

  Habitat : Terrestrial

  Phytography : Small tree; leaves–pinnately compound; flowers in spikes; fruit–pod, flat.

  Fl. & Fr : August–December

  Distribution : Mostly in Indian desert.

  Status : Common, wild

  Uses : Gums principally used in confectionery, chewing gum, *panparag*. Used in burn wounds and wounds of leprosy. It is also used as a hair-setting agent.

- *Acalypha indica* L. (Euphorbiaceae) **(Plate–2)**

  S.-Arittanunjayrie; *H.*-Khokli, *B.*-Muktajari, *E.*-Indian acalypha.

  Habit : Herb

  Habitat : Terrestrial

  Phytography : Herb; leaves–simple, dented; flowers in racemes; fruit-capsule with bracts.

  Fl. & Fr : March–July

  Distribution : Found throughout the plains of India.

  Status : Common, wild.

  Uses : Plant decoction is used as laxative. It possesses the alkaloid acalyphine which is used in Homeopathy against severe cough associated with bleeding from lungs.

- *Aconitum ferox* Wall. ex Ser. (Ranunculaceae) **(Sketch–1)**

  S.-Visha, H.-Bacchang, B.-Katbish, E.-Indian Aconite

  Habit : Herb

  Habitat : Terrestrial

  Phytography : Herb; leaves–simple, sinuted; flowers–racemes, fruit-capsule; root is poisonous.

Fl. & Fr : September–December.

Distribution : Alpine Himalayas, Assam, and Nepal.

Status : Rare, wild.

Uses : Alkaloid pseudaconitine is used in nasal catarrh, tonsillitis, gastric disorder, and debility.

- *Aconitum heterophyllum* Wall. ex Royle (Ranunculaceae)

S.-Ativisha, H.-Atis, E.-Atis root.

Habit : Herb

Habitat : Terrestrial

Phytography : Herb; leaves–simple; flowers–racemes; fruit–capsule; roots–ash gray in colour.

Fl. & Fr : August–December.

Distribution : Found in sub alpine and alpine Himalayas.

Status : In frequent, wild.

Uses : Contains alkaloid like atisine, heteratisine etc. which are used as febrifuse and bitter tonic for abdominal pain and diabetes.

- *Aconitum bisma* (Ham) Rapie. (Ranunculaceae)

Syn : *Aconitum palmatum* G. Don.

H.–Bakhima.

Habit : Herb.

Habitat : Terrestrial

Phytography : Herb; leaves–simple; flowers–racemes; fruit–capsule; roots are non-toxic.

Fl. & Fr. : September-December

Distribution : Alpine Himalayas, Sikkim and Tibet.

Status : Rare and wild.

Uses : Root is used like quinine and also in excessive vomiting.

- *Aconitum nepellus* L. (Ranunculaceae)

S.–Visha, *H.*–Mithazahar, *B.*–Katibish.

Habit : Herb

Habitat : Terrestrial

Phytography : Herb; leaves–simple; flowers–racemes; fruit–capsule.

Fl. & Fr. : August–December.

Distribution : Found mostly in temperate India.

Status : Infrequent, wild.

Uses : It is used as ferox.

- *Acorus calamus* L. (Araceae)

S.-Vacha

H. & B. : Bach. *E.*–Sweet flag

Habit : Herb

Habitat : Marshy or semiaquatic.

Phytography : Semi-aquatic aromatic herb with creeping root stocks; leaves–simple, sessile and linear; flowers in spadix; fruit–berry.

Fl. & Fr. : In rainy season.

Distribution : Throughout India.

Status : In frequent, wild.

Uses : Rhizome is used in epilepsy and other mental ailments. It is also used in chronic diarrhoea, dysentery and glandular and abdominal tumours.

- *Adenanthera pavonina* L. (Caesalpiniaceae)

S.–Kunchandana. *H.*–Barigumchi. *B.*–Rakta kambal. *E.*–Coral wood, red wood.

Habit : Medium sized tree.

Habitat : Terrestrial

Phytography : Tree; leaves–pinnately compound; flowers in racemes; fruit–pod, seeds–scarlet.

Fl. & Fr. : March–July

Distribution : Found in Eastern Himalaya, Western Ghat, Andaman and South India.

Status : Occasional, planted.

Uses : Seed and wood decoction is used in pulmonary infections. Wood is adulterated with red sandalwood.

- *Adhatoda vasica* Ness (Acanthaceae)

Read as *Justicia adhatoda* L.

S.–Vasaka,

H. & B. : Bakas. *E.*–Vasaka

Habit : Shrub.

Habitat : Terrestrial

Phytography : Shrub; leaves–opposite, simple; flowers in racemes, white with pink spots inside; fruit–capsule.

Fl. & Fr. : February–May

Distribution : Found throughout the plains of India.

Status : Abundant, wild.

Uses : Leaf decoction is used in bronchial trouble; leaf juice is used in diarrhoea. Whole plant is used in green manure.

- *Adiantum philippense* L. (Adiantaceae)

Syn. : A. *lunulatum* Burm.

H. & S. : Kalijhant. E.–Walking Maiden hair fern.

Habit : Herb

Habitat : Terrestrial

Phytography : Shrub; leaves–frond, pinnately compound; sori on the dorsal side of the frond.

Fl. & Fr. : In winter season.

Distribution : Found in moist places almost throughout India.

Status : Abundant, wild.

Uses : Used in epileptic fits, it is also used against rabies and elephantiasis.

- *Aegle marmelos* Correa. ex. Koen. (Rutaceae) **(Plate-3)**

S.-Silva

H. & B. : Bel, *E.*-Bale tree

Habit : Tree

Habitat : Terrestrial

Phytography : Trees; leaves–compound, trifoliate; flowers in cymes, greenish; fruit–drupe with hard shell.

Fl. & Fr. : December–June

Distribution : Found throughout India in deciduous forest.

Status : Common, both wild and planted.

Uses : Ripe fruit is astringent, digestive so used in stomach-ache. Fruit pulp is used in dysentery.

- *Aschynomene aspera* L. (Fabaceae)

H. & B. : Sola. *E.*-Sola pith plant

Habit : Shrub

Habitat : Aquatic plant

Phytography : Under shrub; stems with warts and white pith; flowers in racemes, yellow; fruit–pod.

Fl. & Fr. : August–December

Distribution : Throughout India and mostly in moist places

Status : Threatened due to Habitat disturbances, wild.

Uses : Pith used in decorations.

- *Agava sisalana* Perr. (Agavaceae)

Syn : A. *rigida* Mill. var. *Sisalana* Engelm

E.–Sisal

Habit : Shrub

Habitat : Terrestrial

Phytography : Large tufted shrub with stout rhizomes; leaves–long, stout with spines, margin denoted like saw-blade; flowers–campanulate; fruit-capsule and beaked.

Fl. & Fr. : April-August

Distribution : Throughout India.

Status : Common, planted and wild as scape.

Uses : Produce microorganism resistant fibre named ship cordage.

- *Ailanthus excelsa* Roxb. (Simarubiaceae) **(Plate-4)**

S.–Modala.

H. & B. : Maharuk. *K.*–Tree of heaven.

Habit : Big tree

Habitat : Territorial

Phytography : Trees; leaves–pinnately compound; flowers in racemes, small; fruit–winged.

Fl. & Fr. : March–July

Distribution : Found throughout India in deciduous forests.

Status : Common, wild.

Uses : Anthelmintic, febrifuge, expectorant and antispasmodic. Stem bark is used in asthma. The bark is substitute of *kurchi* bark *(Holarrhena pubescens)* source of 'hog gum'.

- *Alangium salviifolium* (L. of.) Wang (Alangiaceae) **(Plate-5)**

Syn : *A. lamarckii* Thw.

S.–Ankola, H.–Akola. B.–Akar-Kanta

Habit : Small tree

Habitat : Terrestrial

Phytography : Medium size tree; branches spinescent; leaves–alternate, simple; flowers in axillary and terminal panicles; fruit-berry

Fl. & Fr. : April-June

Distribution : Found in throughout the drier parts of India.

Status : Abundant, wild.

Uses : Root bark is used in stem diseases. It is astringent, anthelmintic. Children eat fruits; seed oil is used in illumination.

- *Alpina galanga* Willd. (Zingiberaceae)

S.–Sugandha

H. & B. : Kalanjan/Barakulanjan, E.–Greater Galanga

Habit : Herb

Habitat : Terrestrial

Phytography : Herb; leaves–simple, long; flowers in spikes; fruit–small, anilated.

Fl. & Fr. : May–September

Distribution : Eastern Himalayas and South India.

Status : Infrequent, wild.

Uses : Rhizome used in rheumatism and bronchial catarrh. Rhizome is a source of methyl cinnamate. Flowers eaten.

- *Aleurites moluccana* (L.) Willd. (Euphorbiaceae)

H. & B. : Jangli Akhrot, E.–Wood at tree

Habit : Tree

Habitat : Terrestrial.

Phytography : Trees; leaves–simple; flowers in racemes, green; fruit-drupe.

Fl. & Fr. : February–August

Distribution : Wild in South Indian forests.

Status : Not common.

Uses : Seed is the source of '*Lumban* oil'.

- *Anacardium occidentale* L. (Anacardiaceae)

S.–Kaju, H. & B.-Kajubadam, *E.*-Cashew nut tree

Habit : Medium size tree

Habitat : Terrestrial.

Phytography : Tree; leaves–alternate, simple; flowers in racemes; fruit-drupe.

Fl. & Fr. : December–May

Distribution : Naturalised and cultivated in coastal India.

Status : Common, cultivated.

Uses : Kernel is eaten raw and fried. Nutshell yields a oil which is used against termites.

- *Ananas comosus* (L.) Merrill (Bromeliaceae)

  Syn. : A. *sattiva* Schultf.

  H. & B. : Anaras. *E.*–Pineapple.

  Habit : Herb

  Habitat : Terrestrial.

  Phytography : Shrub; leaves–simple, long, margin dented; fruit-aggregate.

  Fl. & Fr. : December–August

  Distribution : Naturalized and cultivated in Assam, West Bengal and Western coast.

  Status : Generally found in cultivated fields. Introduced from Brazil.

  Uses : Ripe fruits are eaten. Aggregate fruit is used in confectionery and for preparation of vinegar. Ripe fruit is diuretic, antiscorbutic. Green fruits are abortive.

- *Amomum subulatum* Roxb. (Zingiberaceae)

  S.–Birhada ela, H. & B.–Bara elach, E.–Nepal cardamom.

  Habit : Herb

  Habitat : Terrestrial.

  Phytography : Herb; leaves–simple long; flowers in racemes; fruit – drupe.

  Another species *Amomum aromaticum* Roxb.-available in forest.

  Fl. & Fr. : March–August

  Distribution : Hilly forest areas of West Bengal, Assam and Sikkim.

  Status : Mostly cultivated.

  Uses : Seeds used as *chotoelachi (Elettaria cardamomum)*. Seed oil is applied to inflammed eyelids.

- *Andrographis paniculata* Wall. ex Ness (Acanthaceae)

  S.–Kirata, H.–Kirayat, B.-Kalmegh, E.–Creat.

  Habit : Herb

Habitat : Terrestrial

Phytography : Herb; stems–4-angled; leaves–simple, opposite; flowers in axillary and terminal racemes; fruit–capsule.

Fl. & Fr. : September–May

Distribution : All over Indian plains

Status : Common, wild.

Uses : Used as bitter tonic, febrifuse and plant astringent. Extract is used in liver diseases and in dysentery.

- *Anemone obtusiloba* D. Don. (Ranunculaceae)

H.-Rattangog.

Habit : Herb

Habitat : Terrestrial

Phytography : Herb; leaves simple; flowers racemes; fruit–capsule.

Fl. & Fr. : September–November

Distribution : Temperate Himalaya, Sikkim and Darjeeling Himalaya

Status : Rare, wild.

Uses : Rootstock is used for concussions. Seed oil is used against rheumatism.

- *Annona squamosa* L. (Annonaceae)

H.–Sharifa, B.–Ata, E.–Custard apple

Habit : Small tree

Habitat : Terrestrial

Phytography : Tree; leaves–simple, opposite; flowers–solitary small, yellowish; fruit–ovoid with projecting convex ovoid areole.

Fl. & Fr. : April–June.

Distribution : Found throughout India mostly in deciduous forest belt.

Status : Infrequent, wild, sometimes cultivated.

Uses : Fruits are edible. Root is used as purgative. Seed is abortifacient.

- *Antiaris toxicaria* Lesch. (Moraceae)

  S.-Valkala, E.-Ulas tree.

  Habit : Tree

  Habitat : Terrestrial

  Phytography : Tree; leaves–simple; flowers–racemes; fruit–berry.

  Fl. & Fr. : May–December

  Distribution : Found in the forests of Western Ghats.

  Status : Infrequent, wild.

  Uses : Latex is used as cardiac and circulatory stimulant. Seed is a febrifuge. Latex is used in arrow poisoning.

- *Antidesma diandrum* Heyne ex Roth (Euphorbiaceae)

  H.–Amari, B.–Mutta

  Read as A. *acidum* Retz.

  Habit : Shrub

  Habitat : Terrestrial

  Phytography : Large bush or small tree; leaves–simple alternate, obovate; flowers–pedicelled, small; fruit–drupe.

  Fl. & Fr. : April-November

  Distribution : All over dry deciduous forest

  Status : Common, wild.

  Uses : Leaves used as vegetable, seeds yield oil and fruits are eaten.

- *Aphanamixis polystachya (Wall.)* Parker (Meliaceae)

  Syn. : *Amoora rohituka* Wight & Arn.

  S.–Rohituka, H.–Harinhara, B.–Pitta-raj.

  Habit : Tree

  Habitat : Terrestrial

  Phytography : Tree; leaves–pinnately compound; fruit–drupe.

  Fl. & Fr. : April–January

  Distribution : Found in the plains in India.

Status : Infrequent, wild.

Uses : Timber is used for boat making. Seeds accredited with laxative and anthelmintic properties. It yields an essential oil used in liver and spleen diseases. Bark dust is effective as poultice on rheumatic swellings.

- *Aquilaria malaccensis* Lamk. (Thymelaeaceae)

Syn. : *A. agallocha* Roxb.

S.-Agaru, *H. & B.*-Agar/Agaru, *E.*-Agarwood

Habit : Tree

Habitat : Terrestrial

Phytography : Large evergreen tree; leaves–alternate, nerves are parallel; flowers–hermaphrodite, pedicelled in. umbellules; fruit -compressed loculicidal capsule.

Fl. & Fr. : May–October

Distribution : Found in Eastern Himalayas, Bhutan, Assam and Tripura.

Status : Critically endangered, wild.

Uses : A good source of *agaru*. *Agaru* is an oleoresin developed in woods of older trees and is considered as a high-class perfumery. Beside perfumery uses it is used as stimulant and cardiac tonic. *Agaru* is obtained from the fungus-infected tree trunk. Aggar *attar* is prepared by using sandalwood oil, vegetable oil or liquid paraffin as bases. Spent wood, obtained after distillation is used for making *agarbatties*.

- *Aristolochia bracteolata* Lamk. (Aristolochiaceae)

Syn. : *A. bracteata* Retz.

S.–Dhumrapatra, *H.*–Kiramar, *B.*–Iswarmul, *E.*–Bracteated birthwort.

Habit : Creeper

Habitat : Terrestrial

Phytography : Herb; branches–glabrous; leaves–alternate, reniform to cordate; flowers–solitary, puduncle with a large orbicular bract; fruit-capsule.

Fl. & Fr. : June-January.

Distribution : Found in Indian plains.

Status : Not common, wild.

Uses : Leaf is used as purgative and anthelmintic. Application of leaf juice with caster oil controls eczema. Root decoction expels intestinal worms. The plant possesses aristolochic acid and nitro phenanthrene compounds with tumor inhibiting properties.

- *Aristolochia indica* L. (Aristolochiaceae) **(Sketch-2)**

S.–Iswarmul

H. & B. : Iswarmul, *E.*–Indian birthwart.

Habit : Creeper

Habitat : Terrestrial

Phytography : Perennial herb; leaves–alternate, entire; flower–solitary, yellowish; fruit–capsule.

Fl. & Fr. : June–December

Distribution : Found in low hills and plains of India

Status : Infrequent, wild.

Uses : Root is used as gastric stimulant and bitter tonic. Seed is used in inflammations and biliousness. The chief active principle is aristolochic acid.

- *Artemisia nilagirica* (C. B. Clarke) Pamp. (Asteraceae)

Syn. : *A. valgaris* auct. non L.

S.–Nagadamini. H.–Nagadrona, B.–Nagdana, E.–Santonic plant.

Habit : Shrub.

Habitat : Terrestrial.

Phytography : Shrub; leaves–simple, palmate, flowers–raceme; fruit-nutlet.

Fl. & Fr. : July-December

Distribution : Found throughout mountainous districts of India.

Status : Infrequent, wild.

Uses : Leaves and flowering tops are used in asthma. Plant is used in brain diseases.

- *Asparagus racemosus* Willd. (Liliaceae) **(Plate-6)**

S. & B.–Satamuli, H.–Satawar.

Habit : Twiner

Habitat : Terrestrial.

Phytography : Perennial under-shrub; stem–greenish, spiny, grooved; leaves–alternate, scale like, flattened axillary cladode; flowers in racemose, greenish white, small; fruit a globose pulpy berry.

Fl. & Fr. : June–November

Distribution : Found throughout tropical and subtropical India.

Status : Not very common, wild.

Uses : Root tonic, diuretic and galactagogue. Fresh root juice with honey is given in dyspepsia. Medicated oil prepared from roots is used in nervous and rheumatic diseases.

- *Asplenium adiantum-nigerum* L. (Aspleniaceae)

H. & B. : Nelapana E.–Black spleenwort.

Habit : Perennial herb

Habitat : Terrestrial.

Phytography : Perennial fern; leaves–fronds; flowers absent, sori has spores, propagated by spores.

Fl. & Fr. : Onset of winter

Distribution : Found almost throughout India

Status : Infrequent wild.

Uses : Laxative and diuretic. Fronds are used in spleen disorders and jaundice. It produces sterility in women. Tribals use the fern as a antifertility drug.

- *Atylosia scrabaeoides* Benth. (Fabaceae)

H. & B. : Bankulatha.

Habit : Twiner

Habitat : Terrestrial

Phytography : Herbaceous, annual, twiner; leaves–pinnately 3-foliolate; flowers–small, in racemes, light yellow in colour; fruit–pod like, with convex faces clothed with fine spreading silky hairs.

Fl. & Fr. : April–December

Distribution : Found throughout India.

Status : Abundant, wild.

Uses : Plant is a good fodder for cattle. Plant decoction is used in dysentery of cattle.

- *Avicennia marina* Vierh. (Avicenniaceae) **(Plate-7)**

H. & B. : Bine.

Habit : Tree

Habitat : Mangrove plant

Phytography : Tree; leaves–simple; flowers–raceme; fruit–drupe.

Fl. & Fr. : March–September.

Distribution : Found in mangrove forest.

Status : Population is declining day-by-day, wild.

Uses : Leaf juice is used as an abortifacient. Bark is used to extract tannin.

- *Azadirachta indica* A. Juss (Meliaceae)

Syn. : *Melia azadirachta* L.

S.–Nimba, H. & B.–Nim, E.–Margosa tree

Habit : Tree

Habitat : Terrestrial

Phytography : Tree with dark wood; leaves–pinnately compound, leaflets–toothed; bitter in test; flowers–in axillary panicle; fruit–fleshy drupe; seeds–enclosed in an aril and winged.

Fl. & Fr. : March–July.

Distribution : Throughout India

Status : Common, wild

Uses : Bark is used in skin diseases. Leaf is considered as antiseptic. Seed oil has also medicinal properties. Berries are used as purgative. Nim cake is used as an anti-termite.

- *Balanites aegyptiaca* (L.) Delile (Balanitaceae)

Syn. : *B. roxburghii* Planch.

S.–Ingudi

H. & B. : Hingan

Habit : Small tree

Habitat : Terrestrial

Phytography : Small spiny tree; leaves–coriacious, bi-foliolate, leaf margin–entire; flowers in small axillary cymes, green in colour; fruit–1-5 seeded samaras with large membranous wing.

Fl. & Fr. : April–September

Distribution : Dry deciduous forest in India.

Status : Threatened, wild.

Uses : Ripe fruits are used in whooping cough and skin troubles. Seed is an expectorant. Seed extract hypotensive.

- *Baliospermum montanum* Mull. Arg. (Euphorbiaceae)

S. & H.–Dauti.

Habit : Shrub

Habitat : Terrestrial

Phytography : Large stout shrub; leaves–alternate, toothed; flowers -monoecious, small and arranged in racemose, fruit–bi-valved, crustaceous cocci.

Fl. & Fr. : March–September

Distribution : Himalayas and plains of India.

Status : Not common, Wild.

Uses : Seed is a good purgative. It is used as substitute of Jamalgota *(Croton tiglium)*

- *Barleria prionites* L. (Acanthaceae)

  S.–Karunta, H.–Jhinti, B.–Kata Jati

  Habit : Shrub

  Habitat : Terrestrial

  Phytography : Spinescent bush; leaves–opposite entire; flowers -solitary, small, sessile and yellow; fruit–an oblong capsule; seed compressed.

  Fl. & Fr. : September–December

  Distribution : Found throughout the deciduous forests of India.

  Status : Common, wild.

  Uses : Leaf juice with honey is used in catarrh of children. Root paste is applied on glandular swellings for cure. Leaves chewed to get relief from toothache.

- *Barringtonia acutangula* (L.) Gaertn. (Barringtoniaceae)

  S.-Dhatripala, H.-Ingar, B.-Hijal

  Habit : Tree

  Habitat : Mangrove associated plant.

  Phytography : Small tree; leaves–alternate, crowded towards end; flowers–small, in spikes, of greenish white colour; fruit–a fibrous berry.

  Fl. & Fr. : March–July

  Distribution : In costal India

  Status : Common in saline area, wild.

  Uses : Bark, root and leaves are used for fish poisoning. Fruit contains saponin. Bark contains 16% tannin.

- *Barringtonia racemosa* Roxb. (Barringtoniaceae)

  Syn. : B. *racemosa* Blume

  S.-Nipa, H.-Ijjul, B.-Samudraphal.

  Habit : Tree

  Habitat : Mangrove

Phytography : Medium size tree; leaves–alternate; flowers in lateral racemes, small and deciduous; fruit–ovoid, berry crowned by the persistent calyx limb.

Fl. & Fr. : March–August

Distribution : Found in western coasts of India.

Status : Rare, wild.

Uses : Tender leaves are eaten. Tribals extract starch from seeds.

- *Bauhinia purpurea* L. (Caesalpiniaceae)

S.-Vanaraja, H.-Khairwal, B.-Raktakanchan

Habit : Small tree

Habitat : Terrestrial

Phytography : Medium size tree; leaves–simple, usually bi-lobed with mucronate sinus; flowers–showy, racemose, purplish-pink and pale-purple in colour; fruit–linear pod.

Fl. & Fr. : April–July

Distribution : Found mainly in sub-Himalayan tract

Status : Infrequent, wild.

Uses : Root is carminative. Bark is used for cordage. Flower buds are eaten. Leaves are used as fodder. Seeds yield lubricant oil.

- *Bauhinia vahlii W.* & A. (Caesalpiniaceae)

Read-*Phonera valii*

H.-Maljan, *B.*-Sihar.

Habit : Small tree

Habitat : Terrestrial

Phytography : Perennial creeper; leaves–simple lobed; flowers-racemes; fruit–pod, flat.

Fl. & Fr. : April-December

Distribution : Found throughout India.

Status : Population is declining, wild

Uses : Bark yields cordage fiber and has 17% tannin in it. Seeds are eaten raw. Leaves are used as plate, for preparing umbrellas.

- *Berberis aristata* DC. (Berberidaceae)

  S.-Daruharidra, *H.*-Rasant, *B.*-Darhaldi. *E.*-Indian barberry.

  Habit : Large shrub.

  Habitat : Terrestrial

  Phytography : Large shrub with yellow wood; leaves–simple, pinnate; flowers–solitary, small, yellow; fruit–berry.

  Fl. & Fr. : March–December

  Distribution : Himalayas and Nilgiri hills.

  Status : Endangered, wild.

  Uses : Dried stem is used as bitter tonic. Dried berries are edible. Root contains an alkaloid called berberine.

- *Betula utilis* D. Don (Betulaceae)

  Syn. : *B. bhojpattra* Wall.

  S. & B.–Bhurjapatra, H.-Bhujpatra, E.-Himalayan silver birch

  Habit : Tree

  Habitat : Terrestrial

  Phytography : Tree; leaves–simple; flowers–racemes.

  Fl. & Fr. : April–December

  Distribution : Found in temperate Himalayas, Sikkim, Kashmir and Bhutan.

  Status : Rare, wild.

  Uses : Ancient manuscripts were written on the bark of this tree. Infusion of the bark is aromatic. Bark contains betulin, which has some antifertility activity.

- *Bixa orellana* L. (Bixaceae)

  S.-Shendri

  H. & B. : Latkan, E.-Annatto tree.

  Habit : Small tree.

  Habitat : Terrestrial

  Phytography : Small evergreen tree; leaves–simple, digitately nerved; flowers–in terminal panicles, whitish; fruit–2-valved loculicidal capsule.

Fl. & Fr. : April-September

Distribution : Occurs in Malabar coasts, West Bengal, Bihar, Orissa and Assam.

Status : Rare, wild.

Uses : Seeds yield an edible dye, which is used for colouring butter, ghee, cheese, chocolate, hair oil etc.

- *Bombax ceiba* L. (Bombacaceae) **(Plate-8)**

S.-Salmali, H.-Semul, B.-Lal simul, E-Silk cotton tree

Habit : Tree

Habitat : Terrestrial

Phytography : Tree; leaves–palmately compound; flowers–big, red; fruit–capsule.

Fl. & Fr. : March–May

Distribution : Found throughout deciduous forest in India.

Status : Plantation is abundant. Natural and wild population is declining in a rapid way.

Uses : Wood is used in match industry, tea chest, ply wood, pencil making etc. Floss used for pillow. Tender leaves are used as fodder. Root is useful in ulceration of urinary bladder and kidney. Seed yields edible oil. The oil is also used for soap making.

- *Borassus flabellifer* L. (Arecaceae)

S. & B.-Tal, H.-Tar, E-Palmyar plam.

Habit : Branchless tree

Habitat : Terrestrial

Phytography : Tree, trunk stout, unarmed; leaves–terminal fan-shaped, plicately multifid; flowers-diocious, peduncle sheathed with open spathes; fruit–large subglobose drupe, seed–oblong a 3-lobed apex.

Fl. & Fr. : March–September

Distribution : Throughout India

Status : Common, wild

Uses : It is a toddy-yielding plant. In ancient times leaves were used as writing material. Leaves are used for thatching, mat, hats, bags, toys and flutes. Juice is a source of jaggery. Seeds and pulp of tender fruits are edible. Leaf stalks yield timber. Stem is used for house building. Leaf stock and fiber is used for hat making.

- *Boswellia serrata* Roxb. (Burseraceae)

S.–Kunduru

H. & B. : Salai, E–Indian olibanum tree.

Habit : Quick growing tree

Habitat : Terrestrial

Phytography : Tree; leaves–deciduous, alternate usually serrete leaflets; flowers–small, white arranged in axillary racemes or panicles; fruit -drupe.

Fl. & Fr. : March–November

Distribution : Found throughout the dry deciduous forest in India.

Status : Infrequent, wild.

Uses : Wood is a source of newsprint, plywood, and toys. Ammunition boxes are made from this wood is better than those from teak. It is a non-coniferous source of turpentine. Oleo-gum-resin is olibanum or Indian Frankincense used as incense. Rasin is used as a substitute of Canada balsam. Bark is used in diarrhoea and skin diseases. Gum is an expectorant, diuretic. Flowers and seeds are eaten.

- *Bruguiera gymnorhiza* Lam. (Rhizophoraceae) **(Plate-9)**

B.-Kankra.

Habit : Small tree

Habitat : Mangrove plant

Phytography : Large tree with terete branches; leaves–opposite, coriacious with entire margin; flowers–solitary or cymose in axillary peduncles; fruit–1-seeded, turbinate.

Fl. & Fr. : April-November

Distribution : Found in mangrove forests in India.

Status : Population is declining in a formidable rate, wild.

Uses : Bark yields 60% tannin. Alkaline extract of bark is used for softening water in boiler; wood is used for house building. Bark is used in diarrhoea. Peeled hypocotyls are eaten and also cooked.

- *Buchanania lanzan* Spreng. (Anacardiaceae) **(Plate-10)**

S.–Piyalaka, H.–Chironji, B.-Piyal, E.-Almondette tree.

Habit : Small tree

Habitat : Terrestrial

Phytography : Tree; leaves–simple, alternate, entire; flowers–small, pubescent, sessile; fruit–slightly fleshy drupe.

Fl. & Fr. : March–June

Distribution : Found throughout deciduous forest in India.

Status : Population is declining in a formidable way, wild.

Uses : Seeds are eaten; nutritious, seeds yield fatty oil substitute of olive oil. It is used on glandular swellings of the neck. Wood is useful in house building. Wood is termite resistant. Gum is used in diarrhoea and intercostal pain.

- *Butea monosperma* (Lam.) Kuntze (Fabaceae) **(Plate-11)**

Syn. : *B. frondosa* Koenig ex Roxb.

S.-Kingsuk, H.-Dhak Pales, B.-Pales, E.-Flame of forest.

Habit : Tree

Habitat : Terrestrial.

Phytography : Erect tree; leaves–pinnately 3-foliate, leaflets–large; flowers–large, showy, arranged densely in axillary or terminal racemes; fruits–widely ligulate follicies; seed–aborate, somewhat compressed.

Fl. & Fr. : February–May

Distribution : Found throughout India and is a keystone species in deciduous forest.

Status : Common, wild.

Uses : Leaves are used for making plates, cups, and bark yields fiber. Bark is astringent and is used for piles and menstrual disorders. Gum is known as 'Bengal-kinogum', which is astringent and used in diarrhoea. Flowers yield dye. Seed powder with lemon juice is used against herpes. Seed powder with common salt is used against intestinal worms.

- *Caesalpinia crista* L. (Caesalpiniaceae)

  Syn. : *C. bonducella* Flem

  S.-Putikaranja, H.-Kat-Karanja, B.-Nata, E.-Fever nut.

  Habit : Scrambler

  Habitat : Terrestrial

  Phytography : Prickly woody rambler, stems–not much thicken than a quill; leaves–alternate, pinnatisect; flowers–small, often scorpiod spikelets; fruits–globose; seed–smooth or pitted.

  Fl. & Fr. : March–September

  Distribution : Throughout Indian forests.

  Status : Population is declining, wild.

  Uses : Young tender shoots are used in medicine. Other species are *C. andamanicus* Kurz, *C. aesius* Blume, *C. extensus* Roxb, *C. flagellum* Criff., *C. latifolius* Roxb., *C. manan* Miq., *C. rotang* L., *C. travancoricus* Bedd. ex Hook.f., *C. viminalis* Willd. All these species are used in various purposes.

- *Calamus spp.* L. (Arecaceae) **(Plate-12)**

  B.–Bet, E.–Rattan Palms or Rattan Canes. About 40 species are found in India of which 38 are wild and 2 have been introduced. Important species are *C. rotang* and *C. erectus.*

  Habit : Shrub or shrubby climber

  Habitat : Terrestrial

  Photography : Mostly climber scrambling with the aid of hooked spines, twinning over trees; Leaves–pinnate and sheathing leaf base; Flowers -Small polygamo-dioecious; Fruit–Globose or ellipsoid.

  Fl. & Fr. : August–October

  Distribution : Throughout evergreen and semi-evergreen forests of Assam and other Northeastern States, West Bengal, Karnataka and Maharashtra.

Status : Moderate to rare.

Uses : The pith is used for making a variety of articles such as handbags and fancy decoration pieces. Buds of some pieces are edible; ripe, cooked fruits of some species are eaten. Slender shoots of C. *rotang* are used as vegetable. The fleshy mucilaginous, sweet and sour pulp surrounding the seed is eaten to quench thirst.

- *Cannabis sativa* L. (Cannabinaceae) **(Plate-13)**

S.-Bhanga

H. & B. : Bhang, E.-Hemp.

Habit : Shrub

Habitat : Terrestrial

Phytography : Tall annual shrub; leaves–opposite below and alternate above; flowers–small, males in axillary racemes; fruit–scrustaceous nut.

Fl. & Fr. : March–September

Distribution : Naturalised in sub-Himalayan tract.

Status : Abundant in wastelands, wild.

Uses : Source of hemp fiber and narcotic like *Bhang, Charas* and *Ganja.* Female flower heads contain 15-20% of resin, cannabinol and essential oil.

- *Carallia brachiata* (Lour.) Merrill. (Rhizophoraceae)

Syn. : *C. integerrima* DC. B.–Kierpa.

Habit : Small tree

Habitat : Mangrove

Phytography : Evergreen small tree with shining leaves; leaves-opposite, petioled, ovate or elliptic; flowers–small, sessile, pedicelled axillary; fruit–1-seeded, globose.

Fl. & Fr. : April–October

Distribution : Mangrove forest in India.

Status : Common, wild.

Uses : Timber used in house building. Fruits are edible. Leaves are used as tealeaf. Seed yields edible oil.

- *Careya arborea* Roxb. (Lecythidaceae)

  S.-Kumbhi

  H. & B. : Kumbi, E.-Kumbi

  Habit : Big tree.

  Habitat : Terrestrial

  Phytography : Large tree; leaves–alternate, crowded towards the ends of branches; flowers–large showy, in racemes; fruit–large, globose, fibrous berry, crowned by the persistent calyx.

  Fl. & Fr. : April-December

  Distribution : Found throughout deciduous forest areas of India.

  Status : Frequent, wild.

  Uses : Wood is used in house building purpose. Seeds are used in diabetes.

- *Carissa spinarum* L. (Apocynaceae)

  S.-Karadika, H.-Karaunda

  Habit : Shrub

  Habitat : Terrestrial

  Phytography : Small sub-erect shrub; leaves–small, glabrous, opposite; flowers–terminal and axillary cymes, small; fruit–globose berry.

  Fl. & Fr. : April-September

  Distribution : Throughout India.

  Status : Infrequent, wild.

  Uses : Fruits are edible. Leaves are rich in tannin 9-15%. Root is purgative. Wood is used for comb making.

- *Caryota urens* L. (Arecaceae)

  S.–Mada, H.–Mari, B.–Salpa, E.–Sagopalm.

  Habit : Unbranched palm

  Habitat : Terrestrial

Phytography : Tall unbranched palm; leaves–compound, few very large, twice pinnatesect; flowers–monoecious and spadix; fruit–globose.

Fl. & Fr. : January-May

Distribution : Found throughout deciduous forests of India.

Status : Infrequent, wild.

Uses : The plant yields sago, toddy, and fiber. Young unfolded leaves are edible. Nuts are acrid.

- *Casearia elliptica* Willd. (Samydaceae)

H.-Chilla, B.-Maslibisa.

Habit : Shrub

Habitat : Terrestrial

Phytography : Perennial shrub; leaves–simple, alternate, elliptic oblong to lanceolate; flowers–axillary, small and greenish yellow; fruit-succulent globose.

Fl. & Fr. : April–September

Distribution : Throughout India.

Status : Infrequent, wild.

Uses : Bark yields tannin 11%. Wood is used for comb making. Fruit is diuretic. Bark powder is applied in dropsy. Fruit is used as fish poison.

- *Cassia fistula* L. (Caesalpiniaceae)

S.–Suvarnaka, H.–Amaltas, B.–Sundali, E.–Indian Labur num.

Habit : Medium size tree

Habitat : Terrestrial

Phytography : Medium size tree; leaves–compound with evenly pinnate; flowers–usually showy, axillary racemes and yellow in colour; fruit -woody lomentum within a flattened pod.

Fl. & Fr. : March-December

Distribution : Throughout India.

Status : Common, both wild and planted.

Uses : Fruit pulp is a purgative, root decoction used in black stool.

- *Castanopsis indica A.* DC. (Fagaceae)

E.-Indian chest nut.

Habit : Tree

Habitat : Terrestrial

Phytography : Small evergreen tree; leaves–simple, alternate, entire; flowers–small monoecious, males in erect covered by the closed fruiting involucre.

Fl. & Fr. : April-December

Distribution : Throughout Himalayan region.

Status : Infrequent, wild.

Uses : Wood is used in construction work. Bark yields tannin 6-12%. Leaves are used for wrapping *beedis.*

- *Catharanthus roseus* (L.) G. Don. (Apocynaceae) **(Plate-14)**

Syn. : *Vinca rosea* L.

H.-Sadabahar, B.-Nayantara

Habit : Shrub

Habitat : Terrestrial

Phytography : Erect shrub; leaves–simple, opposite; flowers–solitary, axillary, pinkish in colour; fruit–2, slender, erect, follicles.

Fl. & Fr. : Almost throughout the year

Status : Common, both wild and cultivation.

Uses : Leaf is used in diabetes. Leaf infusion is used in menorrhagia. Root contains ajmalicine, serpentine and reserpine. The alkaloids obtain from the plant possess hypotensive, sedative and tranquilizing effects.

- *Cedrela toona* Roxb. (Meliaceae)

Syn. : *Toona ciliata* M Roem.

S.-Nandivriksha, H. & B.-Tun, Mahanimba, E.-Toon.

Habit : Tree

Habitat : Terrestrial

Phytography : Tall tree; leaves–compound, opposite, imperipinnate; flowers–white, terminal and subterminal in panicles; fruit–capsule.

Fl. & Fr. : April–September

Distribution : Throughout India

Status : Infrequent, wild.

Uses : Wood is used for furniture and plywood making. Bark is used in dysentery of infants.

- *Cedrus deodara* (Roxb.) Loud. (Pinaceae)

Syn. : *C. libani* Barrel var. *deodara* Hook.f.

S.–Devadaru. Other vernacular names derived from Sanskrit name.

Habit : Tree

Habitat : Terrestrial.

Phytography : Tree; leaves–needle like; flowers–in cones; fruit–dry.

Fl. & Fr. : September–May

Distribution : Found in north-west Himalayas.

Status : Infrequent, wild.

Uses : Wood is used for house building. Plant yields oleoresin and oil. Needle oil is carminative. Bark is used in diarrhoea and dysentery.

- *Ceiba pentandra* (L.) Gaertn. (Bombacaceae) **(Plate-15)**

Syn. : *Eriodendron anfractuosum* DC.

S.-Swet simul, H.-Safed simul, B.-Sada simul, E.-White silk cotton.

Habit : Tree

Habitat : Terrestrial.

Phytography : Trees; leaves–palmately compound; flowers–racemes; fruit–capsule.

Fl. & Fr. : April-September

Distribution : Throughout India

Status : Common, mostly planted.

Uses : Floss is the source of kapok fiber. Wood is used for toys. Seed oil after refining is often used as edible oil. Cake is used as feed for cattle. Gum is used in bowel complaints. Root paste is a diuretic.

- *Celastrus paniculatus* Willd. (Celastraceae)

S.-Jyotishmati, H. & B.-Malkagni

Habit : Creeper.

Habitat : Terrestrial.

Phytography : Scandent shrub; leaves–alternate, entire; flowers-terminal or axillary racemes or panicles; fruit–globose or ovoid.

Fl. & Fr. : May–December

Distribution : Found in the drier parts of India

Status : Rare, wild.

Uses : Bark paste abortifacient. Seed oil is aphrodisiac and nerve stimulant. Seed yields celastrine an alkaloid. Seed oil used as hair oil and also reduces exhaustion when applied on the head.

- *Celosia argentea* L. (Amaranthaceae)

S.-Vitunna, H.-Sufaid Murga, B.-Sweet Murga. Quail grass.

Habit : Herb

Habitat : Terrestrial

Phytography : Erect annual herb; leaves–simple, alternate; flowers-small, white arranged in axillary spikes; fruit–capsule.

Fl. & Fr. : April-December

Distribution : Throughout India.

Status : Common, wild.

Uses : Inflorescence is used in decoration. Flowers are astringent. Seeds demulcent. Young plants are eaten as vegetable.

- *Centella asiatica* (L.) Urban. (Apiaceae)

  S.-Mandukparni, H.-Braham-manduki, B.-Thalkuni, E.-Indian Pennywork.

  Habit : Prostrate herb.

  Habitat : Terrestrial.

  Phytography : Annual herb; leaves–simple; flowers–umbels; fruit-capsule.

  Fl. & Fr. : September–March

  Distribution : Throughout India.

  Status : Common, wild.

  Uses : Whole plant is used as diuretic and tonic. The plant yields glycosides. Plant paste is used in dysentery and diarrhoea.

- *Centratherum anthelminticum* Kuntze (Asteraceae)

  S.–Somaraji, H.–Somraj, B.-Somraj.

  Habit : Shrub

  Habitat : Terrestrial.

  Phytography : Tall, robust, leafy annual shrub; leaves–simple, alternate, toothed; flower–heads homogamous, involucre sub hemispheric; fruit–cypsela.

  Fl. & Fr. : June–December

  Distribution : Throughout India.

  Status : Infrequent, wild.

  Uses : Seed is anthelmintic.

- *Cerbera manghas* L. (Apocynaceae)

  B.-Dabur, E.-Dog-bane.

  Habit : Small tree

  Habitat : Plant is found in saline swamp.

  Phytography : Tree; leaves–simple; flowers–racemes; fruit–berry.

  Fl. & Fr. : April-September

  Distribution : Coastal zones of India.

Status : Infrequent, wild.

Uses : Fruit is narcotic and poisonous. Seed is used for killing wild animals. Fruits are used to stupefy fish. Fruits yield poisonous substances like cereberin and odollin.

- *Ceriops tagla* (Perr.) C.B. Robins. (Rhizophoraceae) **(Plate-16)**

B.-Goran

Habit : Small tree

Habitat : Mangrove plant

Phytography : Tree; leaves–simple; flowers–racemes; fruit–berry.

Fl. & Fr. : April-December

Distribution : Coast zones of India.

Status : Found in muddy shores and tidal creeks of India.

Uses : Bark yields about 43% tannin. It acts as haemostatic. Bark decoction is applied on ulcer. Tannin is also used as dye. Shoot decoction is used as that of quinine in Africa.

- *Ceropegia spp.* L. (Asclepiadaceae)

Almost all the species of the genus are threatened due to its medicinal value. The important species are C. *bulbosa* Roxb.; C. *pusilla* W. & A. ; C. *tuberosa* Roxb. etc.

Habit : Herb

Habitat : Terrestrial.

Phytography : Creeping herb; leaves–simple; flowers–raceme; fruit -follicle.

Fl. & Fr. : April–September

Distribution : Hills and plains of India.

Status : Rare, wild.

Uses : Root is used both in medicine and food. *C. bulbosa* root produces a bitter tonic and contain uropegine, *C. pusilla* has tuberous root which is edible. *C. tuberose* tubers are also edible and is considered as a tonic with digestive properties.

- *Chloroxyion swietenia* D.C. (Rutaceae)

  H.-Bhirra, E.-East Indian Satin wood.

  Habit : Tree

  Habitat : Terrestrial

  Phytography : Medium to large size tree; leaves–compound, evenly pinnate; leaflet–obtuse; flowers–small in axillary and terminal panicles; fruit–leathery, laculicidally capsule.

  Fl. & Fr. : April-September

  Distribution : Found in dry deciduous forest in India.

  Status : Infrequent, wild.

  Uses : Wood is durable and can resist insect and fungal attack. Plant yields an alkaloid chloroxylonine, which causes irritation on skin. Tree yields gum used in rheumatism.

- *Cinnamomum spp.* Blume (Lauraceae)

  Most important species are *C. burmanii* Blume, *C. camphora* Ness and Eberm., *C. assia* Blume, *C. cecidodaphne* Meissn, *C. glanduliferum* Meissn., *C. impressineviem* Meissn., *C. iness* Reinw. *C. macrocarpum* Hooke. F; *C. obtusifolium* Ness., *C. tamala* Ness & Eberm., *C. zeylanicum*, Breyn.

  Habit : Medium size tree; leaves–simple; flowers–panicles; fruit–berry.

  Habitat : Terrestrial.

  Phytography : Under tree; leaves–simple; flowers–racemes; fruit–berry.

  Fl. & Fr. : September–March

  Distribution : Found in hill forest and plains in India.

  Status : Both wild and planted.

  Uses : All the species of *Cinnamomum* are minor forest produces and are well known for their barks, leaves, fruits and extracted oils used for medicine, spices with aroma.

- *Cissampelos pareira* L. (Menispermaceae)

  S.-Ambashtha, H. & 8.-Akanadi, E.-Falesh pareira.

  Habit : Twiner

Habitat : Terrestrial

Phytography : Climbing shrub; leaves–often petiolate; female flowers -racemose, male flowers–cymose; fruit–drupaceous.

Fl. & Fr. : August–January

Distribution : Throughout tropical and sub-tropical India.

Status : Common, wild.

Uses : Root is diuretic and purgative. Root powder is used in urinary trouble dyspepsia and dropsy. Plant yields an alkaloid known as pelosine.

- *Cissus quadrangularis* L. (Vitaceae)

S.–Asthisanhara, H.-Hadjora, B.-Barjora, E.-Edible stemmed vine.

Habit : Climber

Habitat : Terrestrial.

Phytography : Square and succulent stem and climber; climb usually by leaf opposed tendrils and rarely by adventitious root, leaves–simple; flowers–usually small cymes; fruit–ovoid and globose.

Fl. & Fr. : September-February

Distribution : Throughout hotter parts of India.

Status : Infrequent, wild.

Uses : Stem pest is used on bone fracture. Plant juice is used in scurvy.

- *Cissus repanda* Vahl. (Vitaceae)

Syn. : *Vitis repanda* W. & A.

H.-Panibel

Habit : Climber

Habitat : Terrestrial

Phytography : Large climber; leaves–simple, wide cordate, often margin repend; flowers in small cymes, fruit–indehiscent berry like.

Fl. & Fr. : August–December

Distribution : Dry parts of India.

Status : Infrequent, wild.

Uses : Plants yield large quantities of potable water.

- *Citrullus colocynthis* Schrad. (Cucurbitaceae)

  S.-Mahandravaruni

  H. & B. : Indrayan makhal, E.-Colocynth, Bitter apple.

  Habit : Climber

  Habitat : Terrestrial

  Phytography : Climber herb; leaves–simple; flowers–racemes; fruit -pome.

  Fl. & Fr. : March-September

  Distribution : Found in South Indian forests.

  Status : Infrequent, wild.

  Uses : Fruit pulp is a hydragogue and cathartic. Fruit yields citrullin, which is a glycoside and also has an alkaloid citrullol. Root is used as purgative. It is also used in jaundice and rheumatism.

- *Citrus aurantifolia* (Christm.) Swingle (Rutaceae)

  S.-Kaghzi, H.-Kaghzi, B.-Kaghzi nebu, E.-Lime.

  Habit : Shrub

  Habitat : Terrestrial

  Phytography : Annual climber; leaves–simple; flowers–racemes; fruit -pome.

  Fl. & Fr. : April-October

  Distribution : Wild in warm valleys.

  Status : Rare in wild conditions.

  Uses : Fruits stomachic and is a source of vitamin-C.

- *Clausena excavata* Burm. f. (Rutaceae)

  B.-Agnijal.

  Habit : Shrub

  Habitat : Terrestrial

Phytography : Shrub; leaves–compound, imparipinnate; leaflet-membranous; flowers–small in terminal panicles; fruit–dry-fleshy, berry like.

Fl. & Fr : March-September

Distribution : Found in Chotonagpur, Orissa and Sub-Himalayan tract.

Status : Threatened, wild.

Uses : Plant is used as diuretic and is useful in diarrhoea. Root infusion is used in colic.

- *Cleistanthus collinus* (Roxb.) Benth ex Hook.f. (Euphorbiaceae)

H.-Garasi, B.-Parasi

Habit : Shrub

Habitat : Terrestrial

Phytography : Small tree and shrub; leaves–simple, alternate, entire; flowers–small, in axillary clusters; fruit–subglobose capsule.

Fl. & Fr. : September–February

Distribution : Found in dry deciduous forest in India.

Status : Common, wild.

Uses : Leaves are abortifacient. Root is a fish poison. Bark yields tannin.

- *Cocculus hirsutus* (L.) Diels (Menispermaceae)

Syn. : Vasavti tiksa, H.-Jamtikibel, B.-Huyer.

Habit : Climber

Habitat : Terrestrial

Phytography : Climber, very rarely suberect; leaves–alternate, lanceolate; flowers–panicle; fruit–compressed drupe.

Fl. & Fr. : April-December

Distribution : Found in forests of tropical and subtropical India.

Status : Infrequent, wild.

Uses : Leaf extract with water is used as refrigerant. Leaf is applied on eczema. Root is a laxative.

- *Cochlospermum religiosum (l.)* Alston (Cochlospermaceae) **(Plate-17)**

  S.-Swarna simul. H.-Galgal, B.-Sonali simul. E.-Silk cotton tree.

  Habit : Tree

  Habitat : Terrestrial

  Phytography : Small tree; leaves–solitary, very larger, yellow; fruit–3-5 valved capsules with membranous endocarp.

  Fl. & Fr. : April–August

  Distribution : Found throughout deciduous forests in India.

  Status : Infrequent, wild.

  Uses : Plant yields gum used in leather and ice-cream industry.

- *Colebrookea oppositifolia* Sum. (Lamiaceae)

  H.-Pansra.

  Habit : Shrub

  Habitat : Terrestrial.

  Phytography : Densely hoary shrub; leaves–opposite, elliptic-oblong in shape; flowers–small, arranged in panicled spikes; fruit of 4 obovoid, dry nutlet.

  Fl. & Fr. : May–December

  Distribution : Mostly in hilly forests in India.

  Status : Frequent, wild.

  Uses : Leaf paste is applied on wounds. Root is used in epilepsy. Recently it shows its efficacy in cancerous wounds.

- *Combretum decandrum* Roxb. (Combretaceae)

  Syn. : *C. roxburghii* Spreng.

  H. & B.-Atang.

  Habit : Shrub

  Habitat : Terrestrial

Phytography : Small bushy shrub with subscandent branches, leaves-opposite; flowers–small, white and panicled; fruit–dry drupe.

Fl. & Fr. : March–December

Distribution : West Bengal, Assam and Southern India.

Status : Wild population is declining in a rapid rate.

Uses : Stem is used in basket making. Leaf is used as febrifuge.

- *Commiphora wightii* Bhandari (Burseraceae)

Syn. : C. *mukal* (Hook. f. ex Stocks)

S.-Ilkktah, E.-Hill mango.

Habit : Tree

Habitat : Terrestrial

Phytography : Tree; leaves–compound, alternate; leaflet–opposite; flowers–receme; fruit–fleshy drupe.

Fl. & Fr. : April-November

Distribution : Karnataka, rare in other parts of South India.

Status : Rare in wild.

Uses : Root is astringent and used in muscular rheumatism and as a uterine stimulant.

- *Cordia dichotama* Forst. f. (Boraginaceae)

Syn. : *C. obliqua* Willd., *C. myxa* Roxb.

S.-Bahuvaraka, H.-Lasora, B.-Bahodari. E.–Sebesten Palm.

Habit : Medium size tree

Habitat : Terrestrial

Phytography : Tree of 40 ft. (approx) high; leaves–simple, alternate; flowers–arranged in corymbs; fruit–ellipsoid drupe.

Fl. & Fr. : April–December

Distribution : Throughout India.

Status : Infrequent, wild.

Uses : Fruit is edible, astringent, anthelmintic. It is used in diseases of chest and urinary passages. Wood is durable even in water.

- *Coptis tecta* Wall. (Ranunculaceae)

H.-Mamira, B.-Tata.

Habit : Herb.

Habitat : Terrestrial.

Phytography : Herb; leaves–pinnately compound; flowers–racemes; fruit–capsule.

Fl. & Fr. : June–December

Distribution : Endemic to Arunachal Pradesh

Status : Critically threatened, wild.

Uses : Rhizome is a tonic for debility. Root contains berberine.

- *Corypha umbraculifera* L. (Araceae)

S.–Karalika, B.–Bajar-battuler, E.-Fan palm.

Habit : Branchless tree.

Habitat : Terrestrial

Phytography : Tall straight tree, leaves–simple; huge deeply partite; petiole usually spinous; flowers–small, terminal arranged in pyramidally paniculate spadix; fruit–heshy drupe.

Fl. & Fr. : May–October

Distribution : Wild in Malabar coast, Andaman and Nicobar islands.

Status : Wild

Uses : Leaves are used for thatching and making an indigenous umbrella. Sacred Buddhist books were written on the leaves of the plant. Pith yields Sago.

- *Costus speciosus* (Koenig) Sm. (Zingiberaceae) **(Plate-18)**

S.-Kerunka, H. & B.-Keu

Habit : Shrub

Habitat : Terrestrial

Phytography : Shrub with leafy stem; leaves–simple, alternate, with broad sheath; flowers in a dome ovoid terminal head and white in colour; fruit–ovoid capsule.

Fl. & Fr. : August–December.

Distribution : Throughout Indian forest.

Status : Common.

Uses : Rhizome is cooked and eaten. Root is used as a tonic and anthelmintic. Rhizome is a good source of steroid.

- *Crateava nurvala* Buch.-Ham. (Capparidaceae)

  Syn. : *C. religiosa* Hook. f. & Thoms. var. Frost. f.

  S.-Varuna, H.-Baruna, B.-Barun.

  Habit : Small tree

  Habitat : Terrestrial.

  Phytography : Stunted tree; leaves–digitately 3-foliate; flowers–large and creamy white to pale rose in colour; fruit–hard indehiscent.

  Fl. & Fr. : April–August

  Distribution : Throughout India.

  Status : Infrequent.

  Uses : Bark stimulates liver. Bark extract is used to promote appetite.

- *Croton tiglium* L. (Euphorbiaceae)

  S.-Jaypala, H.-Jamaigota, B.-Jayapala, E.-Purging croton.

  Habit : Small tree

  Habitat : Terrestrial.

  Phytography : Small evergreen tree; leaves–3-nerved, ovate to cordate in shape and hairy; flowers–small, monoecious and arranged in racemes; fruit–capsule.

  Fl. & Fr. : March–September

  Distribution : East India.

  Status : Infrequent both wild and cultivated.

Uses : Seed oil is purgative. Fruit is used to stupefy fish.

- *Croton roxburghii* Balajk. (Euphorbiaceae) **(Plate-19)**

  Syn. : *C. oblongifolius* Roxb.

  S.-Bhutankusa, H.-Chucka, B.-Baragach.

  Habit : Big shrub

  Habitat : Terrestrial

  Phytography : Small deciduous tree; leaves – simple, alternate, margin toothed elliptic oblong or elliptic lanceolate in shape; flowers-monoecious, small; fruit–globose capsule.

  Fl. & Fr. : April–September

  Distribution : Sub-Himalayan tracts and Eastern India.

  Status : Common, wild.

  Uses : Seed oil purgative. It is also used as an insecticide.

- *Cryptomeria japonica* (L. f.) D. Don. (Pinaceae)

  H. & B.-Dhupi, E.-Japanese Cedar.

  Habit : Tree

  Habitat : Terrestrial

  Phytography : Tree; leaves–needle like; flowers–like other gymnosperms; fruit–berry.

  Fl. & Fr. : July–January

  Distribution : Hilly forests of Eastern India.

  Status : Naturalized, wild.

  Uses : Wood is used in house building. Leaves are used as incense.

- *Curculigo orchioides* Gaertn. (Amaryllidaceae) **(Plate-20)**

  S.-Mushali, H.-Kali musli, B.-Talmuli

  Habit : Herb

  Habitat : Terrestrial

  Phytography : Slender herb with rhizomatous stem; leaves–large flat and lanceolate; flowers–unisexual and arranged in distichous spikes, scapes hidden by leaf sheaths.

Fl. & Fr. : March–October

Distribution : Throughout India.

Status : Infrequent, wild.

Uses : Root is used in skin diseases. It is considered as demulcent, diuretic. Root paste is applied on piles. Root decoction is used in piles, diarrhoea, jaundice and asthma.

- *Curcuma angustifolia* Roxb. (Zingiberaceae)

S.-Tavakshira, H.-Tikhur, B.-Tikhur, E.-East India Arrowroot.

Habit : Herb

Habitat : Terrestrial

Phytography : Small herb with rhizome; leaves–oblong, narrow; flowers in dense compound spikes; fruit–tardily dehiscent globose capsules.

Fl. & Fr. : August–November

Distribution : Central Himalayas, Bihar, and North Bengal

Status : Infrequent, wild.

Uses : Tubers yield arrowroot used for aged people and children.

- *Curuma aromatica* Salisb. (Zingiberaceae) **(Plate-21)**

S.–Vana-haridra, H.-Jangli haldi, B.-Ban haldi, E.-Yellow zedoary.

Habit : Herb.

Habitat : Terrestrial

Phytography : Small herb with yellowish tuber; leaves–large and broad; flowers–yellowish and arranged in long spikes.

Fl. & Fr. : September-December.

Distribution : Throughout India.

Status : Frequent, wild.

Uses : Rhizome is used in medicine. It is used as a substitute of turmeric.

- *Curcuma caesia* Roxb. (Zingiberaceae)

H.-Black zedoary.

Habit : Herb

Habitat : Terrestrial

Phytography : Large herb with bluish-gray tubers; leaves–large and broad; flowers–arranged in spikes with grayish in colour; fruit–nutlet.

Fl. & Fr. : September–January

Distribution : Endemic to West Bengal.

Status : Infrequent, wild.

Uses : Rhizome is used in sprains and as cosmetics. Rhizome yields oil, which is a good source of camphor.

- *Curcuma zedoaria* Rosc. (Zingiberaceae)

  S.-Sati, H. & B.-Kachura, E.-zedoary.

  Habit : Herb.

  Habitat : Terrestrial.

  Phytography : Large herb with yellowish tuber; leaves–large and broad; flowers–arranged in spikes; fruit–nutlet.

  Fl. & Fr. : September-February

  Distribution : Deciduous forests of India.

  Status : Infrequent, wild.

  Uses : Rhizome is stimulant, carminative; it is a source of starch used as a substitute of arrowroot called zedoary and is also used as a condiment.

- *Cyathea brunoniana* C. B. Clarke & Baker (Cyathaceae) Pakpa.

  Habit : Tree fern

  Habitat : Terrestrial.

  Phytography : Perennial; leaves–pinnately compound; flowers–in the form of sporophylls.

  Fl. & Fr. : June-September

  Distribution : Himalayan forests in India.

  Status : Endangered.

  Uses : Pith used as starch.

- *Cycas bedomi* (Cycadaceae)

  Habit : Shrub

  Habitat : Terrestrial

  Phytography : Perennial cycas; leaves–pinnately compound; flowers–in the form of sporophylls.

  Fl. & Fr. : June–December

  Distribution : Orissa and other parts of South India.

  Status : Critically endangered

  Uses : Fruit is used for starch.

- *Cycas circinalis* L. (Cycadaceae)

  S.-Varaguna, H.-Janglimandan-mast-ka-phul.

  Habit : Tree like.

  Habitat : Terrestrial

  Phytography : Small palm like tree; leaves in terminal crown, clothed with woody bases of petioles.

  Fl. & Fr. : April-November

  Distribution : Throughout South Indian forest.

  Status : Infrequent.

  Uses : Pith is a source of sago. Seeds yield starch. Tender leaf juice is used in flatulence.

- *Cycas rumphii* Miq. (Cycadaceae)

  H.-Kama.

  Habit : Tree like

  Habitat : Terrestrial

  Phytography : Small palm like tree; leaves in terminal crown, clothed with woody bases of petioles; flowers–sporophylls.

  Fl. & Fr. : April-December

  Distribution : South India, West Bengal and Andaman Nicobar islands.

  Status : Rare.

  Uses : Pith yields sago.

- *Daemonorops kurzianus* Hook.f. (Arecaceae)

  Syn. : Aprang, E.–East Indian Dragonis Blood.

  Habit : Branchless palm

  Habitat : Terrestrial

  Phytography : Small tree like palm; leaves–palmately compound; flowers in spadix; fruit–berry.

  Fl. & Fr. : May–December

  Distribution : Eastern Himalayas and Andaman and Nicobar islands.

  Status : Rare

  Uses : Red resin is used as stimulant.

- *Dalbergia spinosa* Roxb. (Fabaceae)

  Chillanki

  Habit : Shrub

  Habitat : Terrestrial

  Phytography : Erect spiny shrub; leaflets–alternate, stipule small; flowers–small numerous and arranged in terminal cymes; fruit -compressed lomentum.

  Fl. & Fr. : May–December

  Distribution : Forests of seashore of West Bengal and Peninsulas.

  Status : Infrequent.

  Uses : Root powder with water is used to allay the effect of alcohol.

- *Dapne papyraca* Wall. ex Stend. (Thymelaeaceae)

  H.-Satpura, Gande

  Habit : Shrub

  Habitat : Terrestrial.

  Phytography : Shurb; leaves–simple; flowers–racemes; fruit–drupe.

  Fl. & Fr. : April–November

Distribution : Eastern Himalayas.

Status : Common.

Uses : Plant is used for producing paper pulp.

- *Delphinium denudatum* Wall. (Ranunculaceae)

S.-Aparisha, H.-Jadwar

Habit : Shrub

Habitat : Terrestrial

Phytography : Perennial shrub; leaves–simple; flowers–racemes; fruit -berry.

Fl. & Fr. : May-January

Distribution : Western Temperate Himalayas.

Status : Frequent

Uses : Root is a used as a tonic. It is used as adulterant of aconite.

- *Desmodium gangeticum* DC. (Fabaceae)

S.-Shalaparni, H.-Sarivan, B.-Salpani

Habit : Shrub

Habitat : Terrestrial

Phytography : Sub-erect under shrub with membranous leaflets; leaflets-alternate; flowers–small and in terminal panicle; fruit–jointed pod.

Fl. & Fr. : March–October

Distribution : Throughout India.

Status : Common, wild.

Uses : Root is used as febrifuge, expectorant and diuretic.

- *Dillenia aurea* Sm. (Dilleniaceae) Dheugr.

Habit : Tree

Habitat : Terrestrial

Phytography : Spreading tree with large simple nerved leave; flowers-large, solitary and off white in color; fruit–globose with persistent calyx.

Fl. & Fr. : May–December

Distribution : Deciduous forest of India.

Status : Infrequent.

Uses : Fruits are eaten. Bark is used in gum infection.

- *Dioscorea spp.* L. (Dioscoreaceae) **(Plate-22)**

All species of the genus *Dioscorea* are available in wild conditions starting from high hill forests to plains. Some of them like *D. alata* L. are cultivated. The species like *D. deltoidea* Wall., *D. floribunda* Mart & Gal., *D. prazeri* Prain & Burkill are the sources of diosgenin.

Habit : Climber

Habitat : Terrestrial

Phytography : Climbing herb; leaves–simple; flowers in racemes; fruit-dry berry.

Fl. & Fr. : July–January

Distribution : Throughout India.

Status : Species like *D. alata, D. deltoidea, D. floribunda* and *D. prazeri* are under ultivation.

Uses : Used both as edible vegetable and medicine.

- *Diospyros spp.* L. (Ebenaceae)

All the species of the genus *Diospyros* are available in wild condition except *D. kaki* L. f. and *D. discolor* Willd. Found in hilly forest or in the forest of plain lands including islands. The species like *D. ebenum* Koeing., *D. tomentosa* Roxb., *D. montana* Roxb., *D. marmorata* Parker are timber yielding plants. The taxa like *D. buxifolia* (Blume) Hiern., *D. chloroxylon* Roxb., *D. ferrea* (Willd.) Bakh., *D. insignis* Th. W., *D. lotus* L., O. *marmorata* Parker., *D. melanoxylon* Roxb., *D. peregrina* (Gaertn.) Curke, *D. racemosa* Roxb. etc. do not produce good timber but produce *'bidi pata'*.

Habit : Mostly tree.

Habitat : Terrestrial.

Phytography : Tree; leaves–simple; flowers–raceme; fruit–drupe.

Fl. & Fr. : April–February

Distribution : Hills, islands, deciduous forests.

Status : Common, wild.

Uses : Fruit is used as vegetable; leaf is used to wrap *bidi* and is traded almost throughout India as an important NTFP.

- *Diptetacanthus prostratus* Ness. (Acanthaceae)

  B.-Chauli.

  Habit : Herb

  Habitat : Terrestrial.

  Phytography : Herb; leaves–simple; flowers–racemes; fruit–capsule.

  Fl. & Fr. : May–December

  Distribution : Throughout India.

  Status : Common.

  Uses : Tuberous roots are used for making tablets used for fermentation and preparation of country liquor.

- *Dregea voloubiles* (L. f.) Benth. ex Hook. f. (Asclepiadaceae)

  B.–Tita kumra

  Habit : Climber

  Habitat : Terrestrial

  Phytography : Stout tall climber; with green branches; leaves–simple, opposite; flowers–in umbellate cymes; fruit–follicle.

  Fl. & Fr. : March–May

  Distribution : Throughout India.

  Status : Infrequent.

  Uses : Flowers are eaten as a vegetable. Root has emetic and purgative property.

- *Edgeworthia tomentosa* Nakai (Thymelaeaceae)

  Syn. : *E. gardeneri* Meissn.

  Habit : Shrub

Habitat : Terrestrial

Phytography : Shrub; leaves–simple; flowers–raceme; fruit–berry.

Fl. & Fr. : May–December

Distribution : Assam, Eastern Himalayas, Sikkim.

Status : Infrequent.

Uses : Rootstock is used as a remedy of buboes. Stem is used for paper pulp.

- *Ehretia laevis* Roxb. (Boraginaceae)

H.–Chamror; B.–Tamboli.

Habit : Small tree

Habitat : Terrestrial

Phytography : Medium size tree; leaves–simple, alternate; flowers-small, white and arranged in axillary corymbs; fruit–subglobose drupe.

Fl. & Fr. : March–September

Distribution : Deciduous forest of India.

Status : Infrequent.

Uses : Bark is used as a masticatory.

- *Elaeocarpus sphaericus* (Gaertn.) K. Schum. (Elaeocarpaceae)

Syn. : *E. ganitrus* Roxb.

S.-Rudraki, B.-Rudrakhya, E.-Utrasum bead tree.

Habit : Tree.

Habitat : Terrestrial.

Phytography : Tree; leaves–simple; flowers–arranged in axillary racemes and bisexual, ovary 5-celled; fruit–drupe.

Fl. & Fr. : April–December.

Distribution : Duars and Eastern Himalayas.

Status : Rare.

Uses : Fruit is used in epilepsy. Stem is used as beads for rosaries.

- *Elephantopus scaber* L. (Asteraceae) **(Plate-23)**

  S.-Gojihcia, H.-Gobhi, B.-Gojialata.

  Habit : Herb

  Habitat : Terrestrial

  Phytography : Rigid, dichotomous branched scabrid herb; leaves-alternate, toothed; flowers–compressed within a involucre of bract; fruit -cypsela.

  Fl. & Fr. : April-December.

  Distribution : Throughout India.

  Status : Common.

  Uses : Root decoction is used in dysuria, diarrhoea. Root paste is used to arrest vomiting. Leaf paste is also applied on eczema and ulcer for relief.

- *Embelia ribes* Burm.f. (Myrsinacenaceae)

  S.-Vidanga, H.-Baberang, B.-Biranga.

  Habit : Creeper.

  Habitat : Terrestrial.

  Phytography : Scandent shrub; leaves–entire; flowers–terminal as well as axillary and gray in colour; fruit–small globose berry.

  Fl. & Fr. : April–November.

  Distribution : Throughout India.

  Status : Infrequent.

  Uses : Fruit is used as stomachic tonic and anthelmintic. Active principle is embelin.

- *Emblica officinalis* Gaertn. (Euphorbiaceae) **(Plate-24)**

  Syn. : *Phyllanthus emblica* L.

  S.-Adiphala, H.-Amla, B.-Amlaki, E.-Embelic myrobalan.

  Habit : Tree.

  Habitat : Terrestrial

  Phytography : Deciduous tree with flaky bark; leaves–very small, closely pinnately set; flowers–small yellowish and axillary; fruit–large fleshy drupe.

Fl. & Fr. : April–November.

Distribution : Throughout India.

Status : Both planted and wild.

Uses : Fruits astringent, cooling laxative. A rich source of vitamin C. Fruit is used in hair dye. Dried fruit is used for shampooing hair.

- *Entada phaselides* Merril (Mimosaceae)

Syn. : *E. scandens* Benth.

H. & B. : Gilaq, E.–Elephant creeper.

Habit : Large creeper.

Habitat : Terrestrial

Phytography : Immense woody climber with bi-pinnate leaves ending in a point tendril; flowers in slender spikes with cream colour; fruit–in a pod.

Fl. & Fr. : May–February.

Distribution : Himalayas, Bihar, Orissa, Assam, West Bengal.

Status : Occasional

Uses : Seed anthelmintic and emetic. It contains two saponins. Bark is used for cordage. Seed is used as soap substitute.

- *Erycibe paniculata* Roxb. (Convolvulaceae)

B.-Karibili

Habit : Creeper

Habitat : Terrestrial

Phytography : Small tree with spreading branches; leaves–alternate, entire and elliptic; flowers–small, yellowish and axillary; fruit–fleshy berry.

Fl. & Fr. : March–December.

Distribution : Found throughout deciduous forest of India.

Status : Common.

Uses : Berries eaten, bark used in cholera.

- *Erythrina variegata* L. var. *orientalis* (L.) Merrill. **(Plate-25)**

  Syn. : *E. indica* Lam.

  S.–Mandar. H.–Dadap. B.–Palita mamdar. E.–Indian coral tree.

  Habit : Tree

  Habitat : Terrestrial

  Phytography : Tree with blackish prickles; leaves–pinnately tri-foliate, petioles prickly; flowers–red and arranged axillary racemes; fruit–linear pod.

  Fl. & Fr. : March–November.

  Distribution : Throughout India.

  Status : Common. White flowered plants are in frequent.

  Uses : Leaves are used as cattle fodder. Leaf paste is diuretic, anthelmintic and galactogogue. Bark contains resins and alkaloids.

- *Euphoria langen* Steud. (Sapindaceae)

  Syn. : *Nephelium longana*

  B.-Ashphal, E.-Longan.

  Habit : Tree

  Habitat : Terrestrial

  Phytography : Tree; leaves alternate : flowers–regular, and in axillary panicles; fruit–indehiscent and globose.

  Fl. & Fr. : February–July

  Distribution : West Bengal, Western Ghats.

  Status : Both planted and wild.

  Uses : Aril of the fruit is eaten. Wood is used for house building. Seed contains saponin.

- *Excoecaria agallocha* L. (Euphorbiaceae)

  B.-Gengwa, E.-Agallocha.

  Habit : Tree

  Habitat : Mangrove.

Phytography : Small evergreen tree, with milky juice; leaves–alternate : flowers–dioecious and in terminal, fruit–capsules.

Fl. & Fr. : March–September

Distribution : Mangrove forests of India.

Status : Mangrove restricted, wild.

Uses : Plant is used for house building. Bark Paste contains tannin. Latex is used in leprosy and paralysis. It is abortifacient. Leaf decoction is used in epilepsy.

- *Ficus cunia* Buch-Ham. (Moraceae)

S.-Kharapatra, H.-Jahrphali, B.-Dumbur.

Habit : Tree

Habitat : Terrestrial

Phytography : Medium size tree with milky juice; leaves–simple, opposite; flowers–within a globular structure called hypanthodium; fruit-hollow cup shaped closed receptacle.

Fl. & Fr. : April–November.

Distribution : Sub-Himalayan tract, Bihar, Satpura Hills, West Bengal.

Status : Infrequent.

Uses : Fruits edible. Fruit juice is used in bladder ailments. Bark yields-fiber.

- *Ficus glomerata* Roxb. (Moraceae)

Syn. : *E. racmosa* L.

S.-Udambara, H.-Gular, B.-Jagya dumur.

Habit : Tree

Habitat : Terrestrial

Phytography : Large tree; leaves–simple alternate with reddish mid vein and glabrous; flowers–within a subglobose structure called hypanthodium; fruit–hollow cup shaped sub globose receptacles.

Fl. & Fr. : April–December

Distribution : Throughout India.

Status : Common

Uses : Leaves are used as fodder. Bark is given to cattle in rinder-pest diseases. Root is used in diarrhoea and diabetes. Latex is applied on piles. Bark yields 14% tannin.

- *Ficus hispida* L. f. (Moraceae)

S.-Kakadumbura. H.-Daduri. B.-Dumoor.

Habit : Tree

Habitat : Terrestrial

Phytography : Tree; leaves–simple rough; flowers–within sub-globose structure; fruit–receptacles.

Fl. & Fr. : April–December.

Distribution : More or less throughout India.

Status : Common, wild.

Uses : Fruits edible. Ripe fruits are lactagogue and emetic. Barks contain tannin.

- *Ficus lucescens* Blume. (Moraceae)

Syn. : *F. infectoria* Roxb., *F. facor* Buch-Ham. S.-Plaksha, H.-Kahimal, B.-Pakur.

Habit : Tree

Habitat : Terrestrial

Phytography : Tree; leaves–simple; flowers–receptacles; fruit-hypanthodium.

Fl. & Fr. : May–November.

Distribution : Plains and lower hill forests of India.

Status : Common, wild.

Uses : Leaves are used as fodder; fruits are edible. Bark decoction is used as gargle and vulnerary. In Chinese medicine bark is used as a diaphoretic.

- *Ficus religiosa* L. (Moraceae)

S.-Ashvatha, H.-Pipal, B.-Ashathwa, E.-Peepal.

Habit : Tree

Habitat : Terrestrial

Phytography : Tree; leaves–simple; flowers–receptacles; fruit-hypanthodium.

Fl. & Fr. : May–December.

Distribution : Sub-Himalayan forests, Bengal and Madhya Pradesh.

Status : Both planted and wild.

Uses : Fruits edible. Leaves are used as fodder. Latex is used to fill up cavities in hollow ornaments. It is the source of birdlime. Bark contains tannin. Bark infusion used in ulcer and applied on skin troubles.

- *Flacourtia ramontchi* L'Herit (Flacourtiaceae)

  S.-Talis, H.-Bilangra, B.-Benchi. E.-Governor's plum.

  Habit : Tree

  Habitat : Terrestrial

  Phytography : A ramblin shrub; leaves–simple, toothed; flowers–small, whitish in colour; fruit–indehiscent with endocarp and cell one seeded.

  Fl. & Fr. : May–January.

  Distribution : More or less throughout India and usually in scrub forest.

  Status : Infrequent, wild.

  Uses : Fruits are edible, appetiser and digestive. It is good in jaundice. Bark contains tannin and acts as astringent.

- *Flacourtia sepiasia* Roxb. (Flacourtiaceae)

  H.-Kondai, B.-Baichi.

  Habit : Shrub

  Habitat : Terrestrial

  Phytography : Spiny shrub; leaves–simple, entire; flowers–medium size, sometime thorn bears leaves or flower; fruit–berry.

  Fl. & Fr. : May–January.

  Distribution : Dry-deciduous forests of India.

  Status : Infrequent, wild.

Uses : Leaves are used as fodder. Bark is used in gout and rheumatism.

- *Garcinia cowa* Roxb. (Clusiaceae)

Syn. : G. *kydia* Roxb.

H.-Cowa, B.-Kau.

Habit : Tree

Habitat : Terrestrial

Phytography : Tall erect tree with yellow juice; leaves–broad, unisexual and arranged in axillary or terminal fascicles; fruit–berry like.

Fl. & Fr. : April–December.

Distribution : Assam, Andaman Islands and West Bengal.

Status : In frequent, both wild and planted.

Uses : Fruits edible. Sun dried fruit slices are used in dysentery. Leaves are cooked and eaten. Bark yields a dye. Plant yields a gambose-like gum resin used in varnish for metallic surface.

- *Garcinia indica* Choisy (Clusiaceae)

H.-Kakom, E.-Mangosteen oil tree.

Habit : Tree

Habitat : Terrestrial

Phytography : Tree; leaves–simple; flowers–racemes; fruit–berry like.

Fl. & Fr. : April-November.

Distribution : Western Ghats and peninsula India.

Status : Infrequent and found both wild and planted.

Uses : Fruit are used in jellies and syrups. Seed oil is anthelmintic. It is the source of kokam butter.

- *Gardenia campanulata* Roxb. (Rubiaceae)

H.–Bitmara.

Habit : Shrub

Habitat : Terrestrial

Phytography : Medium size shrub with membranous an elliptic leaf; flowers–small and white in colour; fruit–subglobose berry.

Fl. & Fr. : January–July.

Distribution : Assam and Sikkim Himalayas. Wild.

Status : Rare

Uses : Fruits and leaves are eaten after cooling. Fruits are cathartic and anthelmintic. It is also used to remove stains from silk.

- *Gardenia gummifera* L. f. (Rubiaceae)

S.-Pindava, H. & B.-Dikamali, Bhuru.

Habit : Shrub

Habitat : Terrestrial

Phytography : Woody bushy shrub with resinous bud; leaves–simple, alternate and shining; flowers–medium, white and axillary or terminal; fruit–ellipsoid berry.

Fl. & Fr. : April–September.

Distribution : Dry deciduous forests of India.

Status : Common, wild.

Uses : The resinous exude or gum called 'cumbi-gum' is used in nervous disorders of children. It also acts as insect repellent. Wood is used for making combs.

- *Gardenia latifolia* Act. (Rubiaceae)

H.-Papra, E.-Box wood-gardenia.

Habit : Tree

Habitat : Terrestrial

Phytography : Small tree; leaves–dull green, broadly elliptic almost sessile; flowers–large terminal or axillary; fruit–subglobose berry.

Fl. & Fr. : April-December.

Distribution : Found throughout deciduous forests of India.

Status : Infrequent, wild.

Uses : Wood used as boxwood. It is used for making combs, tobacco pouches.

- *Gentiana kurroo* Royle (Gentianaceae)

  S.-Kutki

  H. & B. : Karu.

  Habit : Herb

  Habitat : Terrestrial

  Phytography : Herb; leaves–simple; flowers–in racemes; fruit–berry.

  Fl. & Fr. : May–October.

  Distribution : Kashmir and western Himalayas, wild.

  Status : Threatened. Wild.

  Uses : Rhizome and root contain Gentian, which is used as tonic and febrifuge. It is used in urinary troubles. It is a substitute of true Gentian G. *lutea* L.

- *Garuga pinnata* Roxb. (Burseraceae)

  H.-Ghogar, B.-Jum.

  Habit : Tree

  Habitat : Terrestrial

  Phytography : Medium size tree with pubescent young branches; leaves-compound, alternate crowded ear end of branches; flowers–polygamous in much branched panicles; fruit–a globose fleshy drupe.

  Fl. & Fr. April–December.

  Distribution : Found throughout mixed deciduous forests of India.

  Status : Occasional, wild.

  Uses : Wood is used for house building. Gum possesses cooling and digestive property.

- *Gloriosa superba* L. (Liliaceae) **(Plate-26)**

  S.-Langli, H.-Kasihasi, B.-Bishalonguli, E.-Malabar Glory lily.

  Habit : Climber.

Habitat : Terrestrial.

Phytography : Climber with tendrilar leaf apex; leaves–alternate and lanceolate in shape; flowers–large red-yellow in colour, axillary or terminal; fruit–a large septicidal capsule.

Fl. & Fr. : April–November.

Distribution : Found throughout tropical forests of India.

Status : Rare, wild.

Uses : Tuber is poisonous due to presence of alkaloid colchicine. It is used to promote labour pain and against neuralgic pains. Root paste is applied on gout and rheumatism.

- *Grewia sapida* Roxb. (Tiliaceae)

B.-Phalsatenga.

Habit : Shrub

Habitat : Terrestrial

Phytography : Small shrub with woody stock; leaves–rounded or obovate, apex obtuse and stipule present; flowers in axillary or terminal cymes, flower bud ovoid; fruit–sub-globose drupe.

Fl. & Fr. : April–December.

Distribution : Deciduous forests of India.

Status : Wild

Uses : Leaves are used as fodder and fruits are edible.

- *Gymema sylvestre* R. Br. (Asclepiadaceae) **(Sketch-3)**

S.-Meshashringi, H.-Gur-mar, B.-Mera-singi.

Habit : Twiner

Habitat : Terrestrial

Phytography : Twining shrubs; leaves–simple, opposite; flowers–small, white and in umbellate cymes; fruit–2, slender, acuminate, follicles like and with 2 horns.

Fl. & Fr. : April–December

Distribution : Mixed deciduous forests of India.

Status : Wild.

Uses : Leaves when chewed paralyze the sense of taste of sweet. It is used in diabetes. Leaf powder is a cardiac stimulant. Leaves contain gymnemic acid.

- *Gynocordia odorata* R. Br. (Flacourtiaceae)

S.-Kushthapa

H. & B. : Chaulmogra

Habit : Tree

Habitat : Terrestrial

Phytography : Tree; leaves–simple; flowers–racemes; fruit–berry.

Fl. & Fr. : April–February.

Distribution : Sikkim, Assam, Khasia hill forest.

Status : Vulnerable, wild.

Uses : Fruit pulp is a fish poison. Seeds possess insecticidal properties. Seed oil is used for skin ailments.

- *Hedychium spicatum* Buch.-Ham. (Zingiberaceae)

S.-Karpurakachali

H. & B. : Gandhashati, E.-Spiked ginger lily.

Habit : Shrub.

Habitat : Terrestrial.

Phytography : Shrub; leaves–simple; flowers in spike; fruit–nutlet.

Fl. & Fr. : June–September.

Distribution : Sub-tropical forests of India.

Status : Infrequent, both wild and planted.

Uses : Aromatic rhizome powder is used for preparation of coloured *holi* powder. It is stomachic and carminative. Oil is used in dyspepsia and for making soap and hair oil.

- *Helicteres isora* L. (Starculiaceae)

S.-Avartani, H.-Mavorphalli, B.-Almora/Petmochra, E.-East India screw tree.

Habit : Shrub

Habitat : Terrestrial

Phytography : Bushy shrub; leaves–simple, alternate; flowers–axillary or solitary and red in colour; fruit–spirally twisted follicle.

Fl. & Fr. : April–December.

Distribution : Mixed deciduous forests of India.

Status : In frequent, wild.

Uses : Bark yields a fiber used for cordage. Fruits used in intestinal complains-like colic and dysentery.

- *Hemidesmus indicus* R. Br. (Asclepiadaceae) **(Sketch-4)**

S.-Anantamul, H.-Anantamuli, B.-Anantamul, E.-Indian sarsapasilla.

Habit : Twiner

Habitat : Terrestrial

Phytography : Twining shrub; leaves–simple opposite, shining and mid rib is white in colour; flowers-small in sessile, cyme and greenish purple in colour; fruit–follicles.

Fl. & Fr. : June–January.

Distribution : Forests of the Gangetic plains.

Status : Common, wild.

Uses : Dried roots are used as demulcent, diaphoretic. Medicated oil with root is applied in rheumatism and urinary diseases.

- *Hibiscus tilliaceus* L. (Malvaceae)

H. & B.-Bola, E.-Coast cotton tree.

Habit : Shrub

Habitat : Terrestrial

Phytography : Large shrub; leaves–palmetly 3-lobed; flowers–medium size, axillary and yellowish color; fruit–loculicidal capsule.

Fl. & Fr. : April–December.

Distribution : Coastal forests of India.

Status : Common, wild.

Uses : Bark yields fiber used for cordage. Bark mucilage is used in dysentery.

- *Holarrhena pubescens* (Buch-Ham.) Wall. ex. DC.

  (Apocynaceae) **(Plate-27)**

  Syn. : *H. antidysenterica* (L.) Wall.

  S.-Kutaj, H.-Kurchi, B.-Kurchi.

  Habit : Small tree

  Habitat : Terrestrial

  Phytography : Small deciduous tree with broad membranous opposite leaves; flowers–white, in terminal corymbose; fruit–elongated 2-terete follicles, seed–compressed tipped with deciduous coma (Silky hair).

  Fl. & Fr. : April-December.

  Distribution : Throughout India.

  Status : Common.

  Uses : Bark astringent, anthelmintic, tonic, antidysenteric. Bark contains alkaloid conessins.

- *Hydnocarpus kurzii* (King) Warb. (Flacourtiaceae)

  H. & B. : Chaulmurga.

  Habit : Tree

  Habitat : Terrestrial

  Phytography : Tree; leaves–simple; flowers–racemes; fruit–dry berry.

  Fl. & Fr. : April–January.

  Distribution : Forest of upper Assam.

  Status : Infrequent, wild.

  Uses : Kernels yield true *chaulmoogra* oil used on leprosy. Bark rich in tannin.

- *Hyoscyamus niger* L. (Solanaceae)

  S.-Dipya, H.-Khurasania jvayam, B.-Khorasania jawan, E.-Henbane.

Habit : Shrub.

Habitat : Terrestrial

Phytography : Shrub; leaves–simple; flowers–raceme; fruit–berry.

Fl. & Fr. : July–February.

Distribution : Kashmir to Garwal.

Status : Vulnerable, wild.

Uses : Dried leaves and flower buds are the sources of drug Henbane, Hyoscyamus and Hyoscyamin. It contains alkaloids Hyoscyamine and Hyoscine and Scopolamine. All these are narcotic and used in asthma and whooping cough. It is also a substitute for opium.

- *Ichnocarpus frutescens* R. Br. (Apocynaceae)

S.-Syamalata, H.-Kalidudhi, B.-Dorhia lata.

Habit : Twiner/Climber

Habitat : Terrestrial

Phytography : Climber; leaves–opposite; flowers–white, small and axillary and terminal; fruit–divaricated follicle.

Fl. & Fr. : July–February.

Distribution : Throughout India.

Status : Common, wild.

Uses : Root demulcent, tonic, and diaphoretic. Root paste is applied on dyspepsia and skin troubles. Root powder is used in bladder stone. Stem is used for making baskets and fish traps.

- *Illicium griffithii* Hk. F.T. (Magnoliaceae)

H.-Anasphal, E.-Star Anise.

Habit : Tree

Habitat : Terrestrial.

Phytography : Tree; leaves–simple; flowers–raceme; fruit–drupe.

Fl. & Fr. : May–December.

Distribution : Bhutan and Khasia hills.

Status : Infrequent, wild.

Uses : Fruit is stimulant and carminative. It has an odour like that of anise. It is used for flavouring liquors and in cough mixture. Fruit paste is applied on scabies.

- *Indigofera linifolia* Retz. (Fabaceae)

  H.–Torki, B.–Bhangra.

  Habit : Shrub

  Habitat : Terrestrial

  Phytography : Erect herb; leaves–simple; flowers–reddish purple and bract present; fruit–one seeded pod.

  Fl. & Fr. : April–December.

  Distribution : Throughout India

  Status : Common, wild.

  Uses : Plants are used as vermifuge. Seeds are reported to be used as famine food. It is also a fodder.

- *Indigofera tinctoria* L. (Fabaceae)

  S.-Nilla

  H. & B. : Nil, E.-Common Indigo.

  Habit : Shrub

  Habitat : Terrestrial

  Phytography : Twiggy shrub with compound leaves; leaflets-membranous; flowers–small and arranged in racemes; fruit–several seeded recurved green pod.

  Fl. & Fr. : April–December.

  Distribution : Throughout India.

  Status : Both cultivated and wild.

  Uses : Seeds are astringent, aphrodisiac and used in leucorrhoea. It is the source of natural 'indigo' and was cultivated in India and China.

- *Ipomoea digitata* L. (Convolvulaceae)

  Syn. : *I. paniculata* R. Br. Non Burm.

S.-Bhumikushmanda, H.-Bilaikand, B.-Bhuikumra.

Habit : Creeper

Habitat : Terrestrial

Phytography : Large creeper with compound leaves; leaves–palmately lobed; flowers–pink-purple and wide-campanulate in shape; fruit-capsule.

Fl. & Fr. : July–January.

Distribution : Found throughout tropical forests of India.

Status : Vulnerable, wild.

Uses : Root tuber contains resin used as aphrodisiac, demulcent, and galactagogue. Root powder is used in spleen and liver debility.

- *Ixora arborea* Roxb. (Rubiaceae)

Syn. : *I. parviffora* Vahl

S.-Nevali, H.-Kotagamdhal, B.-Swet Rangan, E.-Torch wood Ixora.

Habit : Shrub

Habitat : Terrestrial

Phytography : Large shrub or small evergreen tree; leaves–opposite, stipulate; flowers–in terminal cymes; fruit–globose berry.

Fl. & Fr. : July–February.

Distribution : Forests of Gangetic plains.

Status : Common, wild.

Uses : Bark decoction is used in debility and anaemia. Leaves are used as fodder.

- *Jesminum arborescens* Roxb. (Oleaceae)

S.-Saptala, H.-Bela, B.-Barakunda, E.-Tree jasmine.

Habit : Shrub

Habitat : Terrestrial

Phytography : Large shrub; leaves–simple distinctly petiolate and cordate in shape; flowers–white and arranged in cymes; fruit–berry.

Fl. & Fr. : April–December

Distribution : Gangetic plains, Deccan plains.

Status : Infrequent, wild.

Uses : Juice is used in bronchial obstruction. Berries are used in tonic.

- *Juglans regia* L. (Juglandaceae)

S.-Akschota, H. & B.-Akhrot, E.-Common Walnut.

Habit : Tree

Habitat : Terrestrial

Phytography : Tree; leaves–simple; flowers–racemes; fruit–berry.

Fl. & Fr. : April-January.

Distribution : Temperate Himalayas

Status : Both wild and cultivated.

Uses : Kernel is akhrot of commerce. Wood is used for furniture making.

- *Juniperus communalis* L. (Pinaceae)

S.-Vapusha, H.-Oil, B.-Havushu, E.-Common juniper.

Habit : Tree

Habitat : Terrestrial

Phytography : Tree; leaves–needle like; flowers and fruits are like other gymnosperms.

Fl. & Fr. : May–March.

Distribution : Western Himalayas.

Status : Wild.

Uses : Aromatic fruits are used for liqueurs. Wood is used for house polls.

- *Kandelia candel* (L.) Druce (Rhizophoraceae)

Syn. : *K. rheedii W.* & A.

B.-Goria.

Habit : Small tree

Habitat : Mangrove plant.

Phytography : Small tree; leaves–opposite coriacious and stipulate; flowers in axillary cymes; fruit–1-celled indehiscent ovoid with persistent calyx teeth.

Fl. & Fr. : April-December.

Distribution : Tidal forests of India.

Status : Common mangrove, wild.

Uses : Bark is rich in tannin used in diabetes. Wood is used for making charcoal.

- *Kydia calycina* Roxb. (Malvaceae)

H.-Pula, B.-Pola.

Habit : Tree

Habitat : Terrestrial

Phytography : Tree is with lobed leaves; flowers–in close panicles; fruit-a loculicidal 3-valved capsules.

Fl. & Fr. : February-September

Distribution : Tropical Himalayan forests. Peninsular Indian forests. Gangetic plains.

Status : Infrequent. Wild.

Uses : Bark is used as raincoat. Wood is used for making shoe heels.

- *Lannea coromandelica* (Houtt.) Merrill. (Anacardiaceae)

Syn. : *L. grandis* Engh., *Odina wodier* Roxb. H.-Jhingan, B.-Jiyal

Habit : Tree.

Habitat : Terrestrial

Phytography : Tree; leaves–simple; flowers–racemes; fruit–berry.

Fl. & Fr. : December–March.

Distribution : Found throughout deciduous forests of India.

Status : Infrequent, wild.

Uses : Wood is used for packing cases and is source of 'jhingan gum'. Bark decoction is used as dye.

- *Laportea crenulata* Gaudich (Urticaceae)

H.-Utigun, B.-Chorpata, E.-Fever nettle.

Habit : Shrub

Habitat : Terrestrial

Phytography : Shrub; leaves–simple; flowers–spike like racemes; fruit-nutlet.

Fl. & Fr. : June–December

Status : Common, wild.

Distribution : Duars, Sikkim, Assam.

Uses : Bark is used for cordage. Tribals use the fiber for cloth making. Flowers are eaten in curries. Root paste is used in chronic fever.

- *Lasia spinosa* Thw. (Araceae)

Syn. : *L. heterophylla* Schott., *L. aculeata* Lour.

B.-Kanta-kachu, H.-Kanta Saru.

Habit : Shrub

Habitat : Aquatic

Phytography : Stout intensely prickly shrub with branched rootstock; leaf-blade lobed; petiole–prickly; flowers–long fleshy and in spadix; fruit-berry.

Fl. & Fr. : November-May.

Distribution : Tropical forests of India.

Status : Vulnerable, wild.

Uses : Rhizome and young leaves are used as vegetable. Rhizome is used against rheumatism. Root stock paste is applied on piles.

- *Lowsonia inermis* L. (Lythraceae)

Syn. : *L. alba* L.

S.-Mendika, H.-Mehndi, B.-Mendi, E.-Henna.

Habit : Shrub

Habitat : Terrestrial

Phytography : Glabrous shrub, with spinous branches; leaves–small, opposite; flowers–small arranged in panicled cymes and white in colour; fruit–globose and irregularly dehiscing.

Fl. & Fr. : April-November.

Distribution : Throughout India.

Status : Both naturalised and cultivated.

Uses : Leaf-dye is used for coloring hands, hairs, bread, and eyebrows.

- *Leea asiatica* (L.) Ridsdala (Vitaceae)

Syn. : *L. crispa* L.

B.-Banchalita.

Habit : Shrub

Habitat : Terrestrial

Phytography : Rigid shrub; leaves–simple pinnate and petioles and rachis of ten winged; flowers–on leaf-opposed peduncles, in corymbose cymes; fruit–berry like.

Fl. & Fr. : May–December.

Distribution : Western Ghats, Western Himalayas.

Status : Infrequent.

Uses : Root tuber is used against guinea worm infection.

- *Leea indica* (Burm.) Merril. (Vitaceae)

H. & B.-Kurhurjiwah

Habit : Shrub

Habitat : Terrestrial

Phytography : Shrub; leaves–pinnately compound; flowers–corymbose; fruit–berry.

Fl. & Fr. : May–January.

Distribution : Throughout India.

Status : Infrequent, wild.

Uses : Roots is used in diarrhoea, dysentery and as sudorific.

- *Leea macrophylla* Roxb. ex. Hornem (Vitaceae)

  S.-Dholasamudrika, H. & B.-Dholsamudra

  Habit : Shrub.

  Habitat : Terrestrial

  Phytography : Herb; the lower leaf–often 2 feet across the upper ones .5-1; simple, white beneath with early pubescence; flowers–small and greenish white in colour; fruit–sub-globose.

  Fl. & Fr. : May–December.

  Distribution : Deciduous forest of India.

  Status : Vulnerable, wild.

  Uses : Leaves and fruits are edible. Root-mucilage is used as anodyne.

- *Leucas cephalotes* Spreng (Lamiaceae)

  S.-Dronapusphi, H.-Dhurpi sag, B.-Barahalkusa.

  Habit : Herb

  Habitat : Terrestrial

  Phytography : Annual herb; leaves–linear; flowers white with globose head, Stamens didynamous; fruit–dry nutlet.

  Fl. & Fr. : April–December.

  Distribution : Throughout India

  Status : Common, wild.

  Uses : Leaves are laxative and anthelmintic. Syrup prepared from flowers is used for cough and cold.

- *Litsea monopetala* (Roxb.) Pers. (Lauraceae)

  Syn. : *L. polyantha* Juss.

  H.-Meda, B.-Bara Kukur Chita.

  Habit : Tree

  Habitat : Terrestrial

  Phytography : Evergreen tree; leaves–simple alternate, long petioled, broad and tomentose beneath; flowers–small and umbellate; fruit-drupe.

Fl. & Fr. : May–December.

Distribution : Found in foothill forests of Himalayas, Assam and Satpura.

Status : Infrequent.

Uses : Wood is used for making agricultural implements. Bark is astringent. It contains tannin. Leaves are used to feed *muga* silk larvae. Seed oil is applied on rheumatic swelling.

- *Livistonia jenkinsiana* Griff. (Arecaceae)

Tokapata.

Habit : Tall palm

Habitat : Terrestrial

Phytography : Tree like palm; leaves–palmately compound; flowers in spadix; fruit–berry.

Fl. & Fr. : April-September.

Distribution : Found in foothills of Himalayas, Assam and Nagaland.

Status : Infrequent, wild.

Uses : Leaves are used for thatching, covering tops of palanquins as hats also.

- *Lobelia nicotianaefolia* Hegne (Campanulaceae)

S.-Devanala, H.-Naja, B.-Badenala.

Habit : Shrub

Habitat : Terrestrial

Phytography : Shrub; leaves–simple; flowers–racemes; fruit–berry.

Fl. & Fr. : June–December

Distribution : Western Ghats, Peninsular India.

Status : Infrequent, wild.

Uses : Plant has antiseptic properties.

- *Lycopodium clavatum* L. (Lycopodiaceae)

Nagbeli.

Habit : Shrub

Habitat : Terrestrial, prostrate.

Phytography : Shrub; leaves–scaly; flowers and fruit–absent. Produces spores in sporangia which has reproductive unit.

Fl. & Fr. : June–February.

Distribution : Himalayan forest areas.

Status : Vulnerable.

Uses : Spores–constitute Lycopodium of commerce. It is used in dyspepsia, flatulence. In homoeopathy it is used against diseases of chest and urinary passage.

- *Lygodium flexuosum* (L.) Sw. (Schizaeaceae)

B.-Bhut-raj

Habit : Shrub

Habitat : Terrestrial, climber.

Phytography : Graceful climbing fern with wide-twinig stem; fronds (leaves) scandent, palmately lobed; sporangia paired in the axils of clasping inducia.

Fl. & Fr. : Spore formation in winter season.

Distribution : Deciduous forests of India.

Status : Common, wild.

Uses : Plant is used as expectorant. Root is applied on rheumatism, sprain, scabies, and eczema.

- *Madhuca indica* J.F. Ganbel. (Sapotaceae)

Syn. : *M. latifolia* Macbr., *Bassia latifolia* Roxb.

H.-Mahua, B.-Mahwa, E.-Butter tree.

Habit : Tree

Habitat : Terrestrial.

Phytography : Large deciduous tree; leaves–simple, petioled and coriaceous; flowers–axillary pedicels, small with yellowish in colour and with nice scent; fruit is ellipsoid berry.

Fl. & Fr. : March–December.

Distribution : Deciduous forests of India.

Status : Common, wild.

Uses : All parts of the plant are useful. Fruit is used as vegetable. Flowers yield a popular and indigenous liquor called *mahul*. Timber is good for house building. Tree has religious value for some tribals.

- *Madhuca longifolia* (Koenig) Macb. (Sapotaceae)

Syn. : *Bassia longifolia* Koenig.

H. & B. : Mahua, E.-Mowra Butter tree.

Habit : Tree

Habitat : Terrestrial

Phytography : Tree; leaves–simple; flowers–racemes; fruit–berry.

Fl. & Fr. : April–December.

Distribution : Throughout South Indian forests, at low elevations.

Status : Infrequent, wild.

Uses : All parts of the plant are useful. It also has the same usefulness as *Madhuca indica*.

- *Magnolia grandiflora* L. (Magnoliaceae)

S.-Champaka. H. & B.-Him Champa.

Habit : Tree

Habitat : Terrestrial

Phytography : Tree; leaves–simple; flowers–single; fruit–aggregated.

Fl. & Fr. : April–January

Distribution : Found in foothills of Himalayas.

Status : Infrequent, both wild and planted.

Uses : Bark stimulant. Bark decoction is used in malarial fever. Flowers contain essential oil.

- *Mahonia napaulensis* DC. (Berberidaceae)

Syn. : *Berberis nepalansis* Spreng in part.

Chetri

Habit : Shrub

Habitat : Terrestrial

Phytography : Shrub; leaves–compound; flowers–spike like racemes; fruit–pod.

Fl. & Fr. : April-November

Distribution : Himalayan forests.

Status : Vulnerable, wild.

Uses : Plant contains many useful alkaloids. Roots yield a yellow dye. Berries are diuretic and demulcent. "*Rasant*' is prepared from the plant.

- *Mallotus philippensis* Muell.-Arg.

H.-Sindur, B.-Kamala, E.-Kamala tree

Habit : Tree

Habitat : Terrestrial

Phytography : Small evergreen tree; leaves–simple, alternate, beneath subglobose (waxy covered) and covered with red, resinous glands, above glabrous; flowers–small in axillary and terminal racemes; fruit-capsule.

Fl. & Fr. : June–March

Distribution : Found throughout deciduous forests of India.

Status : Infrequent, wild.

Uses : Glandular fruit hairs are the source of Kamala powder. It produces rich golden red colour for silk colouring. Powder is anthelmintic. Seed oil is used as hair fixer. Leaves are used as fodder.

- *Mangifera sylvatica* Roxb. (Anacardiaceae)

B.-Ban am.

Habit : Tree

Habitat : Terrestrial

Phytography : Large tree; leaves–simple, alternate; flowers–small arranged in panicles and greyesh in colours; fruit–drupe with a compressed fibrous stone.

Fl. & Fr. : December–February.

Distribution : Assam and Andaman and Nicobar Islands.

Status : Vulnerable, wild.

Uses : Fruits are edible. Ash made after burning leaves are applied on scabies.

- *Malinkara litoralis* (Kurz.) Dubard. (Sapotaceae)

  Syn. : *Mimusops littoralis* Kurz.

  Habit : Tree

  Habitat : Terrestrial, mangrove.

  Phytography : Tree; leaves–simple; flowers–racemes; fruit–berry.

  Fl. & Fr. : March–June.

  Distribution : Sundarban and Andaman and Nicobar Islands.

  Status : Threatened, wild.

  Uses : Wood used for house building. Barks yield a red vegetable dye.

- *Marsdenia tenacissima* W. & A. (Asclepiadaceae)

  H. & B. : Jiti

  Habit : Shrub

  Habitat : Terrestrial, twiner.

  Phytography : Stout twining shrub; leaves–simple, opposite; flowers -medium, fleshy; fruit–dagger shaped follicle.

  Fl. & Fr. : March-May.

  Distribution : Deciduous forest of India.

  Status : Infrequent, wild.

  Uses : Stems yields fiber used for making fishing net, cordage. Roots yields *"safed nisoth"* used as purgative.

- *Martynia annua* L. (Martyniaceae) **(Plate-28)**

  Syn. : *M. diandra* Glox.

  H.-Bichu, B.-Bagnok, E.-Tiger claw.

  Habit : Shrub

Habitat : Terrestrial

Phytography : Shrubs; leaves–simple; flowers–raceme; fruit–drupe.

Fl. & Fr. : March–December.

Distribution : West lands in forests.

Status : Frequent, wild.

Uses : Leaves are used for epilepsy. Leaf paste is applied on inflammation of tubercular glands. Fruits used alexeteric in inflammation.

- *Melothria heterophylla* (Lour) Cogn. (Cucurbitaceae)

Read-*Solena amplexicaulis* (Lamk.) Gandhi

H. & B. : Kudasi

Habit : Herb

Habitat : Climber.

Phytography : Climber; leaves–simple; flowers–racemes; fruit–pome.

Fl. & Fr. : May–August.

Distribution : Throughout India in deciduous forests.

Status : Vulnerable, wild.

Uses : Tuber is used as food.

- *Mesua ferrea* L. (Clusiaceae)

S.-Nag Keshara, B.-Nagesar.

Habit : Tree

Habitat : Terrestrial.

Phytography : Tree; leaves–simple or compound; flowers–racemes; fruit–capsule.

Fl. & Fr. : March–November.

Distribution : Foothill forests of Himalayas, Assam, West Bengal, Western Ghats.

Status : Frequent

Uses : Wood is used for construction. Seed oil is used for soap making; flowers are used in cough and in making an *Ayurvedic* tonic *Chavanprash*'.

- *Michelia champaca* L. (Magnoliaceae)

  S.-Champaka

  H. & B. : Champa, B.-Champak.

  Habit : Tree

  Habitat : Terrestrial

  Phytography : Large evergreen tree; leaves–bud enclosed within the connate caduceus stipule and simple; flowers–solitary, yellow and very fragrant; fruit–dense elongated spike of coriacious follicle.

  Fl. & Fr. : September-March.

  Distribution : Wild in Eastern Himalayas.

  Status : Infrequent

  Uses : Flower is a source of Champaca oil used in perfumery, hair oils. Seeds contain fat and bark contains tannins.

- *Miliusa velutina* Hook. f. & Thoms. (Annonaceae)

  H. & B. : Domsal.

  Habit : Tree

  Habitat : Terrestrial

  Phytography : Large tree; leaves–simple, tomentose on both surfaces; flowers–racemose, and greenish in colour; fruit–globose.

  Fl. & Fr. : December-May.

  Distribution : Deciduous forests of India.

  Status : Frequent, wild.

  Uses : Wood is used for house building and making match-boxes. Fruit is edible; leaves are used as fodder. Bark is a purgative.

- *Millettia pachycarpa* Benth. Baker in part (Fabaceae)

  B.–Bishloti

  Habit : Shrub

  Habitat : Terrestrial, twiner.

Phytography : Shrubs; leaves–pinnately compound; flowers–racemes; fruit–pod.

Fl. & Fr. : September–March.

Distribution : Deciduous forest of India.

Status : Common, wild.

Uses : Seed is poisonous to fish and birds. It is used as an insecticide due to presence of rotenone.

- *Mitragyna parvifolia* (Roxb.) Kuntze (Rubiaceae)

Syn. : *Stephegyne parvifolia* Korth.

H.-Kaim, B.-Gulikadam.

Habit : Tree

Habitat : Terrestrial

Phytography : Small or medium size tree; leaves–simple, petioled, stipulate; flowers–globose, terminal; fruit–follicles.

Fl. & Fr. : September–April.

Distribution : Deciduous forests of India.

Status : Frequent, wild.

Uses : Wood is used for making matchboxes. Leaves are used as fodder. Root bark used as fabrifuge. Bark is rich in tannin.

- *Moghania strobilifera* J. St. Hilaire (Fabaceae)

Syn. : *Flemingia strobilifera* R. Br.

H.-Kussunt.

Habit : Shrub

Habitat : Terrestrial

Phytography : Erect shrub; leaves–simple; flowers–arranged in cymes, which are hidden by a broad membranous bract; fruit–like pod.

Fl. & Fr. : August–January.

Distribution : Deciduous forests of India.

Status : Common, wild.

Uses : Roots is used in epilepsy and hysteria. Root paste induces sleep and relieves pain. Leaves are used as vermifuge.

- *Momordica cochinchinensis* Spreng. (Cucurbitaceae)

S.-Karkataka, H.-Gulkakra, B.-Go-Kakararol.

Habit : Shrub

Habitat : Climber

Phytography : Long herbaceous climber; leaves–simple, cordate; flowers–dioecious; fruit–spherical rough many-seeded berry.

Fl. & Fr. : June–September.

Distribution : Deciduous forests of India.

Status : Rare in wild. Both wild and cultivated.

Uses : Fruits are eaten as vegetable. Seeds are used in ulcer. Leaf paste is applied on lumbago.

- *Morinda citrifolia* L. (Rubiaceae)

S.-Ashyuka

H. & B. : Guj.

Habit : Small tree

Habitat : Terrestrial

Phytography : Climbing shrub; leaves–glabrous, opposite and stipulate; flowers–axillary and terminal, flower head distinctly peduncled; fruit-whitish, aggregate.

Fl. & Fr. : April-November.

Distribution : Deciduous forests of India.

Status : Infrequent, wild.

Uses : Roots yields morindone used as "Al dye". Bark contains glucoside morindin. Fruits are eaten by silk-worms which are reared on the leaves. Root is cathartic and febrifuge. Leaf juice is applied on gout. Fruits are given in leucorrhoea and sapraemia.

- *Mucuna prurita* Hook. (Fabaceae)

  Syn. : *M. pruriens* Baker. Non DC. Some authors consider *M. pruriens* DC. as Syn. of *M. prurita.*

  S.-Atmagupta, H.-Kiwach, B.-Alkushi, E.-Cowhage.

  Habit : Shrub

  Habitat : Terrestrial, climber

  Phytography : Annual climber of twining shrub; leaves–pinnately tri-foliate; flowers–large purple red arranged in axillary racemes; fruit -thickly leathery linear pod with stinging hair.

  Fl. & Fr. : July–March.

  Distribution : Deciduous forests of India.

  Status : Rare in wild. Sometime cultivated.

  Uses : Roots are diuretic, purgative and emnenagogue. Seed powder is used in diseases of nervous system, renal affections. Ointment is prepared from the roots and applied on elephantiasis. Seed is aphrodisiac.

- *Murraya koenigii* (L.) Spreng. (Rutaceae)

  S.-Surabhininib, H.-Kathnim, B.-Karipata, E.-Curry leaf tree

  Habit : Shrub

  Habitat : Terrestrial

  Phytography : Small spreading tree; leaves–inperipinnate, leaflets -alternate; flowers–arranged in axillary cymes small and white in colour; fruit–berry.

  Fl. & Fr. : March-July

  Distribution : Peninsular Indian forests.

  Status : Frequent, wild.

  Uses : Leaves are used in curry for flavour. Fruit is edible. Root is carminative. Leaf paste is used in dysentery and for checking vomiting. Root decoction gives relief to renal pain.

- *Nardostachya grandiflora* DC. (Valerianaceae)

  Syn. : *Nardostachys jatamansi* DC.

S.-Jata mansi

H. & B. : Jatamansi, E.-Spikenard.

Habit : Herb

Habitat : Terrestrial

Phytography : Herb; leaves–simple; flowers–racemes; fruit–capsule.

Fl. & Fr. : July–January.

Distribution : Alpine Himalayas.

Status : Threatened, wild.

Uses : Root is the source of Nardus. It is substitute of valerian (A stimulant and antispasmodic drug found from a Mediterranean plant known as *Valeriana officinalis)*. Root oil is used as hypotensive. Root infusion is used in leprosy and hysteria. Spikenard oil improves hair growth and darkens the hair.

- *Nepenthes khasiana* Hook.f. (Nepenthaceae)

Pitcher plant

Habit : Shrub

Habitat : Terrestrial, climber.

Phytography : Climber; leaves–simple; flowers–racemes; fruit–nutlet.

Fl. & Fr. : December-April.

Distribution : Assam, Meghalaya.

Status : Critically endangered, wild.

Uses : Pitcher extract is used as eye drop.

- *Nypa fruticans* Wrmb. (Arecaceae) **(Plate-29)**

B.–Goal pata, E.–Nipa palm

Habit : Shrub

Habitat : Mangrove palm.

Phytography : Stemless shruby palm with branching root stock; leaves are compound and pinnatisect; flowers–monoecious on terminal, branched erect spadices; fruit–

large, globose, hexagonal, one seeded with fleshy pericarp and pyramidal tips.

Fl. & Fr. : September–May.

Distribution : Sundarbans and Andaman and Nicobar Islands.

Status : Rare, wild.

Uses : Leaves are used for thatching. Fruit kernel is edible. Inflorescence is tapped for its sweet sap.

- *Nyctanthes arbor-tristis* L. (Oleaceae)

S.-Parijata, H.-Harsinghar, B.-Sedi, E.-Night Jasmine

Habit : Tree

Habitat : Terrestrial

Phytography : Small tree with simple small cordate shape with dentate leaf margin; flowers-small, sessile with white and orange in colour; fruit-capsule.

Fl. & Fr. : September–December

Distribution : Deciduous forests of India.

Status : Common, both wild and planted.

Uses : Corolla tubes contain coloring matter, nyctanthin, and seeds yield a fixed oil. Bark yields tannin. Leaf juice is an expectorant. Leaf extract is also as laxative and anthelmintic.

- *Ochna pumila* Buch-Ham.ex Don. (Ochnaceae)

Bhui Champa.

Habit : Herb

Habitat : Terrestrial.

Phytography : Medium size herb with coarsely toothed larger leaves; flowers–conspicuous, radish and arranged in umbel inflorescence; fruit -drupe and seated on the broad disk.

Fl. & Fr. : March-July

Distribution : Foothills forests and deciduous forests of Bihar, West Bengal, Orissa.

Status : Rare, wild.

Uses : Root is used in epilepsy. Root decoction is used in asthma and against menstrual complaints.

- *Olax scandens* Roxb. (Olacaceae)

H.-Dheniani, B.-Koko-ara

Habit : Shrub

Habitat : Terrestrial, scrambler.

Phytography : Large climber with woody twig and prickly; leaves–simple and alternate; flowers small arranged in cymes; fruit–small drupe and covered by fleshy calyx.

Fl. & Fr. : April–November.

Distribution : Western Ghats and other deciduous forests of India.

Status : Frequent, wild.

Uses : Leaves are used as vegetable. Fruit is eaten. Bark decoction is used in anaemia.

- *Ophiopogon intermedius* D. Don. (Liliaceae)

B.-Piyajimurba.

Habit : Herb

Habitat : Terrestrial

Phytography : Herb; leaves–simple; flowers–raceme; fruit–nutlet.

Fl. & Fr. : March–September

Distribution : Foothill forests of India.

Status : Rare, wild.

Uses : Tuber is used in dropsy. Tuber mucilage is used as substitute of *Ginseng*. Tuber decoction is used to bring down inflammation.

- *Orchis latifolia* L. (Orchidaceae)

H.-Salap

Habit : Herb

Habitat : Terrestrial and epiphytic.

Phytography : Herb; leaves–simple; flowers–spikes; fruit–capsule.

Fl. & Fr. : September–February

Distribution : Western temperate Himalayas

Status : Frequent

Uses : Tuber is the source of *salap* (a drug used in *Ayurveda) used* in dysentery and chronic fever.

- *Oroxylurn indicurn* Vent. (Bignoniaceae)

S.-Shyonaka, H.-Ullu, B.-Sona.

Habit : Tree

Habitat : Terrestrial

Phytography : Glabrous tree; leaves–large, opposite and two-three pinnate; flowers–long terminal racemes with purplish or whitish corolla; fruit–large, linear capsule, compressed parallel to the septum, two valves open septifragally.

Fl. & Fr. : March-October

Distribution : Deciduous forests of India.

Status : Infrequent, wild.

Uses : Bark is astringent and used in diarrhoea and dysentery. Seed is purgative. Tender fruits are eaten. Plant is also used as antiseptic.

- *Orthosiphon rubicundus* Benth. (Lamiaceae)

Jikapota.

Habit : Herb

Habitat : Terrestrial

Phytography : Small herb with simple sessile leaf; flowers–small, campanulate and whitish in colour; fruit–ovoid nutlet.

Fl. & Fr. : April-September.

Distribution : Deciduous forests of India.

Status : Infrequent, wild.

Uses : Tubers are eaten. Root powder used in colic pain.

- *Oxystelma secamone* (L.) Karst. (Asclepiadaceae)

Syn. : *O. esculentum* R. Br.

S.-Dugdhika

H. & B. : Dudhialata

Habit : Twiner

Habitat : Semiaquatic

Phytography : Slender, glabrous, climbing perennial; Flower–large and corolla with corona; fruit–thick, oblique or curved, smooth, two winged follicle.

Fl. & Fr. : September–January.

Distribution : Water bodies in deciduous forests

Status : Frequent.

Uses : Plant is eaten. Latex is antiseptic and galctagogue. Root is good for jaundice.

- *Paederia foetida* L. (Rubiaceae)

S.-Prasarani

H. & B. : Gandhali

Habit : Shrub.

Habitat : Creeper

Phytography : Slender twining shrub with terete branches; leaves-simple, opposite, membranous with odorus smell; flowers–small and in axillary and terminal; fruits are much compressed and globose.

Fl. & Fr. : May–September.

Distribution : Dry deciduous forests of India.

Status : Infrequent. Both wild and planted.

Uses : Leaves are used in dysentery. Leaf paste is applied on herpes. Roots and bark is antithelemetic.

- *Panax schinseng* Nees (Araliaceae)

Syn. : *P. ginseng* Ney

Habit : Herb

Habitat : Terrestrial

Phytography : Herb; leaves–simple; flowers–spadix; fruit–berry.

Fl. & Fr. : May–September.

Distribution : Eastern Himalayas.

Status : Threatened, wild.

Uses : Tuber is the source of *Ginseng*.

- *Pandanus andamanensium* Kurz. (Pandanaceae) **(Plate-30)**

  H.-Keora

  Habit : Shrub

  Habitat : Terrestrial moist loving plant

  Phytography : Shrub; leaves–simple, long, dented; flowers–spadix; fruit–aggregate.

  Fl. & Fr. : June–January

  Distribution : Andaman and Nicobar Island.

  Status : Infrequent, wild.

  Uses : Leaves yield fiber. Fruits are edible.

- *Pandanus leram* Jones (Pandanaceae)

  E.-Nicobar bread fruit.

  Habit : Shrub

  Habitat : Terrestrial but loves marshy lands.

  Phytography : Shrubs; leaves–simple; flowers–spadix; fruit–aggregate.

  Fl. & Fr. : May–February.

  Distribution : Andaman and Nicobar Islands.

  Status : Vulnerable, wild.

  Uses : Fleshy lower parts of the fruits are used for making flour for making *chapati* and bread.

- *Pandanus odoratissimus* L. (Pandnaceae)

  Syn. : P. *tectorius* Solanol ex Parkinson, *P. fascicularis* Lam., *P. laevis* Kunth, *P. variegatus* Miq., *P. latifolius* Hassk., *P. amaryllifolis* Roxb., *P. utilis* Bary.

  S.-Ketaki, H.-Keura, B.-Keya, E.-Screw pine.

  Habit : Shrub

Habitat : Terrestrial, marshy plant.

Phytography : Much branched bushy shrub with stilted root; leaves -narrow, coreacious with spinously toothed; flowers–dioecious small and arranged in axillary spadices which are clothed with leafy spathes; fruit-oblong woody drupe.

Fl. & Fr. : January–February.

Distribution : Seacoast of Indian Peninsula and Andaman Islands.

Status : Common, wild.

Uses : Male flowers possess fragrance used for hair decoration and extraction of Kewda water. Young buds are eaten. Seeds are edible. Leaves are used for thatching, mat making. Sap obtained from inflorescence is used in rheumatic arthritis and in veterinary medicine.

- *Pavetta indica* L. (Rubiaceae)

Syn. : P. *crassicaulis* Bremex.

S.-Papata, H.-Papari, B.-Kuhurachura, E.-White Pavetta.

Habit : Shrub

Habitat : Terrestrial

Phytography : Shrub with terete branches; leaves–opposite, petioled and membranous; flowers–axillary or terminal and medium in size; fruit -small fleshy berry.

Fl. & Fr. : March–December

Distribution : Throughout India. Some botanists state that *P. indica* is available only in South India.

Status : Infrequent, wild.

Uses : Fruits are eaten. Root decoction is a purgative and diuretic and also used in urinary diseases.

- *Pedalium murex* L. (Pedaliaceae)

H.-Bara-gokhru, B.-Bara-ghokru

Habit : Herb

Habitat : Terrestrial

Phytography : Herb; leaves–simple; flowers–racemes; fruit–berry.

Fl. & Fr. : April–October

Distribution : Gujrat and Deccan Peninsula.

Status : Infrequent, wild.

Uses : Viscid mucilage resembles gum arabic. It is diuretic. It is also used in dysuria. Fruits are aphrodisiac.

- *Pergularia daemia* (Forsk.) Chiov. (Asclepiadaceae)

Syn. : *P. extensa* N.E. Br., *Daemia extensa* R. Br.

S.-Phala, H.-Sagovani, B.-Chagulbati.

Habit : Climber

Habitat : Terrestrial

Phytography : Twining glabrous under shrub; leaves–opposite, membranous and ovate in shape; flowers–small, pale yellow, arranged in subumbellate cymes, fruit–2-lanceolate follicles.

Fl. & Fr. : April-December

Distribution : Deciduous forests of India.

Status : Frequent, wild.

Uses : Plant is antiemetic, expectorant, anthelmintic. Fruits juice with limejuice applied on rheumatic swellings.

- *Peucedanum dhana* Buch.-Ham ex C.B. Clarke. (Apiaceae) Bhojraj

Habit : Herb

Habitat : Terrestrial

Phytography : Big glabrous perennial herb; leaves–compound, ovate in shape; flowers–yellowish white and arranged in compound umbels; fruit–sub-quadrate dorsal furrows 1-vittate, lateral 2-vittate.

Fl. & Fr. : September–April

Distribution : North Bengal and Western Himalayas, Madhya Pradesh.

Status : Vulnerable

Uses : Carrot like roots are edible. Seeds are carminative.

- *Phlogacanthus thyrsiflorus* Nees (Acanthaceae)

  Syn. : *Phlogacanthus thyrsiformis* (Hordus) Mabberby

  B.-Rambasak.

  Habit : Shrub

  Habitat : Terrestrial

  Phytography : Shrub; leaves–simple; flowers–racemes; fruit–capsule.

  Fl. & Fr. : August–April

  Distribution : Bihar, West Bengal and upper Gangetic plain.

  Status : Vulnerable, wild.

  Uses : Plant is used as a substitute of *Adhatoda vasica.*

- *Phoenix acaulis* Roxb. (Arecaceae)

  H.-Khajur

  Habit : Shrub

  Habitat : Terrestrial

  Phytography : Slender dwarf palm of lateritic West Bengal; leaves-pinatisect, segments lanceolate with induplicate margins; flowers-dioecious, small yellowish on several branched spadices; fruit–oblong terete, 1-seeded with a fleshy pericarp and membranous endocarp.

  Fl. & Fr. : October–April

  Distribution : Dry deciduous forests of India.

  Status : Infrequent, wild.

  Uses : Plant pith is eaten. Pith yields sago. Leaves are used for thatching and for making mats known as *sitalpati.*

- *Phoenix paludosa* Roxb. (Arecaceae)

  H. & B. : Hetal

  Habit : Shrub

  Habitat : Mangrove palm.

  Phytography : Gregarious tall soboliferous palm of Sundarbans; leaves-pinnatisect; flowers–dioecious small on drooping spadices; fruit-oblong, one seeded.

Fl. & Fr. : September–April

Distribution : Mangrove forests of India-specially Sandarbans of West Bengal.

Status : Frequent, wild.

Uses : Leaves are used for thatching. Stem is used as walking sticks. Pith and fruits are edible.

- *Phoenix sylvestris* Roxb. (Arecaceae)

H.-Khajur, B.-Khajur, E-Wild date palm.

Habit : Tree

Habitat : Terrestrial

Phytography : Tall graceful tree with elongated stem and clothed with persistent leaf bases; leaves one pinnatisect and log segments lanceolate; flowers–small dioecious white and arranged in several branched spadices; fruit–oblong, one seeded and orange in colour when ripe.

Fl. & Fr. : November-April

Distribution : Throughout India.

Status : Common, wild.

Uses : Leaves are used for thatching and for making mats, brooms, jaggery. Sap is also used as *nira,* which is a source of vitamin-B and vitamin-C. Fruit is edible. Wood is used for construction. Root is used in toothache.

- *Phyllanthus acidus* Skeels (Euphorbiaceae)

Syn. : *Cicca acida* (L.) Merrill., *C. disticha* L,. *P. distichus* Muell.-Arg.

B.-Nauli.

Habit : Tree

Habitat : Terrestrial

Phytography : Tree; leaves–pinnately compound; fruit–berry.

Fl. & Fr. : April–September

Distribution : North-East Indian forest.

Status : Vulnerable, both wild and planted.

Uses : Fruits eaten. Roots and seeds are cathartic. Root barks yield 18% tannin.

- *Phyllanthus fartemus* Webster (Euphorbiaceae)

  Syn. : *P. niruri Hook.f.non* L

  S.-Bhmyamalaki, H.-Jaramla, B.-Bhui-amla.

  Habit : Herb

  Habitat : Terrestrial

  Phytography : Annual small herb found abundantly; leaves–alternate, entire; flowers–axillary from node and small pedicillate; fruit–drupe.

  Fl. & Fr. : February–August.

  Distribution : Found throughout hotter parts of India.

  Status : Common, wild.

  Uses : Plant decoction is used in jaundice. Whole plant has astringent, diuretic and febrifuge properties. Leaves are used in dropsy and in urino-genetal system. Root galactogogue. Plant decoction is used to dye cotton black.

- *Picrorhiza kurroa* Royle ex Benth. (Scrophulariaceae)

  Syn. : *P. kurroa* Hook.f. in part non Royle ex Benth.

  S.-Katuka.

  H. & B. : Kuru.

  Habit : Herb

  Habitat : Terrestrial

  Phytography : Herb; leaves–simple; flowers–racemes; fruit–berry.

  Fl. & Fr. : June–January

  Distribution : Alpine Himalayas from Kashmir to Sikkim.

  Status : Vulnerable, wild.

  Uses : Dried root is used as a substitute or adulterant of *Gentiana kurroo* Royle. It is cholagogue, stomachic and laxative. It yields picrorhizin, kutkin etc.

- *Pinus roxburghii* Sarg. (Pinaceae)

  Syn. : *P. longifolia* Roxb.

  S.-Sarala, H.-Chir, E.-Chir pine.

Habit : Tree

Habitat : Terrestrial

Phytography : Herb; leaves–needle; flowers cones; fruit–with wing seeded.

Fl. & Fr. : April–January

Distribution : Alpine Himalayas

Status : Infrequent, both wild and planted.

Uses : Natural source of turpentine used in paints and in pharmaceutical preparation. Plant yields oleoresin used in chronic bronchitis and as expectorant. It is applied in rheumatic afflictions. Turpentine is used in variety of industries.

- *Pinus wallichiana* A.B. Jaackson (Pinaceae)

  Syn. : *P. excelsa* Wall. ex D. Don., *P. griffithi* Mc Clelland H.-Kil, E.-Blue pine.

  Habit : Tree

  Habitat : Terrestrial

  Phytography : Tree; leaves–needle; flowers–cones; seed–winged.

  Fl. & Fr. : April–December

  Distribution : Throughout Alpine Himalayas.

  Status : Infrequent, wild.

  Uses : Source of oleoresin for good quality of turpentine. Wood used for matchbox.

- *Piper retrofractum* Vahl. (Piperaceae)

  Syn. : *P. chaba* Hunter non Blume P. *officina rum* DC.

  H.-Chab, B.-Chai, E.-Java long peppers.

  Habit : Creeper

  Habitat : Terrestrial

  Phytography : Stout climber with flexuous branches; leaves–simple with very short petiole; flowers–dioecious small and in spike; fruit–small globose berry found in dense cylindrical spike.

Fl. & Fr. : October–March

Distribution : Foothills and plains of India, Orissa, and West Bengal.

Status : Both wild and planted.

Uses : It has weak aroma. It is used in pickles. Fruits stimulant and carminative. Root is chewed after cooking. Stem powder is used in dyspepsia and gastralgia.

- *Piper longum* L. (Piperaceae)

S.-Pippali, H.-Pipal. B.-Piplamor. E.-Indian long pepper.

Habit : Creeper.

Habitat : Terrestrial

Phytography : Slender creeper shrub with erect flowering shoots; leaves-simple with long petioled; flowers–small in the axil of bract; fruit–small berry.

Fl. & Fr. : July–December

Distribution : Deciduous forests of India.

Status : Infrequent.

Uses : Roots and fruits are used for diseases of respiratory tract. Fruit powder is used to get relief of muscular pain. Fruit decoction is carminative and is used as sedative in insomnia and epilepsy.

- *Piper mulfesna* Don. (Piperaceae)

Syn. : *P. brachystachyum* Wall. ex Hook.f.

E.–Pahari piper; B.–Pahari pan.

Habit : Herb

Habitat : Terrestrial

Phytography : Herb; leaves–simple; flowers–spike; fruit–berry.

Fl. & Fr. : June–December

Distribution : Eastern Himalayas

Status : Vulnerable.

Uses : Leaves yield volatile oil.

- *Plumbago indica* L. (Plumbaginaceae)

  Syn. : Chitraka, H.-Chitra, B.-Lal chita

  Habit : Herb

  Habitat : Terrestrial

  Phytography : Herb; green branches; leaves–oblong to ovate in shape; flowers–small and red in color; fruit–capsule.

  Fl. & Fr. : April–October

  Distribution : Throughout India probably native of Sikkim.

  Status : Both wild and planted in garden.

  Uses : Root abortifacient. Root paste is applied on leucoderma. Plumbagin is the active principle. It acts also on central nervous system and relieves muscle pain.

- *Plumbago zeylanica* L. (Plumbaginaceae)

  S.-Chitraka

  H. & B. : Chita

  Habit : Herb

  Habitat : Terrestrial

  Phytography : Herb with rambling green branches; leaves–oblong to ovate with base suddenly narrowed into a short amplexicaul petiole; flowers–arranged in terminal panicled spikes of very pretty long-tubed white flowers; fruit–capsule.

  Fl. & Fr. : April–October

  Distribution : Throughout India.

  Status : Vulnerable, wild.

  Uses : Root abotifacient. Root paste used is in dyspepsia and piles. Root infusion is used in influenza. Active principle is plumbagin.

- *Podophyllum hexandrum* Royle. (Berberidaceae)

  Syn. : *P. emodi Wall.* ex. Hook.f. & Thoms.

  H.-Bakrachimaka. B.-Papra, E.-Indian podophyllum.

  Habit : Herb;

Habitat : Terrestrial

Phytography : Herb; leaves–simple; flowers–racemes; fruit–berry.

Fl. & Fr. : July–January

Distribution : Himalayan forests

Status : Threatened, wild.

Uses : Source of podophyllin used as purgative and vermifuge. Podophyllin is the active principle. It controls some forms of cancer.

- *Polyalthia suberosa* Thw. (Annonaceae)

H.-Cham-khirni, B.-Brachali.

Habit : Shrub

Habitat : Terrestrial

Phytography : Shrub with oblong to obtuse leaves; flowers–many or few in axils; stamens–numerous; fruit–berry like.

Fl. & Fr. : April–October

Distribution : Deciduous forests of India.

Status : Infrequent.

Uses : Wood is used for general carpentry work. Fresh root decoction is used as an abortifacient.

- *Polygala chinensis* L. (Polygalaceae)

Syn. : *P. telephioides* Willd., *P. brachystachya* DC.

H.-Meradu, B.-Meradu

Habit : Herb

Habitat : Terrestrial

Phytography : Annual herb of usually in everywhere and on roadside; leaves–simple alternate; flowers–conspicuous arranged in axillary cymes; fruit–2 celled loculicidal capsule.

Fl. & Fr. : May–September

Distribution : Throughout India.

Status : Infrequent, wild.

Uses : Tender leaves are eaten. Leaf infusion is used in asthma.

- *Polygonum barbatum* L. (Polygonaceae)

  Syn. : P. *stagninum* Buch.-Ham. ex Meissn.

  B.-Bekh-unjubar.

  Habit : Shrub

  Habitat : Terrestrial

  Phytography : Stout annual herb; leaves–alternate, entire, lanceolate; stipules usually sheathing the stem; flowers–small white and arranged in terminal cymes, fruit–small nuts.

  Fl. & Fr. : April-October

  Distribution : Found throughout hotter parts of India.

  Status : Infrequent, wild.

  Uses : Plant is used in curries. Shoot decoction is stimulant. Seed is a purgative drug.

- *Polygonum hydropiper* L. (Polygonaceae)

  Syn. : *P. flaccidum* Meissn.

  B.-Packur mul, E.-Pepper wort.

  Habit : Herb

  Habitat : Semiaquatic

  Phytography : Stoutish annual herb of wet places; leaves–subsessile, gland-dotted and linear to lanceolate in shape; flowers–small, pink in colour and arranged racemes; fruit–nut type.

  Fl. & Fr. : June–December

  Distribution : Both hilly and plain forest areas.

  Status : Infrequent, wild.

  Uses : Plant is stimulant, diuretic and emmenagogue. Liquid extract of plant is used as contraceptive.

- *Polygonum molle* D. Don (Polygonaceae)

  Syn. : *P. paniculatum* Blume., *P. rude* Meissn., *P. froudosum* Meissn.

  Habit : Herb

Habitat : Terrestrial

Phytography : Herb; leaves–simple; flowers–recemes; fruit–nut like.

Fl. & Fr. : May–January

Distribution : Himalayan forests.

Status : Frequent.

Uses : Shoot is eaten. The whole plant is used in diarrhoea.

- *Pongamia pinnata* Pierre (Fabaceae)

  Syn. : *P. glabra* Vent.

  S., H. & 8.-Karanij, E.-Pongam oil tree.

  Habit : Tree

  Habitat : Terrestrial

  Phytography : Large tree with odd-pinnate leaves. Leaves–ovate to oblong with shining upper surface; flowers–medium size, pinkish white colour and in axillary raceme; fruit–pod, 1-seeded.

  Fl. & Fr. : April–October

  Distribution : All over Indian forests, mostly on the river-banks.

  Status : Frequent, wild.

  Uses : Seeds yield oil, which is used for soap making. It is also used in tanning industry. Oil is applied on scabies; leaf juice is used in dyspepsia, diarrhoea and cough. Root and leaf powders are used as fish poison. Seed cake is a good manure.

- *Premna bengalensis* C.B. Clarke (Verbenaceae)

  B.-Dauli

  Habit : Small tree

  Habitat : Terrestrial

  Phytography : Tree; leaves–simple, elliptic to oblong or ovate; flowers-yellowish white or greenish white in large terminal panicles; fruit–drupe.

  Fl. & Fr. : April–September

Distribution : Deciduous forests of India.

Status : Frequent, wild.

Uses : Bark edible. Wood is used for house building.

- *Premna herbacea* Roxb. (Verbenaceae)

Syn. : *Pygmaeopremna herbacea* Moldenke

H.-Bharangi, B.-Bhui Jam

Habit : Herb.

Habitat : Terrestrial, prostrate.

Phytography : Small herb; leaves–sessile obovate and closely appressed to the ground; flowers–very small white and arranged in small corymb; fruit–drupe.

Fl. & Fr. : April–September

Distribution : Sub-tropical Himalayas, Bihar, Orissa, Western Ghats.

Status : Vulnerable, wild.

Uses : Root powder is used in asthma, dropsy. Leaves are generally prescribed in cough and rheumatism.

- *Premna latifolia* Roxb. (Verbenaceae)

H.-Bakar, B.-Gohara

Habit : Tree

Habitat : Terrestrial

Phytography : Small scraggy tree or shrubby with usually ovate sometimes elliptic leaves; flowers–arranged in hairy panicles; fruit -globose drupe.

Fl. & Fr. : March–October

Distribution : West Bengal, Orissa, Bihar and in dry forests of India.

Status : Infrequent.

Uses : Leaves are used in curries. Leaf decoction is diuretic. It is used in dropsy.

- *Pouzolzia zeylanica* (L.) Benn. (Urticaceae)

Syn. : *P. indica* Gaudich., *P. tuberosa* Wight.

Habit : Herb

Habitat : Terrestrial

Phytography : Perennial herb with alternate leaves; flowers–small and sessile in axils of leaves; fruit–small achene with brittle, shining pericarp.

Fl. & Fr. : July–October

Distribution : Throughout Indian forests.

Status : Common.

Uses : Tuberous roots are edible. Leaves are eaten as vegetable. Consumption of leaves as vegetable expells intestinal worms. Leaf juice is used as galactagogue.

- *Psoralea corylifolia* L. (Fabaceae)

S.-Bakuchi, H.-Babchi, B.-Bavchi.

Habit : Shrub

Habitat : Terrestrial

Phytography : Small shrub with sparsely hairy striate branches; leaves-simple broadly ovate and long petioled; flowers–heads of small purple flowers on peduncles as long or exceeding the leaves; fruit–pod ovate and indehiscent.

Fl. & Fr. : May–November

Distribution : Throughout India.

Status : Endangered in wild condition. Cultivated.

Uses : Seeds are laxative and diaphoretic. Seed oil is applied on leucodema, leprosy and psoriasis. Seed yields psoralen and iso-psoralen as active principle. Seed cake is used both as feed and manure.

- *Pterospermum acerifolium* Willd. (Sterculiaceae)

S.-Karnikara, H. & B.-Kanak Champa.

Habit : Tree

Habitat : Terrestrial

Phytography : Large tree with simple, broad, large and leathery leaves; flowers–large, axillary or terminal and yellow in colour; fruit–woody and loculicidally 5-valved capsules.

Fl. & Fr. : April & July

Distribution : Deciduous forest of India.

Status : Infrequent.

Uses : Wood is used for house building, toy making and walking sticks. Flowers are used for inflammation and tumour. Leaves are used for thatching huts and making plates.

- *Pueraria tuberosa* DC. (Fabaceae)

H.-Sural, B.-Shimia batraji, E.-Indian kudzu

Habit : Herb

Habitat : Terrestrial

Phytography : Very large woody climber; flowering when leaf less then very handsome; leaves–are pinnately tri-foliate and stipulate; flowers -bluish purple and terminate on the nodes of racemes; fruit–pod.

Fl. & Fr. : May–November

Distribution : Western Himalayas, West Bengal, Orissa, Bihar and Punjab.

Status : Infrequent, wild.

Uses : Tuberous roots are eaten. Root is demulcent. Root paste is applied on swollen joints.

- *Randia spinosa* Poir (Rubiaceae)

Syn. : *R. dumetorum* Poir., *R. brandisii* Gamble; *R. longispina* W. & A., *R. tomentosa* W. & A. non Blume., *Xeromphis spinosa* Keay.

S.-Madana, H. & B.-Mainphal, E.-Common emetic nut.

Habit : Shrub

Habitat : Terrestrial

Phytography : Large shrub or small tree usually armed with straight axillary thorns; leaves–mostly clustered on short branchlets and ovate in shape; flowers–white and in cymes; fruit–globose berry.

Fl. & Fr. : April–November

Distribution : Deciduous forests of India.

Status : Infrequent.

Uses : Fruits are edible after boiling. Unripe fruits are used in diarrhoea and dysentery during pregnancy. Leaves are used as cattle fodder.

- *Ranunculus sceleratus* L. (Ranunculaceae)

Jaldhania, E.-Celery-leaved crow foot.

Habit : Herb

Habitat : Terrestrial

Phytography : Annual or perennial herb with alternate, lobed and stipulate leaves; flowers–small, white or yellow and in panicles; fruit is a head or spikelet of beaked achene.

Fl. & Fr. : June–December

Distribution : Himalayas, West Bengal.

Status : Vulnerable

Uses : Tuber causes blisters. It is toxic to animals. Plant is stimulant and diuretic. Plant juice is used in dysuria and asthma. Root drug is used as vermifuge. Seeds are used in kidney troubles.

- *Rauvolfia serpentina* Benth. ex Kurz. (Apocynaceae) **(Plate-31)**

S.-Sarpagandha, H.-Chandrabhaga, B.-Chandra, E.-Serpentine.

Habit : Herb

Habitat : Terrestrial

Phytography : Pretty glabrous herb with shining leaves; leaves–opposite, oblong or ovate, and whorled; flowers–small, reddish white and in cymes; fruit–drupe.

Fl. & Fr. : April-September

Distribution : Throughout India.

Status : Rare in wild state, cultivated.

Uses : Roots are used in nervous disorders, in schizophrenia insanity, insomnia and epilepsy. Root extract stimulate uterine contraction. Root decoction is used in treatment of hypertension and as sedative and tranquillizer. Active principles are alkaloids like serpentine, ajmaline, ajmalinine etc.

- *Rauvolfia tetraphylla* L. (Apocynaceae) **(Plate-32)**

  Syn. : *R. canescebs* L., *R. heterophylla* Roem. & Schult. H.-Barachandrika.

  Habit : Shrub

  Habitat : Terrestrial

  Phytography : Small spreading shrub; leaves–usually in two unequal pairs in a whorl; flowers–small and in terminal or axillary cymes; fruit-globose berry of red in colour.

  Fl. & Fr. : April–November

  Distribution : Forests of moist and hot regions of India.

  Status : Infrequent.

  Uses : Roots are used as substitute of *R. serpentina*. Alkaloid reserpine content is low in comparison to *R. serpentina*. Plant extract with castor oil is applied on skin diseases.

- *Rheum emodi Wall.* ex Melssn. (Polygonaceae)

  Syn. : *R. australe* D. Don.

  S.-Gandhini, H.-Hindirevandchini, B.-Bangalarevanchini, E.-Indian rhubarb.

  Habit : Herb

  Habitat : Terrestrial

  Phytography : Herb; leaves–simple; flowers–spadix; fruit–berry.

  Fl. & Fr. : May–October

  Distribution : Sub-Alpine and Alpine Himalayas.

  Status : Vulnerable, wild.

  Uses : Source of Indian rhubarb-used as purgative. It is good in dyspepsia. Leaves are eaten after boiling. The active principle 'rutin' is present in leaves.

- *Rhizophora mucronata* Lam. (Rhizophoraceae) **(Plate-33)**

  Syn. : *R. mangle* Roxb. Non. L.

  B.-Bhora Kamo, E.-True mangrove.

  Habit : Tree

  Habitat : Mangrove–a halophyte.

  Phytography : Handsome small evergreen tree standing on stilt roots with elliptic or elliptic coriacious leaves; flowers in long pedicelled few-fid cymes; fruit–brown at time of germination and obpyriform.

  Fl. & Fr. : April–September

  Distribution : Along muddy shore and tidal creeks of India.

  Status : Infrequent, wild.

  Uses : Bark is richest source of tannins used in leather industry. Bark decoction is used in hemorrhages and angina. Fruits are eaten. Wood yields good charcoal.

- *Rhododendron arboreum* Sm. (Ericaceae) **(Plate-34)**

  H.-Burans, B.-Baras, E.-Tree Rhododendron

  Habit : Tree

  Habitat : Terrestrial

  Phytography : Tree; leaves–simple; flowers–racemes; fruit–berry.

  Fl. & Fr. : March–September

  Distribution : Temperate Himalayas, Khasia hills, Nilgiris.

  Status : Infrequent, wild.

  Uses : Wood is used in making tool handles. Flowers are eaten. Flowers are used in liquor. Petals used in diarrhoea and dysentery.

- *Rhododendron campanulatum* D. Don. (Ericaceae)

  H.–Cherailer, B.–Ghentaboras.

  Habit : Small tree

  Habitat : Terrestrial

  Phytography : Tree; leaves–simple; flowers–racemes; fruit–berry.

Fl. & Fr. : April–October

Distribution : Alpine Himalayas

Status : Abundant, wild.

Uses : Leaves used in syphilis and sciatica. It is poisonous to livestock. Flowers yield some essential oil.

- *Rhus griftithii* Hook. f. (Anacardiaceae)

Bhalayo; Darmi-asing.

Habit : Tree

Habitat : Terrestrial

Phytography : Tree; leaves–simple; flowers–raceme; fruit–drupe.

Fl. & Fr. : December–May

Distribution : Himalayan forests.

Status : Abundant, wild.

Uses : Leaves are eaten as vegetable. Wax is obtained from fruits and is used for lipsticks. Galls on branches are used as expectorant, stimulant, diarrhoea and dysentery. Leaf extract shows anti cancer and antiviral activities.

- *Rhus succedanea* L. (Anacardiaceae)

H. & B. : Kakra singi, E-Wax tree.

Habit : Tree

Habitat : Terrestrial

Phytography : Tree; leaves–simple; flowers–racemes; fruit–drupe.

Fl. & Fr. : December–June

Distribution : Temperate Himalayas and Khasia hills.

Status : Abundant.

Uses : Leaf juice creates blisters on skin. Kernel yields fatty oil. Thorn like out growth on branches is used diarrhoea and dysentery of children.

- *Rivea hypocrateriformis* Choisy (Convolvulaceae)

H.–Phong, B.–Ban pui. E.–Mindapur creeper.

Habit : Creeper.

Habitat : Terrestrial

Phytography : Creeper with silky branches; cordate leaves often broader than long and white while with appressed silky hairs beneath; flowers -axillary or terminal and white in color; fruit–globose.

Fl. & Fr. : June–January

Distribution : Deciduous forests of India.

Status : Infrequent, wild.

Uses : Leaves are eaten as vegetable. Roots are given to a mother after parturition.

- *Rubia cordifolia* L. Sensu Hook. f. (Rubiaceae)

  S.-Manjistha, H.-Manjit, B.-Manjistha, E.-Indian madder.

  Habit : Creeper.

  Habitat : Terrestrial

  Phytography : Creeper scrambling over bushes by means of its scabrid or minutely aculeate stems; leaves–petiolate, with hooked prickles, ovate or cordate and in whorled; flowers–yellowish white, small and in umbel; fruit–fleshy globose.

  Fl. & Fr. : April–November

  Distribution : All the hilly districts in India.

  Status : Abundant, wild.

  Uses : Roots are the source of Indian madder. Plant yields a yellow dye, a combination of purpurin and munjistin. Root tonic is antidysenteric, antiseptic. Active principle is septilin used in rhinosinal infection.

- *Rubus ellipticus* Sm. (Rosaceae)

  H.–Hinsalu.

  Habit : Shrub

  Habitat : Terrestrial

  Phytography : Long sarmentose shrub with long branches and red hairs, compound, petioled, trifoliate leaves with small prickles on mid rib; flowers in long panicles with short pedicel; fruit–yellow and raspberry.

Fl. & Fr. : September–March

Distribution : Hilly districts in India.

Status : Abundant, wild.

Uses : Fruits are edible.

- *Ruellia tuberosa* L. (Acanthaceae)

B.–Chauli

Habit : Herb

Habitat : Terrestrial

Phytography : Small herb, which is branched from a rootstock of fleshy tubers; leaves–simple, elliptic or ovate and petioled; flowers–medium in size, axillary and deep blue in colour; fruit–linear capsule.

Fl. & Fr. : April-September

Distribution : Throughout India.

Status : Common, wild.

Uses : Root is emetic, used as substitute for Ipecacuanha. Tuberous roots are used for fermentation of rice beer.

- *Rumex masitimus* L. (Polygonaceae)

H.-Jub-palum, B.-Bun-palung.

Habit : Herb

Habitat : Terrestrial, marshy.

Phytography : Short, stout, erect and annual with deeply grooved stems; leaves–simple and lanceolate in shape; flowers–green and in whorl panicle, perianth persistent and brown in small nut type fruit.

Fl. & Fr. : June–October

Distribution : Assam, West Bengal.

Status : Abundant, both wild and planted.

Uses : Plant is eaten and known to have cooling effect. Leaf paste is applied on burns. Root is purgative.

- *Ruta graveolens* L. (Rutaceae)

S.–Somalata, H.–Sadab, B.–Ermul, E.–Garden rue.

Habit : Shrub

Habitat : Terrestrial

Phytography : Shrub; leaves–simple; flowers–racemes; fruit–berry.

Fl. & Fr. : February–October

Distribution : Deciduous forests of India.

Status : Abundant, wild.

Uses : Fruit is expectorant, used in epilepsy and as fish poison. Seed powder is insecticidal. Fruits bear saponins used as natural detergent. Wood is used for making charcoal.

- *Sapindus trifoliatus* L. (Sapindaceae)

Syn. : S. *laurifoliatus* Vahl.

S.-Arishta, H.-Rectha, B.-Bara ritha, E.-South Indian Soap nut tree.

Habit : Tree

Habitat : Terrestrial

Phytography : Large evergreen trees with compound leaves; flowers-regular, small and arranged in panicles; fruit–subglobose.

Fl. & Fr. : April–September

Distribution : South Indian forests

Status : Infrequent, both wild and planted.

Uses : Fruit is emetic, anthelmintic, used in asthma; seeds yield a black dye. Wood is used for cartwheel.

- *Saraca asoca* (Roxb.) De Wilde. (Caesalpiniaceae)

Syn. : S. *indica* Auct. non. L.

S., H. & B.-Asok, E.-Ashok.

Habit : Tree

Habitat : Terrestrial

Phytography : A strikingly beautiful tree when in flowers; leaves-compound, leaflets–evenly pinnate; flowers–red in colour and arranged in corymb; fruit–flattened, oblong, firmly coriaceous or almost woody pod.

Fl. & Fr. : March–October

Distribution : Foothills of Himalayas, West Bengal and Western Peninsula.

Status : Abundant.

Uses : Bark is astringent, used in uterine diseases. Flowers are used as uterine tonic. Seeds are used as areca nut.

- *Saussurea costus* C. B. Clarke (Asteraceae)

S.-Kushtra

H. & B. : Kut. E.-Kuth.

Habit : Herb

Habitat : Terrestrial

Phytography : Herb; leaves–simple; flowers–heads; fruit–achene.

Fl. & Fr. : June–December

Distribution : Western Himalayas, Kashmir.

Status : Endangered, wild.

Uses : Roots contain essential oil. The alkaloid saussurine and bitter resin obtained from the root is used in asthma, cough and in rheumatism. The oil is very much effective in bronchial asthma.

- *Semecarpus anacardium* L. f. (Anacardiaceae) **(Plate-35)**

H.-Bhela, B.-Bhela, E.-Marking nut.

Habit : Tree

Habitat : Terrestrial

Phytography : Large tree with simple, alternate leaves; leaves–leathery, blunt, densely pubescent and strongly veined beneath; flowers–small and usually in terminal panicles; fruit–oblong or subglobose drupe with pendulous seed.

Fl. & Fr. : April–December

Distribution : Sub-Himalayan tract. Found in deciduous forests of India.

Status : Abundant, wild.

Uses : Ripe hypocarp of the fruit is eaten; plant latex and fruit juice is vesicant; kernel is edible; gum and resin is applied in leprosy; fruit extract is effective against human epidermoid carcinoma of the nasopharynx in tissue culture.

- *Sesbania grandiflora* Pers. (Fabaceae)

S.-Agati, H. & B.-Bak, E.-Swamp pea.

Habit : Tree

Habitat : Terrestrial

Phytography : Soft wooded tree; leaves–compound and even, pinnate; flowers in lax axillary racemes with slender panicles; fruit–linear flattened pod.

Fl. & Fr. : May–January

Distribution : Plains of Himalayas to Sri Lanka.

Status : Abundant, wild. Red variety is endangered.

Uses : Plant is used to provide support to betel vine, as shade plant for coconut seedlings and as windbreak in banana plantations. Tender, leaves, pods, flowers are eaten as vegetable. Wood is used for toys and gunpowder charcoal. Bark paste is applied on scabies. Flower juice used as an eye drop.

- *Sida spinosa* L. (Malvaceae)

S.-Nagbala, H.-Gulsakari, B.-Gorak chaulia, E.-Prickly sida.

Habit : Shrub

Habitat : Terrestrial

Phytography : Small shrub; leaves–simple, only sometimes cordate at base; flowers–solitary or fascicled; fruits consisting of separating ripe carpels generally 2-awned at the tip and opening irregularly to admit of the fall of the seed.

Fl. & Fr. : April-September

Distribution : Found throughout deciduous forests of India.

Status : Common, wild.

Uses : Root bark decoction is used in gonorrhoea. Root decoction is tonic and diaphoretic.

- *Smilax ovalifolia* Roxb. (Liliaceae)

  Syn. : *S. macrophylla* Roxb. Non-willd.

  H.-Jangli-aushbah, B.-Kumarika.

  Habit : Climber.

  Habitat : Terrestrial

  Phytography : Stuot prickly climber; flowering branches mostly unarmed; leaves–broadly elliptic or orbicular; flowers in umbels on long peduncles and white in colour; fruit–globose berry.

  Fl. & Fr. : April–October

  Distribution : Throughout India.

  Status : Infrequent, wild.

  Uses : Root used as sarsaparilla applied in treatment of veneral diseases.

- *Smilax perfoliata* Lour. (Liliaceae)

  Syn. : *S. prolifera* Roxb.

  H.-Ramdatun.

  Habit : Climber

  Habitat : Terrestrial

  Phytography : Stout prickly climber and flowering branches mostly unarmed; leaves–broadly elliptic or orbicular; flowers–in umbels on long peduncles and white in colour; fruit–globose berry.

  Fl. & Fr. : April–December

  Distribution : Deciduous forests of India.

  Status : Abundant, wild.

  Uses : Root decoction is used in female diseases.

- *Smilax zeylanica* L. (Liliaceae)

  Habit : Climber

  Habitat : Terrestrial

  Phytography : Shrub; Leaves–simple; flowers–umbels; fruit–berry.

Fl. & Fr. : April–November

Distribution : Sub-Himalayan tract, deciduous forests of India.

Status : Abundant, wild.

Uses : Root is eaten in veneral and skin diseases. Leaves are also taken as vegetable.

- *Solanum erianthum* D. Don. (Solanaceae)

Syn. : *S. verbascifolium* Auct. non L.

S.-Vidari, E.-Potato tree.

Habit : Shrub

Habitat : Terrestrial

Phytography : A large shrub or small tree with white or buff stellate tomentose unarmed stems and branches; leaves–large lanceolate or elliptic; flowers–white in long peduncled cymes; fruit–yellowish berry.

Fl. & Fr. : April-October.

Distribution : Sub-Himalayan tract.

Status : Infrequent, wild.

Uses : Root decoction is used in body pain and urinary troubles. Leaf decoction is used in vaginal discharges.

- *Solanum nigrum* L. (Solanaceae)

S.-Kakamachi, H.-Makoi, B.-Gurkamal, E.-Black night shade.

Habit : Small shrub

Habitat : Terrestrial

Phytography : An erect branched herb with entire or toothed, lanceolate-ovate leaves; flowers–small, white in extra axillary umbel; fruit–globose berry.

Fl. & Fr. : February–June.

Distribution : Deciduous forests of India.

Status : Common.

Uses : Plant is antiseptic and antidysenteric. Its infusion is used in anthrax. Berries produce tonic, diuretic and used to combat heat diseases. Leaves are used in eye trouble. They are used as an adulterant of belladonna.

- *Solanum surattense* Burm.f. (Solanaceae)

  Syn. : *S. xanthocarpum* Schrad. & Wendl.

  S.-Kantakari, H.-Ketali, B.-Kantikari, E.-Yellow-berried night shade.

  Habit : Shrub

  Habitat : Terrestrial, prostrate.

  Phytography : A diffuse prostrate small shrub, sometimes woody at the base with copiously armed all over with straight yellow very sharp prickles; leaves–deeply lobed, oblong or elliptic with scattered stellate hairs; flowers–blue in extra axillary cymes; fruit-globose.

  Fl. & Fr. : April–November

  Distribution : Throughout India.

  Status : Abundant, wild.

  Uses : Root is used in cough, asthma and pain in chest. Stem is carminative. Berry is used in sore throat. Plant extract shows antiviral activity.

- *Solanum trilobatum* L. (Solanaceae)

  S.-Alarka.

  Habit : Herb

  Habitat : Terrestrial, prostrate.

  Phytography : Very prickly climbing much-branched herb with sharp recurved short compressed prickles, more or less deltoid leaves; flowers–purple blue with erect yellow stamens; fruit–globose berry.

  Fl. & Fr. : April–October

  Distribution : Coastal forests of India.

  Status : Abundant, wild.

  Uses : Berries and flowers are used in cough. Leaves are cooked and eaten as vegetable.

- *Sonneratia caseolaris* Engl. (Sonneratiaceae) **(Plate-36)**

  Syn. : *S. acida* L.f.

  B.-Orcha.

  Habit : Tree

  Habitat : Mangrove plant.

  Phytography : Small tree of tidal forest; leaves–broadly obovate or oblong with short petiole; flowers–with large calyx and linear petioles; fruit–globose much depressed capsules.

  Fl. & Fr. : May-December

  Distribution : Mangrove forests of India, Sundarbans.

  Status : Vulnerable, wild.

  Uses : Fruits are eaten both raw and cooked. Pericarp is used as vermifuge. Wood is used for making musical instruments.

- *Sonneratia apetala* Buch-Ham. (Sonneratiaceae) **(Plate-37)**

  B.-Keora.

  Habit : Tree

  Habitat : Mangrove plant.

  Phytography : Large tree with pale bark and numerous conical pneumatophores; leaves–oblong or narrow elliptic; flowers both lateral and terminal with numerous stamens; fruits are embraced by the persistent calyx.

  Fl. & Fr. : April–December

  Distribution : Mangrove forests of India, Sundarbans.

  Status : Vulnerable, wild.

  Uses : Wood is used in house building. Barks and fruits are used for extraction of tannin. Fruits are eaten as vegetables.

- *Soymida febrifuga* A. Juss. (Meliaceae) **(Plate-38)**

  H.-Rohun, B.-Rohan, E.-Indian red wood.

  Habit : Tree

  Habitat : Terrestrial

Phytography : Large tree with thick wrinkled branchlets; leaves-compound, leaflets–peripinnate; flowers–white in terminal panicles; fruit–from the large ellipsoid or ovoid pendent woody septifugal capsule.

Fl. & Fr. : November–April

Distribution : Deciduous forests of India.

Status : Infrequent.

Uses : Bank decoction is used in treatment of diarrhoea, dysentery and fever. It is applied an vaginal infections. Bark yields 17-41% tannin. Wood is used for furniture and house building.

- *Stephania hernandifolia* Walp. (Menispermaceae)

Syn. : *S. japonica* Miers., *S. rotunda* Hook. f. & Thoms. S.-Vanatiktika, B.-Akandi, E-Taperine.

Habit : Twinner

Habitat : Terrestrial

Phytography : Slender twinner with peltate leaves; flowers–axillary cymes; fruit–drupe, glabrous, endocarp compressed and horse shoe shaped.

Fl. & Fr. : June–February

Distribution : Deciduous forests of India.

Status : Infrequent, wild.

Uses : Root decoction is used in fever, diarrhoea, dyspepsia and urinary diseases. Alkaloid akandine is extracted from the root, which is useful in obstetrical practice.

- *Sterculia foetida* L. (Sterculiaceae) **(Plate-39)**

H. & B. : Janglibadam.

Habit : Tree

Habitat : Terrestrial

Phytography : Tall tree with compound digitate leaves; flowers–in terminal panicles and campannulate in shape; fruit–suppression of woody carpels forming a 5 rayed star which are not open till the seeds are ripen.

Fl. & Fr. : March-July.

Distribution : Dry deciduous forests of India.

Status : Abundant, wild.

Uses : Rosted seeds are eaten. Seed oil is used in soap making and for illuminating. Seed oil is laxative.

- *Sterculia urens* Roxb. (Sterculiaceae)

H.-Gular, B.-Kathila.

Habit : Tree

Habitat : Terrestrial

Phytography : Soft wooded tall tree with papery outer bark; leaves-simple, palmately nerved and palmately lobed; fruit–cluster of distinct coriacious carpels.

Fl. & Fr. : March-June.

Distribution : Dry deciduous forests of India.

Status : Vulnerable, wild.

Uses : Gum is used in medicine as karaya gum. It is often adulterated with the gum of *Cochlospermum religiosum.* Wood is used for toy making. Bark yields cordage. Pulverized bark is given to women to facilitate delivery.

- *Sterculia villosa* Roxb. (Sterculiaceae)

H.–Udal.

Habit : Tree

Habitat : Terrestrial

Photography : White barked tall tree; leaves–simple, palmately compound; flowers–in axially or terminal panicles; fruits are cluster of sessile membranous ripe carpels.

Fl. & Fr. : February–May.

Distribution : Deciduous forests of India.

Status : Frequent

Uses : Wood used for making matchsticks and tea-boxes. Bark yields fibre for cordage. Seeds are eaten after roasting. Gum is used in veterinary medicine.

- *Stereospermum suaveolens* DC. (Bignoniaceae)

  S.-Palala. H.-Paral, B.-Parul.

  Habit : Tree

  Habitat : Terrestrial

  Phytography : Handsome large tree with young parts tomentose and glandular; leaves–broadly elliptic or oblong; flowers–dull crimson with very fragrant and in large panicles; fruit–capsule with purple spot.

  Fl. & Fr. : March-August.

  Distribution : Dry deciduous forests of India.

  Status : Abundant, both wild and planted.

  Uses : Wood is used for house building and making agricultural implements and in charcoal preparation. Root decoction is used in asthma and cough. Leaf paste is used in chronic dyspepsia.

- *Streblus asper* Lour. (Moraceae)

  S.-Shakhotaka, H.-Siora, E.-Siamese rough bush.

  Habit : Tree

  Habitat : Terrestrial

  Phytography : Much branched tree with very short petioled rhombic-elliptic leaves; flowers–small, axillary, dioecious, very rarely monoecious; fruit–sub-globose berry, yellowish and covered by the persistent calyx.

  Fl. & Fr. : December-April

  Distribution : Deciduous forests of India.

  Status : Frequent.

  Uses : Leaves are used as fodder; berries are eaten. Leaves are used for polishing ivory and horns. Wood powder is used in Burmese cheroots. Leaf is used as tea. Latex is astringent.

- *Slryschnos nux-vomica* L. (Loganiaceae)

  S.-Karaskara, H.-Bailewa, B.-Kuchila, E.-Nux vomica.

  Habit : Tree

Habitat : Terrestrial

Phytography : Large tall tree with ovate or sub orbicular shining leaves; flowers–white or greenish-white in colour and arranged in short peduncled cymes; fruit–globose bright orange with persistent calyx.

Fl. & Fr. : May–March

Distribution : Dry deciduous forests of India.

Status : Vulnerable, wild.

Uses : Seed is used as nux-vomica in fever and nervous disorder. It is stimulant and tonic. Seeds yield brucine and strychnin. Leaves are poultice. Bark decoction used in epilepsy.

- *Strychnos potatorum* L. (Loganiaceae)

S.-Kataka, H.-Nirmali, B.-Nirmali, E.-Clearing nut tree.

Habit : Tree

Habitat : Terrestrial

Phytography : Small densely branched tree with blackish cracked bark and elliptic obovate leaves; flowers–white in dense glabrous cymes mostly at the axis of fallen leaves with campanulate corolla; fruit-globose drupe.

Fl. & Fr. : May–January

Distribution : Deciduous forests of India.

Status : Rare in wild state, sometimes planted.

Uses : Seed decoction stomachic, demulcent and emetic. Seed powder is used in diabetes and gonorrhoea. Tender fruits are used for purifying water.

- *Suaeda maritima* Dum. (Chenopodiaceae)

Syn. : S. *nudiflora* Meq.

H.-Kharilani, E.-Common Indian saltwort.

Habit : Herb

Habitat : Mangrove weeds.

Phytography : Slender herb; leaves–fleshy and slender; flowers–minute, axillary and running out into spikes, fruit–small membranous spongy utricle.

Fl. & Fr. : April–September

Distribution : Sundarbans and other coastal areas of India.

Status : Common.

Uses : Plant is eaten as vegetable during scarcity. Good saline soil binder and is used for reclamation of saline soil.

- *Symplocos racemosa* Roxb. (Symplocaceae)

S.-Lodhra, H.-Lodh, B.-Lodh.

Habit : Small tree.

Habitat : Terrestrial

Phytography : Small tree with oblong to elliptic leaves; flowers–whitish yellow and in axillary racemes; fruit–drupe, crowned by the calyx.

Fl. & Fr. : September–February.

Distribution : Dry deciduous forests of India.

Status : Abundant, wild.

Uses : Bark yields dye commonly known as *lodh*. Bark decoction is used in diarrhoea and dysentery and dropsy. Bark infusion is used as eye drop in conjunctivitis. Bark paste is used in uterine disorder.

- *Syzygium cuminii* (L.) Skeels (Myrtaceae)

Syn. : *Eugenia jambolana* Lam., *E. cuminii* Druce.

H.-Jam, B.-Jam, E.-Black plum.

Habit : Tree

Habitat : Terrestrial

Phytography : Large tree; leaves–oblong or elliptic; flowers–white, sessile in panicles; fruit–ellipsoid berry.

Fl. & Fr. : December-May.

Distribution : Throughout India.

Status : Common, both planted and wild

Uses : Wood is used for house building. Fruits are edible. Seed is used in diabetes.

- *Swertia chirayita* (Roxb. ex Flem.) Karst. (Gentianaceae)

  Syn. : *S. chirata* Buch. Ham. ex. CB. Clarke.

  S.-Kirata-tikla

  H. & B. : Chirayita, E.-Chiretta.

  Habit : Herb

  Habitat : Terrestrial

  Phytography : Herb; leaves–simple; flowers–racemes; fruit–nutlet.

  Fl. & Fr. : July–January

  Distribution : Eastern Himalayas.

  Status : Threatened.

  Uses : The drug chirata is used in fever, against asthma and liver disorder. It is an important drug of commerce.

- *Tamarindus indica* L. (Caesalpiniaceae) **(Plate-40)**

  H.-Imli, B.-Tentul, E.-Tamarind tree.

  Habit : Tree

  Habitat : Terrestrial

  Phytography : Large tree with even pinnate leaves; leaflets–small close, linear or oblong; flowers–reddish yellow in lax racemes; fruit–linear, compressed, curved pod.

  Fl. & Fr. : July-March.

  Distribution : Throughout India.

  Status : Common, both planted and wild.

  Uses : Fruit pulp is a good source of sauces and beverages; fruit pulp refrigerant, carminative and laxative. Kernels is used as food in scarcity. Flowers are good source of honey.

- *Tamarix dioica* Roxb. (Tamaricaceae)

  S.-Pisula, B.-Guj.

  Habit : Shrub

  Habitat : Terrestrial, marshy

Phytography : Gregarious shrub or small tree, leaves–scaie like and sheathing, flowers–in dense short racemes and whitish in colour.

Fl. & Fr. : January–May

Distribution : Throughout India and Himalayan valleys.

Status : Common, wild.

Uses : Good fuel and the stem is used for making polo stick. Gal is used for extraction of tannin. It yields tannin (50%). It is a rich source of pollen for bee feeds.

- *Tamarix indica* Roxb. (Tamaricaceae)

Syn. : *T gallica* Dyer.

B.-Jhau.

Habit : Shrub

Habitat : Terrestrial, marshy.

Phytography : Shrub; leaves–scale like but not sheathing; flowers–in long racemes and pinkish in colour.

Fl. & Fr. : January–May

Distribution : Coastal plains of India.

Status : Common, wild.

Uses : Bark yields 50% tannin. Good fuel. Galls yield 40-50% tannin.

- *Taxus baccata* L. (Taxaceae) **(Plate-41)**

S.-Mandu parni, H.-Thuno, B.-Buomie, E.-Common yew.

Habit : Tree

Habitat : Terrestrial

Phytography : Tree; leaves–flat needle like; flowers in cones.

Fl. & Fr. : August–March

Distribution : Eastern Himalayas.

Status : Threatened, wild and planted.

Uses : Source of alkaloid taxol is used in treatment of cancer. Leaves are antiseptic and emmenagogue. It is used in hysteria and epilepsy.

- *Terminalia arjuna* (Roxb.) W. & A. (Combretaceae) **(Plate 42)**

  S. & H.-Arjuna, B.-Arjun, E.-Arjun.

  Habit : Tree

  Habitat : Terrestrial

  Phytography : Large tree with pale greenish or gray bark; leaves-opposite, sub-sessile, elliptic; flowers–white and in shortly panicled spikes; fruit–hard drupe with long 5-angled wings.

  Fl. & Fr. : March-September.

  Distribution : Throughout India.

  Status : Common, both wild and planted.

  Uses : Wood is used for making cartwheels and agricultural implements. Bark is used for extraction of tannin. Bark powder is good for heart diseases. Bark decoction is used in fever and as tonic. Leaf juice is used in treating some ear disease.

- *Terminalia bellirica* Roxb. (Combretaceae) **(Plate-43)**

  S.-Bahira H.-Bahera, B.-Bhairah, E.-Belliric myrobalan.

  Habit : Tree

  Habitat : Terrestrial

  Phytography : Large tree with straight and tall trunk; leaves–broadly elliptic and obovate and leaves are long clustered at the end of the branchlets; flowers–greenish white and yellow in solitary axillary or extra axillary spikes, fruit–subglobose berry, gray tomentose in colour.

  Fl. & Fr. : February-July.

  Distribution : Deciduous forests of India.

  Status : Abundant, both wild and planted.

  Uses : Fruits are used as belliric myrobalan. Wood is used as cartwheels. Kernel is edible. Ripe fruit has purgative property. Seed oil is applied on rheumatic swelling.

- *Terminalia catappa* L. (Combretaceae)

  S.-Grahadruma, H.-Deshi badam, B.-Kat badam, E.-Indian almond tree.

Habit : Tree

Habitat : Terrestrial

Phytography : Large tree with wide-spreading crown and buttressed trunk; leaves–clustered towards the end of the branches, obovate are in spikes which are terminal or axillary; fruit–elliptic slightly compressed drupe.

Fl. & Fr. : March-October

Distribution : Throughout India.

Status : Common, both wild and planted.

Uses : Wood is used for house building. Kernel is edible. Leaf paste is applied on rheumatic swellings on joints. Silkworms feed on young leaves.

- *Terminalia chebula* Retz.; CB. Clark. **(Sketch-5)**

S. & H.-Harra, B.-Haritaki, E.-Chebulic myrobalan.

Habit : Tree

Habitat : Terrestrial

Phytography : Small or medium size tree with a rounded crown; leaves-usually sub-opposite ovate or elliptic in shape; flowers–whitish and in spikes; fruit–elliptic 5 ribbed drupe.

Fl. & Fr. : March-October.

Distribution : Deciduous forests of India.

Status : Infrequent, wild.

Uses : Fruits yield 30-32% tannin. Fruit is a myrobalan. It is laxative, stomachic. The active principle is a glycoside.

- *Thespesia populaea* Soland. ex Correa. (Malvaceae)

Syn. : *T. lampas* (Cav.) Dalz & Gibs., *Azanza lampas* (Cav.) Alef.

S.-Gardha-bhanda, H.-Parsipu, 8.-Dumbla, E-Portia tree.

Habit : Tree

Habitat : Terrestrial coastal areas.

Phytography : Tree with glabrous leaves; flowers–large, showy and axillary; fruit–loculicidal capsules covered with rusty coloured scale when unripe.

Fl. & Fr. : Throughout the year.

Distribution : Coastal plains of India.

Status : Common, both wild and planted.

Uses : Wood is used for making agricultural implements and musical instruments. Seed oil is applied on skin diseases. Bark is used in dysentery and hemorrhoids. Leaf is a fodder.

- *Tinospora cordifolia* (Willd.) Miers. ex Hook.f. & Thoms. (Menispermaceae)

S.-Amrita, H.-Amrita, B.-Golancha, E.-Gulancha Tinospora.

Habit : Shrub

Habitat : Climber

Phytography : Climbing shrub with cordate and membranous leaves; flowers–axillary or terminal racemes; fruit–are drupe and red in colour.

Fl. & Fr. : February–September.

Distribution : Dry deciduous forests of India.

Status : Vulnerable, wild.

Uses : Plant decoction is used in general debility, dyspepsia and urinary diseases. Bitter principle in the plant body shows antipyretic activities.

Stem extract is used in treatment of blood sugar. Fruit powder is used in jaundice.

- *Tinospora crisp* (L.) Niss ex Hook. f. & Thoms. (Menispermaceae)

Syn. : *T. rumphii* Boerl., *T. tuberculata* (Lam.) Beumee ex K. Heyne.

Habit : Shrub

Habitat : Climber

Phytography : Shrub; leave–simple; flowers–racemes; fruit–drupe.

Fl. & Fr. : February–September

Distribution : Deciduous forests of India.

Status : Vulnerable, wild.

Uses : Stem is used in fever. Starch extracted from the plant used as a tonic. It has effect on endogenous insulin secretion, glucose uptake and inhibition of peripheral glucose release.

- *Trema orientalis* Blume (Ulmaceae) **(Plate-44)**

Syn. : *T. amboinensis* Auct. non Blume., *Celtis orientalis* L.

S.-Jivanti, H.-Gio, B.-Chikan, E.-Charcoal tree.

Habit : Small tree

Habitat : Terrestrial

Phytography : Small tree with the trunk rigged with stipular scars; leaves–ovate-lanceolate, more or less white or silvery beneath with fine silky hairs; flowers–small, white and in cymes; fruit–drupe, yellow in colour.

Fl. & Fr. : May–December

Distribution : Throughout India.

Status : Common, wild.

Uses : Wood is used as paper pulp. Bark yields fiber used for rope. Root is used for treatment of blood in urine. Root bark is used in epilepsy.

- *Trewia nudiflora* L. (Euphorbiaceae)

S.-Pindara, H.-Bhillaura, B.-Pitali, E.-False white teak.

Habit : Tree

Habitat : Terrestrial

Phytography : Large soft wooded deciduous tree; leaves–opposite, broadly ovate or orbicular; flowers develop before the leaves and in racemose, fruit–fleshy indehiscent or dehiscing loculicidally.

Fl. & Fr. : April–October

Distribution : Found throughout hotter parts of India.

Status : Frequent, both wild and planted.

Uses : Wood used for agricultural implements and for making toys and drums and photo-frames. Root paste is applied on gout.

Status : Frequent, wild.

Uses : It is known as Indian squill and is used as expectorant. Bulb paste is applied on epidermoid carcinoma. Externally bulb paste is used to remove warts and corns.

- *Valeriana jatamansi* Jones (Valerianaceae)

Syn. : *V. wallichii* DC.

H.-Mushkbala, B.-Mushkbala, E.-Indian valerina.

Habit : Herb

Habitat : Terrestrial

Phytography : Herb; leaves–simple; flowers–racemes; fruit–capsule.

Fl. & Fr. : June–January

Distribution : Alpine Himalayas

Status : Threatened, wild.

Uses : Rhizome is used in hysteria, nervous unrest. Rhizome powder is used in urinary troubles. Plant yields valepotriates, a monoterpenic derivative.

- *Vanda tessellata* Lodd. ex Loud. (Orchidaceae) **(Plate-45)**

Syn. : *V. roxburghii* R. Br.

S.-Atirasa, Gandha-nakuli, H.-Banda.

Habit : Herb

Habitat : Epiphyte

Phytography : Epiphyte; stem with narrow complicate premrose and tessellated flowers, flowers–suberect and in racemes with bluish brown colour.

Fl. & Fr. : March-June.

Distribution : Deciduous forests of India.

Status : Infrequent, wild.

Uses : Leaf juice is used in otitis. Root paste is taken in dyspepsia and fever. Medicated oil is prepared from root paste and is applied on rheumatic swellings.

- *Ventilago denticulata* Willd. (Rhamnaceae)

  Syn. : *V. calyculata* Tul.

  B.-Ruktapita, H.-Raidhani.

  Habit : Climber

  Habitat : Terrestrial

  Phytography : Climber of long branches with elliptic leaves and yellowish tomentose in it; flowers–small, yellow-green and in dense panicles; winged fruit.

  Fl. & Fr. : July–January

  Distribution : Forests of hotter parts of India.

  Status : Frequent, wild.

  Uses : Bark paste is applied on sprains. Stem sap is used as a cure for deafness. Seed oil is used for cooking.

- *Ventilago madraspatana* Gaertn. (Rhamnaceae)

  S.-Rakta valli, H.-Pitti, B.-Raktapita.

  Habit : Climber

  Habitat : Terrestrial

  Phytography : Widely scandent shrub with long tomentose branches; leaves–oblong to elliptic in shape with yellow tomentose in it; flowers-small, greenish and arranged in interrupted panicles; winged fruits seated on disc like calyx.

  Fl. & Fr. : August–February

  Distribution : Forests of Deccan Peninsula, Western Ghats.

  Status : Infrequent. Wild.

  Uses : Root a source of red dye-ventilag in used for colouring. Root decoction is used in debility and dyspepsia. Seeds are eaten after cooking.

- *Viscum articulatum* Burm. f. (Viscaceae)

  Syn. : *V. nepalense* Spreng.

  S.-Gandhamadini, H-Budu, B-Mandada.

  Habit : Herb

  Habitat : Semi parasite.

Phytography : Herb with flattened branches, the leaves are reduced to scales; flowers–minute fasciculated at nodes; fruit–yellow and subglobose.

Fl. & Fr. : May–December

Distribution : Himalayas, Assam, West Bengal, Madhya Pradesh, Western Paninsula.

Status : Frequent, wild.

Uses : Plant paste is applied on piles, lumbago. It is also used as febrifuge.

- *Vitex negundo* L. (Verbenacaee)

S-Nirgundi, H-Sambhalu, B-Nisinda

Habit : Small tree

Habitat : Terrestrial

Phytography : Strongly scented large shrub or small tree; leaves are tri to pentafoliate; flowers–white in oblong panicles; fruit–globose drupe.

Fl. & Fr. : April–September

Distribution : Deciduous forests of India.

Status : Frequent, wild.

Uses : Leaves are used in tonic and as vermifuge. Leaf decoction is used as medicinal baths for catarrhal and rheumatic attacks. Root possesses tranquilizing effect. Leaf extract shows anticancerous activity. Flowers are used in diarrhoea and liver complaints.

- *Vitis peduncularis* Wall. ex Schauer (Verbenaceae)

H.-Minjurgorwa, B.-Boruna.

Habit : Small tree

Habitat : Terrestrial

Phytography : Tree; leaves–compound; flowers–in panicles; fruit-drupe.

Fl. & Fr. : April–September

Distribution : Assam, Tripura, Bihar and West Bengal.

Status : Rare, wild.

Uses : Wood is used for house construction. Infusion of barks and leaves are used in malarial fever. Leaf powder is used as antibacterial agent.

- *Wigandia speciosissima* Merrill (Hydrophyllaceae)

Ka-Slang.

Habit : Tree

Habitat : Terrestrial

Phytography : Tree; leaves–simple; flowers–raceme; fruit–capsule.

Fl. & Fr. : June-September

Distribution : Eastern Himalayas.

Status : Infrequent, wild.

Uses : Wood is used by Lamas for making idols of Buddha.

- *Withania somnifera* Dunal (Solanaceae)

S.-Ashwagandha, H.-Asgandh, B.-Asvaganda.

Habit : Small shrub

Habitat : Terrestrial

Phytography : Shrub; sub-tomentose branches with ovate leaves; flowers–greenish or yellowish in the axils; fruit–berry.

Fl. & Fr. : September–April

Distribution : Dry deciduous areas in India.

Status : Rare in wild condition.

Uses : Root powder is used in hiccup, cough, dropsy and gynaecological disorders. It is sedative in case of senile debility. It possesses many alkaloids of which withanolide is most important.

- *Woodfordia fruticosa* Kurz. (Lythraceae) **(Plate-46)**

Syn. : W. *floribunda* Salisb

S.-Dhataki, H.-Dhai, E.-Fire flame bush.

Habit : Shrub

Habitat : Terrestrial

Phytography : Much branched beautiful shrub; leaves–sessile or subsessile, linear to lanceolate with silvery dotted beneath; flowers -brilliant scarlet in colour and arranged in axillary cyme.

Fl. & Fr. : April–December

Distribution : Deciduous forests of India.

Status : Common, wild.

Uses : Flowers yield red dye. Leaves contain 12-20% tannin. Bark contains 20-27% tannin used for washing septic wounds. Flower and leaf powder are used in bowel complaints. Flowers are used for preparing cold drinks.

- *Xylia xylocarpa* (Roxb.) Taub. (Mimosaceae) **(Plate-47)**

S.-Scimsapa, H.-Jambu, B.-Boja.

Habit : Tree

Habitat : Mangrove plant.

Phytography : Tree; leaves–pinnately compound; flowers–racemes; fruit–woody pod.

Fl. & Fr. : June–January.

Distribution : Mangrove forest. Sundarbans.

Status : Vulnerable, wild.

Uses : Substitute of Sal and Teak. Wood lasts well under water. Bark decoction is used in gonorrhoea. Fruits yield a oil used for lamping.

- *Zanthoxylum armatum* DC. (Rutaceae)

Syn. : *Z. alatum* Roxb., *Z. planispinum* Sieb. and Zucc.

S.-Tumburu, H.-Darmar, B.-Gaira.

Habit : Shrub

Habitat : Terrestrial

Phytography : Shrub; leaves–pinnately compound; flowers–racemes; fruit–berry.

Fl. & Fr. : May–January

Distribution : Sub-tropical Himalayan forests. Khasia hills, Kumaon.

Status : Vulnerable, wild.

Uses : Bark and seeds are used as carminative and anthelmintic. Fruits are used in dyspepsia. It yields oil used as antiseptic. Barks are used to toxicate fish. Natural source of Iinalool-a chemical much used in perfumery. Wood is used for walking sticks. Leaves are used as feeds for *muga* silk worms. *Z. oxyphyllum* is used as a substitute of this species.

- *Zizyphus mauritiana* Lam. (Rhamnaceae) **(Plate-48)**

Syn. : *J. jujuba* Lam. non Mill

S.-Ajapriya, H.-Baer, B.-Kool. E.-Common jujube.

Habit : Tree

Habitat : Terrestrial

Phytography : Small tree with alternate ovate leaves, leaves–stipules transformed into sharp hooked or straight prickle; flowers–small, white and in peduncled cymes; fruit–globose drupe.

Fl. & Fr. : September–December

Distribution : Deciduous forests of India.

Status : Common, wild.

Uses : Leaves are good fodder. Fruits are edible. Kernel is sedative. It is used against aconite-poisoning. Wood is used for making agricultural implements.

- *Zizyphus oenoplia* Mill. (Rhamnaceae)

S.-Srigalakoli, H.-Makai, B.-Siakul, E.-Jackie jujube.

Habit : Shrub

Habitat : Terrestrial

Phytography : A prickly shrub with alternate, ovate leaves; flowers-small, greyish white and in cymes; fruit–globose drupe.

Fl. & Fr : September–December

Distribution : Dry deciduous forests in India.

Status : Common, wild.

Uses : Fruits is edible. Bark yields 12% tannin. Fruits are used in stomach-ache. Root contains cyclopeptide. Alkaloids zizyphine-A and zizyphine-B are found in the fruit.

- *Zizyphus xylopyra* Willd. (Rhamnaceae)

  Syn. : Z. *glaberrima* Santapau

  H.-Kat-ber.

  Habit : Shrub

  Habitat : Terrestrial

  Phytography : Straggling and thorny shrub; leaves–ovate in shape hairs; flowers–green in axillary cymes; fruit is globose drupe.

  Fl. & Fr. : September–April.

  Distribution : Deciduous forests of India.

  Status : Infrequent, wild.

  Uses : Leaves are used as good fodder. Fruits possess 12% catechol type of tannin. Wood is used for making agricultural implements and charcoal.

**Abbreviations* : H = Hindi, B = Bengali, E = English, S = Sanskrit, Fl. = Flowering, Fr. = Fruiting, Syn. = Synonym.

CHAPTER 35

KOKUM

A Unique Spice and Medicinal Tree

Kokum *(Garcinia indica choisy)* is one of the most important indigenous spice and medicinal plant of Konkan region. Its cultivation is confined in the coastal hilly region of Maharashtra and Goa states and is popularly know as "Ratamba". Some Kokum trees are also found in western ghat forests of Karnataka, Kerala as well as in parts of West Bengal, Assam and Gujarat.

Kokum belongs to family Crucifereae (previously Guttifereae) (2n = 54). There are approximately 400 species of *Garcinia* identified so far, of which 40 are known to be edible. About 30 species occur in India. Amongst them *G. cambogia, G. morella* and *G. indica* yield nondrying edible fat. *G. mangostina* is considered to be the most delicious among the tropical fruits.

The fruit is used as dessert. It can be made into various preserves. A full grown tree of Kokum attains a height of about 16 to 20 meters. Tree is a pyramid shaped evergreen dioecious having drooping branches. It's leaves are ovate to oblong, lanceolate; broad, dark green above and pale beneath. The flowers of Kokum are unisexual, axillary or terminal and solitary. The fruits are generally globose or spherical dark red when ripe, enclosing 5 to 8 large seeds. Kokum tree starts flowering during the month of November-December and fruits are harvested during April-May. It takes about 8 to 10 years for commercial bearing when grown by seeds. It has characteristic slow growth but very long life and propagated by root suckers and seed. The grafted plants remain dwarf due to strong plagiotropic behaviour.

Medicinal, Food and Industrial Uses

Kokum is valued for its numerous medicinal properties. It is an anthelmentic and cardiotonic. It is also useful in curing piles dysentery, tumors pain and cardial problems. It is widely used for garnishing fish curries, coconut curries and is a good source of acid for coagulating rubber. The fruit of Kokum is of high commercial importance, the rind of which is used to import acid taste to curries like tamarind and is locally known as 'Amsul'. A very popular product Amrit Kokum, which is a syrup, is usually prepared from the rind of Kokum. Besides Kokum Nector, Squash, AgaI are also prepared from the rind. The Kokum fruit contains 'Hydroxy citric acid' which has great medicinal value for reducing the obesity without side effects. Kokum seed is also an important constituent.

The kernel in the seed contains about 33 to 44 per cent oil which is popularly known as Kokum butter. It is white in colour with yellowish tinge. It remains solid at room temperature. Importantly it has good export potential and is used in preparation of ointments, cosmetics and suppositories and many pharmaceutical preparations. Ointment is used for local application to ulceration fissure of lips, hands and feet etc. Due to high content of disaturated and monosaturated glycerides, it is in demand as a substitute for cocoa and extender in chocolate and confectionery products.

Package of Practices for Kokum

Soil and Climate : Being a tropical crop, it grows well within a temperature range of 25° to 30°C and thrives best in an area receiving annual rainfall between 2000 to 3000 mm. It is hardy crop and can be grown well in all types of soils with acidic reaction, especially lateritic soils, with poor fertility. The only prerequisite is well drainage of soil. Luxurious growth and high yield potential is realized under Konkan condition having warm and humid climate.

Propagation : Earlier, it was a common practice to propagate Kokum on commercial scale by seed, which shows a lot of variation in yield potential and quality of fruits due to cross pollination. On account of its dioecious nature, seeding progeny gives 50% of each sex. The sex of the seeding can be determined only when it starts flowering *i.e.* after 6 10 8 years. To overcome this problem, vegetative propagation has been standardized by Dr. Balasaheb Sawant Konkam Krishi Vidyapeeth, Dapoli. Inarch grafting with 10 to 18 month old seeding gives 90% success, when done in the month of Dec.-Jan. Recently, softwood grafting has been found to be successful and easier than inarching. The mature scion of 5 to 6 month old can be preferred for softwood grafting. Defoliation of scion as well as root stock is not needed. Kokum seeding at the age of 22 weeks and more can be successfully used as a rootstock for softwood grafting and October is the best season for softwood grafting. The graft could be successfully maintained either in polyshed or under open sun after grafting.

The existence of tropism has been noticed in this crop. The grafts prepared by using plagiotrophic branches do not get the normal erect shape as that of mother tree. It becomes dwarf, irregular bushy or viny in shape. The internodal length is substantially reduced. To overcome this problem trials are in progress at the R.F.R.S.; Vengurle, Dist. Sindhudurg. However, given proper training and pruning to such plagiotrophic dwarf grafted plants for 4 to 5 years, it is possible to do high density planting of Kokum with ease of harvesting and other management practices. Exposed roots also give rise to orthotrophic erect plants naturally and extensive research needs to be done to exploit this property.

Preparation of Land and Planting : Before planting, the land should be cleared by removing unwanted bushes and trees. In the month of April-May pits of 60×60×60 cm are dug at 6×6 m with spacing and are filled with good soil mixed with about 10 kg F.Y.M. In each pit 100 gm of linden powder may be mixed to protect seedlings from termite attack. At the onset of monsoon, one year old vigorously growing grafts can be planted per pit after milling 1 kg single super phosphate in the pit. At the time of planting rootball should be kept intact and graft union should be kept above the soil surface. Planting 10% wale plants/grafts improves fruit set.

After planting, grafts/seedling are required to be protected from stray cattle. In the first year yound graft should be protected from scorching heat by providing overhead shade from October onward. For initial two years, about 15 litres of water be given per week per plant during winter and summer months. Mulching by dry grass may be done in the basin to conserve the soil moisture and prevent weed growth. Most of the time water sprouts are seen below the graft joint, if graft are planted. The water shoots should be removed as and when observed and plants should be staked erect for 3-4 years.

Manures and Fertilizers : Since Kokum is a perennial crop bearing fruits every year, it is necessary to give manures and fertilizers for higher yield, even though natural plants grow without fertilizer. In the first year, each plant should be applied with 2 kg F.Y.M. 50g N, 25 g P_2O_5 and 25 g K_2O.

This dose is increased every year and from tenth year onward each plant may be given 20 kg F.Y.M. 500g N. 250g P_2O_5 and 250g K_2O. Manures and fertilizer should be applied in a circular trench prepared under the canopy of the plant (at 50% distance of canopy line) during the month of August, after heavy rains are over. During early 10 to 12 years after planting, rainfed crops like sweet potato, vegetables, flowering annuals can be grown as inter crops.

Harvesting : In Kokum, flowering starts from October-November and continues upto February. Fruits are ready for harvest during the month of March to June. When fruits turn from green to reddish in colour, they are plucked carefully by hand. The well grown Kokum tree from 15 years onward yields about 30 to 50 kg fruits per year. Raw fruits are sometimes harvested and used for brining.

Pest and Disease : Kokum, being a hardy crop, is seldom affected by pests and diseases. Many times leaf minor is noticed which can be easily controlled by spraying Dimethode 10 ml per 10 lit of water. Dieback disease can be controlled by removing deadwood, followed by spray of 1% Bordeaux mixture.

Economics of Establishing Kokum Plantation : For establishing the Kokum garden, the cost for five years comes to Rs. 62,884/ha excluding cost of land. In the first year the maximum cost of establishment is required (Rs. 20,780). This expenditure is incurred in the first year and involves maximum contribution from labour (29%). The economical payback from plantation starts from 7th year.

The plantation required production cost of Rs. 17960 whereas total return realized is Rs. 21180. It is found that the Kokum plantation gives net profit 9 years onwards. For recovery of establishment cost of Kokum plantation 8 years are required. After 20 years approximately farmer can obtain net profit of Rs. 43,226 per hectare *i.e.* Rs. 156 plant (277 plant/ha).

Research Contribution

In spite of favourable agro-climatic condition in the coastal zone and amazing spice and medicinal properties, systematic Kokum cultivation is not done so far in the region. Mostly, the Kokum trees are found scattered on roadsides, in jungle, backyard of houses, waste land and in coconut and Arecanut plantation. This is because of certain specific features of this crop.

1. Dioecionouses of the plant.
2. Fifty per cent of plants being male and are unproductive.
3. Large variation in quality and quantity of fruits in seeding progeny.
4. Late bearing, harvesting season extending in rainy period leading to loss of fruits.
5. Grafted plants are unable to take normal shape and architecture.
6. In view of several constraints as above and future importance of the crop Dr. B.S. Konkan Krishi Vidyapeeth initiated research on this precious crop.

Table 1 : Characteristics of Kokum accessions

S. No.	*Accession No.*	*Character*
1.	kv-18	Big size fruit with attractive colour
2.	kv-19	Early type with big size and attractive colour fruit.
3.	kv-20	Red attractive colour
4.	kv-21	Attractive blood red colour
5.	kv-22	Red with white strips on fruit
6.	kv-23	Early type
7.	kv-27	Early type with heavy yielder
8.	kv-29	Early bearing type
9.	kv-32	Early flowering male plant
10.	kv-35	Very early male type
11.	kv-38	Very early male type

After studying and evaluating the local types collected from all over the Konkan for several years, the variety, 'Kokan Amrita' which is the first variety of Kokum in India was released in 1998.

For development of superior genotypes the germplasm of Kokum is being strengthened further for the extraordinary Kokum types from the natural bio-diversity present at the regional level for following characteristics.

1. Earliness
2. Big size fruit with high juice content
3. Better processing characters
4. Attractive colour and shape of fruit

Some of the accession collected recently at the Regional Fruit Research Station, Vengurla for these peculiarities are described in (Table 1).

First national seminar and exhibition on Kokum was organized at RFRS, Vengurla in May 2000. The exhibition provided opportunity to identify elite Kokum genotypes from the existing biodiversity in Western Ghats.

In all 108 elite type of Kokum have been collected and assessed for fruit colour, fruit weight, pulp weight, seed weight, pulp to fruit ratio and seed to fruit ratio.

On the basis of these character ten promising types have been identified (Table 2). These types will be studied for their physio–chemical characters subsequently.

Table 2 : Physical characters of promising Kokum types

Accession No.	*Weight of the fruit (gm)*	*Rind weight per (gm)*	*Seed weight per (gm)*	*Rind/Fruit ratio*	*Seed/Fruit ratio*
VGL 01133	83.33	46.67	36.00	0.560	0.432
VGL 01021	83.20	44.80	36.60	0.538	0.452
VGL 01072	67.20	30.40	36.80	0.452	0.548
VGL 01136	59.60	37.60	20.00	0.631	0.336
VGL 01189	54.40	21.20	32.40	0.389	0.596
VGL 0110	58.40	30.00	27.20	0.514	0.466
VGL 01058	48.80	18.80	28.80	0.385	0.590
VGL 01130	57.00	27.50	29.50	0.482	0.517
VGL 01005	49.60	28.80	20.30	0.580	0.410
VGL 01162	35.20	25.60	9.20	0.727	0.261

The data indicated that the fruit weight was maximum in accession No. 01133 and 01021 *i.e.*, 83.33g and 83.20g per fruit, respectively. Similarly, rind weight was also maximum in accession No. 01133 and 01021 *i.e.*, 46.67g and 44.80g per fruit, respectively.

The maximum rind/fruit ratio was 0.727 in accession No. 01162 while for seed purpose, the accession No. 01189 was promising with seed/fruit ratio of 0.596.

Post Harvest Technology

Kokum fruits require processing and several products can be prepared from it.

1. **Dried Green Kokum Rind :** This product is prepared from the mature green Kokum fruits. They are cut into four longitudinal pieces and kept in solution of 2500 ppm potassium metabisulphide for two hours and then dried at 50°C temperature either in dryer or in sun. Other products made from ripe fruit are as follows :

1. Kokum syrup
2. Dried ripe Kokum rind
3. Kokum agal (salted syrup)

The dry Kokum rind fetches good price in market. However, hygienic drying of rind for late maturing fruits in June is an urgent need to mitigate loss due to rainfall.

CHAPTER 36

ALOE BARBADENSIS

A Useful Plant

The word, "Aloe" is derived from the Arabic word "alloeh", which means "bitter and shiny substance". Aloe describes more than three hundred fifty closely related plants that grow all over the world. Aloe has been used in folk remedies around the world for over 2000 years. The ancient Chinese used it to relieve stomach ailments.

The South and Central American Indians use aloe gel to treat kidney and bladder problems and to increase longevity. In India, Aloe medicines are used for a variety of conditions, particularly for their cathartic, stomachic, emmergagogic and anthelminthic properties. Cultivated throughout India in many varieties some of which run wild as on the coasts of Mumbai, Gujarat and South India. *A.barbadensis* have become completely naturalised in India especially in the hot dry valleys of north-western Himalayas and throughout the central table-land extending as far as Cape Comorin.

The Aloe plant grows in different sizes. It can be 10 inches to 30 feet. It grows in a rosette formation. This shape helps the Aloe plant catch as much rain as possible. It also helps shade itself from the tremendous heat of the desert.

The Aloe plant varies in colour from light to dark green. It usually is darker tnan regular cactus. The plant has long flat leaves. The edges have points but they are not sharp. It has a waxy coating to keep the moisture inside.

Cultivation

(1) **Preparation of Land** : The field should be ploughed through two to three times before cultivation.

(2) **Propagation** : Aloe barbadensis propagation should be conducted at any specified center.

(3) **Sowing** : Sowing of Aloe plant should be done in the month of July–August, so that initially plant will get sufficient water. The plant should be in a line, distance between line should be 1.0 to 1.5 ft., distance between consecutive plant in a line should be I ft. For one hectare field, it requires about 40-50 thousand plant.

(4) **Irrigation** : Not required

(5) **Diseases & their control** : Not required.

(6) **Harvesting & Storage** : Aloe plant can be harvested of the grown leaves every three months with the help of sharp knife. The remaining part will again develop in next 3 months time and so on.

(7) **Production and Yield** : 500 q. per hectare.

(8) **Market Price** : Aloe plant fresh leaves cost @ Rs. 2 to Rs. 5 per kg.

Uses

Historically, we find that Aloe plant has been used to treat human and medical problems from A to Z.

A. Allergies, abscesses, abrasions, asthma, acne, acid indigestion, allergic reaction, anemia, arthritis, athlete's foot, AIDS.

B. Bad breath, burns, boils bursitis, baldness, blisters, bruises, bronchitis, bloddy scours in calves, body cleanser, bladder infections, blood pressure.

C. Corneal, ulcers, contusions, cuts, cataracts, chapped skin, coughs, colds, colitis, carbuncles, colics, cradle, cap, cystitis, candida, chemotherapy, constipation.

D. Dermatitis, dandruff, dry skin, denture sores, diaper rash, dishpan hands, dysentery, diabetes, depression.

E. Edema, erysipelas, epidermis, Epstein-Barr virus, exanthema, enteritis in fowl, eyes, ear ache.

F. Favus, fissured nipples, fever blisters, fungus.

G. Genital Herpes, gingivitis, glaucoma, gangrene.

H. Heat rash, headache of all kinds, hemorrhoid, heart burn, high blood pressure.

L. Impetigo, inflammed joints, insomnia, ingrown toenails, infertility, due to anovulatory cycles, indigestion, insect bites.

J. Jaundice, joints.

K. Keratosis follicularis, kidney infections.

L. Laxation, leprosy, laryngitis, lupus, liver ailments, leukemia

M. Multiple sclerosis, mastitis in diary cattle, mouth irritations, muscle cramps strains, moles.

N. Nausea of all kinds.

O. Onycholysis, odor control of chronic ulcers, oral disorders.

P. Pin worms, psoriasis, prostatitis, poison, pancreas.

R. Razor burn, radiation burns, radiation dermatitis, rashes.

S. Stings, styes, sprains, senile moles, sores of all kind, seborrhea, stretch marks, sore throat, shingles, staph infection, sunburns, sciatic nerve, sicklecell disease.

T. Tonsilitis, tendinitis, trachoma, tuberculosis.

U. Ulcerations of all kinds, urticaria, ulcers.

V. Vaginitis, venereal sores, venous stasis, varicose veins.

W. Wind burns, wheal, wounds of all kinds, warts.

X. X-ray burns.

Y. Yeast infections.

Z. Zoster.

Aloe is known to contain well over 100 separate ingredients or constituents between those found in the leaf and those found in the mucilaginous gel inside the leaf. According to Indian Herbal Pharmacopoeia, the chemical constituents are :

Major : Hydroxyanthraquinone derivatives (25-40%), aloin and 7-hydroxyaloin isomers.

Minor : Include aloe emodin, chrysophanol, aloeresin B with its p-coumaryl derivatives and aglycone aloesone.

Bacteria and Fungi known to be eliminated

Bacteria	*Causes*
Streptococcus pyogenes	Rheumatic fever, strep throat and scarlet fever
Serratia marcescens	Endocarditis, pneumonia
Klebsiella pneumoniae	Pneumonia
Escherichia coli	Diarrhoea and fatal food poisoning
Staphylococcus aureus	Food poisoning and toxic shock syndrome
Pseudomonas aeruginosa	Severe and fatal blood poisoning
Candida albicans	Vaginal, respiratory and skin infection, thrush and endocarditis
Trichophyton	Fungal infection of the skin, nails.
Mycobacterium tuberculosis	Tuborculosis, lupus and erythematosus

The gel of Aloe plant has been known to have healing powers for centuries. Aloe extracts can eliminate dozen of harmful bacteria. They are :

1. Antibacterial findings : Leaves of Aloe plant were cut and allowed the juice to drain out. This juice inhibited Staphylococcus aureus when fresh, but the activity was unstable unless the extract was refrigerated or heated and then freeze dried. The freeze dried extract inhibited some species but the other parts did have these properties.

It was found to have antibacterial activity in the pericyclic juice, particularly against gram-positive bacteria. It was also reported immunostimulant properties in an extract of Aloe plant.

2. Anti inflammatory findings : A. berbadensis gel has some kind of anti-inflammatory effect. A material having anti-brady kinin activity on isolated guinea pig ileum was partially purified from the nondialysate of the pulp of Aloe by repetition of gel chromatography using a hydrophilic polyvinyl gel and dextran gels.

If such a material was present in Aloe gel, then it would cause breakdown of bracykinin and activation of angiotensin to reduce pain and inflammation.

The lectin Aloctin A was found in Aloe gel and this Aloctin A had anticancer effects. Some found that Aloe gel had antiprostagladin effects. Prostaglandins and thromboxanes are compounds involved in the long term inflammatory response in damaged tissue.

They had number of different physiological effects including promotion of fever and pain. There is some evidence that Aloe gel inhibits the synthesis of prostaglandins from arachidonic acid. The % inhibition was found to increase according to the concentration of commercial product. Thus in clinical application *to* burns and similar lesions Aloe product could cause recent studies supports this theory. It was suggested that Aloe gel contains anthraquinone and related compounds which helps as healing agent on skin.

3. Anaesthetic Findings : Several biologically active substances such as aloenin, magnesium lactate, aloe emodin, barbaloin and succinic acid were found to be contained in the leaf juice of Aloe barbadensis which has widely been used in domestic medicines. Aloenin and magnesium lactate were elucidated to exhibit an inhibitory action on the gastric juice secretion.

Pain relieving properties to Aloe gel and found salicylate, lactate and magnesium responsible for the anaesthetic property could either be due to an aspirin like effect or the high magnesium ion content or possibly both acting synergistically. Also it was possible that anthraquinone type compounds such as emodin and barbaloin could be broken down by the kolbe reaction to solicylates.

4. Analgesic Findings : The analgesic effect of evaporated juice from the leaves of Aloe barbadensis was analysed using a rat tail hot wire method. It was found that Aloe extract increased the latent period to 15 s. 60 min after injection. Compared to a latent period of 12 s before the treatment.

The analgesic quality was not as strong as that of the pain relieving drug Novalgin.

5. Thermal Burns : Aloe gel on thermal burns tested in standard guinea pig experiment. The depth of dermal ischaemia was measured by perfusion with Indian Ink. The control animals were burned and untreated while other groups were burned and treated every 8 h with methyl prednisolone, methimazole or a commercial 70% Aloe gel cream. It was found that Aloe cream had similar effect to the methyl prednisolone and methimazole giving improved perfusion of capillaries.

6. Anti Fungal Findings : Different extracts of both fresh and dry leaves of Aloe were screened for their antifungal activity against Aspergillus nigher, Cladosporium herbarum and Fusarium moniliforme. The toxicity of the isolated pure components were evaluated on the tested fungi.

Immuno-stimulatory Findings : The present study together with previous observations show that Aloe gel contain small molecular weight immunomodulators that prevent UVB induced immune suppression in the skin by restoration of UVB induced damages on epidermal cell.

Superoxide Dismutases : Extracts from the parenchymatous leaf gel and the rind of the Aloe plant were shown to contain seven electrophoretically identifiable superoxide dismutases.

7. Opthalmic Solution : The dry eye syndrome is a very common disorder that affects mainly contact lens users and post menopause women. It is also known as keratoconjuntivitis sicca.

Current available treatments for the dry eye are not fully effective because they do not induce re-eipthelization of the cornea and repairing of the physiology of the eye. Aloe solution which have been successfully demonstrated in the treatment of internal and external wounds, could be applied to disorders of the eye involving inflammation, ulceration and infection.

Final Composition

Ingredients	*Minimum*	*Maximum*
Aloe barbadensis	25	75
Citric Acid	0.2	2
Potassium sorbate	0.1	2
Sodium benzoate	0.1	2
Cornstarch	10	70
Albumin	1	10
Hydroxypropylmethyl cellulose	0.5	10
Allantoin	0.5	5
Vitamin A	0.1	5
Vitamin D	0.1	5
Vitamin E	0.1	5

Preparation of the Aloe gel eye drops : The Aloe leaves were processed as soon as possible after cutting from the plant.

The whole extraction process takes no more than 2 hrs. The leaf removed from the plant was washed with a chlorinated solution rinsed with sterilized water and dried. Then the gel is separated from the leaf peeling away the outer green cortex of the leaf.

Carefully but consistently close, not to lose the region just below the skin that contains substantial quantities of a highly active mucopolysaccharide composition. The separated gel matrix is immediately stabilized with 0.1% citric acid and 0.2% potassium sorbate.

Homogenization was performed using a speed blender. The preparation was then fractionated and readily freeze at 18°C. For the preparation of final Aloe gel solution the frozen homogenate was towed, clarified, centrifuged at 1 rpm in continues flow centrifuge and filtered through a 0.2 mu pore filtration system.

The Final 40 : 1 Aloe solution was sterile, aloin free, pulp free and mucilage free. The gel for the use with an Aloe solution was prepared with Natrosol, 0.75% in deionized water with 0.2% of potassium sorbate as a preservative.

8. Aloe Solution for Stimulating Hair Growth : The possession of a healthy and profuse head of hair throughout life is the ambition of most human beings. Various kinds of hair dressings including hair treatment compositions, have been used for slowing down or stopping epillation or depilation.

In this present study a 100% saturated sodium chloride dissolved in Aloe solution was prepared for stimulating hair growth and for decreasing hair loss.

Preparation of Extract : An Aloe leaf was washed, sliced open and filleted. The clean inner gel was retained while the green rind and the latex material were discarded. The filleted material was homogenized and extensively filtered, to remove the pulp.

The clear, viscous gel was acidified to a pH of approximately 3.20 with dill HCl to solubilize the oxalates and lactaes of calcium and magnesium that are usually present to their corresponding water soluble acids. The acid treated gel was then extracted for 4-5 hrs with four volumes of 95% ethanol at ambient temperature.

Floating fibers were removed, then the alcohol/water mixture was siphoned off while the solid precipitated was collected by centrifugation. Most alcohol/water soluble substances were eliminated in the process. The solid was then washed with fresh alcohol, centrifuged, freeze dried and ground to a white amorphous powder.

9. Aloe Gel Toothpaste : It is known that chlorophyll possesses positive antibacterial actions and will stimulate growth. But the use of Aloe gel helps in relieving pain and in promoting healing of topical and other lesions.

10. Aloe Ointment : Fresh aloe gel. *i.e.,* within five hours of removal from the plant and preferably within two hours of its removal, is mixed with Aquaphor in a ratio of 8% of the gel 92% Aquaphor. This ointment is useful in treating skin irritation and promoting the healing of wounds. This ointment is suited for external use only, and must not be taken internally.

CHAPTER 37

SAFED MUSALI – THE MUCH SOUGHT AFTER MEDICINAL CROP

Safed Musali, also known as Dholi Musali in Gujarati, Khairwa in Uttar Pradesh, Sufed or Sufeta Musali in Marathi, Shedeville in Malyalam, Tanirvitang in Tamil and Shaqaqule Hindi in Arabic, is an indigenous medicinal plant of Central India that naturally grows in large quantity in the forests of Madhya Pradesh, Gujarat, Maharashtra, Uttar Pradesh, etc. This small tuberous plant measuring 1 to 2 ft.; has a life span of 3 to 8 months and it surfaces once in a year during the rainy season. Belonging to the "Liliaceae" family of plants, its tubers are used for medicinal purposes.

The tubers of Safed Musali find wide application indifferent Ayurvedic medicines. The Biochemistry of Safed Musali is as following :

- Carbohydrates – 40 to 65%
- Proteins – 10-30%
- Sepongeans – 2-3%
- Steroids – Hezexymin
- Minerals – Ca, Cu, phosphorus, Asparagin, cellulose

Thus it is a rich source of over 25 alkaloids vitamins, minerals, proteins, carbohydrates, steroid saponins, and polysaccharides, etc. Presently a number of health tonics for general and sexual weakness are prepared from it. Its decoction is used in the case of impotency as it is quite rich in glycosides and has spermologic properties.

The traditional use of Musali that everybody must be familiar with is as an essential part of the diet of mothers (after delivering) in the form of *"Laddoos"*. Apart from this, there are a number of other uses of Musali which has made it a much sought after medicinal plant and has taken it to the stage of near extinction from most of the forests. Efforts in countries like U.S.A. and England are also on to convert it into chips/flakes to use it as a nutritious breakfast.

Species of Safed Musali

Although the literature reports about nine species of Safed Musali, but in the commercial parlance only three are quite known. These are :

(*a*) Chlorophytum borivillianum

(*b*) Chlorophytum arundinaceum

(*c*) Chlorophytum tuberocum

The Procurement of Musali

As mentioned earlier, Musali grows widely in forests. With the first rains of the monsoons, it starts appearing. Since it is in great demand, so whosoever finds it just picks it up from the forests and sells it to the local traders or pharmacies. But since this Musali is not fully matured, it lacks most of the medicinal elements. Hence the collectors of Musali get very low price for their produce (ranging between Rs. 5 to 50 per kg only).

If the Musali continues to fall prey to the novices at the same pace, the day is not far when it will totally disappear from the forests. So the only way out is the proper cultivation of this age old natural treasure of our country.

Is the Cultivation of Musali Possible?

During the last decade some systematic efforts have been made to popularise the cultivation of Musali which show that its cultivation is much more profitable than many of the traditional crops. Presently there are more than 100 farmers in the country who have taken the cultivation of Musali in a big way.

In this respect M/s Mittal Musali Farm of Jalgaon Jamod (Maharashtra), where Musali is being cultivated in over 40 acres of land have found that in a year Musali can fetch a net income of Rs. 3-5 lakhs/acre.

Agro Technology for Musali Cultivation

Condition Facilitative for Musali

As Musali grows naturally in most parts of Central region of our country, the normal climate of the Central region suits the crop most and it can grow successfully in the wide range of temperature and rainfall. Sandy loam soil with proper drainage system facilitates its growth.

The Preparation of Land

As the sowing season of Musali starts with the rainy season (during June-July), the land is required to be prepared in the month of April-May. After deep ploughing of the fields, 10 to 15 trolleys (20 to 25 tonnes) per acre of FYM is mixed in the soil. If the soil is of clay nature then soil conditioner at the rate of one tonne per acre could be used to condition it. Green manure can also be used to enrich the soil. For this purpose, the seeds of Sutiboru @60 kg per acre could be sown in the field (proposed for the cultivation of Musali) during March-April and before they start flowering they are cut and mixed into the soil. Sometimes the field is irrigated also for tile easy decomposition of green manure. If the soil gets clods, then 3-1 tillers are given for better pulverization of soil.

After the preparation of land, beds are prepared. For the proper growth of the tubers of Musali and to facilitate the proper drainage, raised beds are prepared. Although the size of these beds could be as per facility and ease, but generally beds with height upto 1 ft and a width of 3.5 ft is recommended for the proper growth and development of Musali tubers.

Sowing of Musali

Being a Kharif crop, the sowing of Musali starts with the first shower of the monsoons (June-July). Although its seeds could also be used for sowing, but for better results, the tubers suit the most. Thus from the old bunch of tubers (which are properly preserved from the old crop or are procured from the forests) their fingers are separated.

Having separated the individual fingers, it is ensured that some part of the crown/disc remains intact with all the fingers which are to be used for swing. For good crop yield those tubers are considered good that carry an average weight of 8 to 12 grams. Generally these fingers are planted at a distance of 6×6", and a total of 60,000 fingers are required per acre with an approximate weight of 400 to 500 kgs.

It is advisable to treat the fingers or tubers (planting material, either with the urine of cow (proportion of cow urine and water should be 1:10) or with Bovestinand Streptocycline (50 gm of Bovestine and 3 gm of Streptocycline to be mixed in 15 litres of water), before planting in order to keep it free from fungus. The fingers are sown in the soil at a depth of approximate 1/2 to 1".

The crop is generally free from most of the pests and diseases. Sometimes water logging causes tuber rot which can be cured by stopping further irrigation.

For better germination, the moisture level in the field must be maintained. As far as possible the sprinkler system should be utilised to irrigate the crop. The sprouting starts taking place within a week of sowing.

Controlling of Weeds

Generally two weedings of the crop are required to free it from the weeds—the first within 20-25 days of sowing and the other 20-25 days later. The fields must be kept free from weeds to obtain the proper yield for the crop.

Fertilisers and Pesticides Required for the Crop

Apart from the FYM used during the preparation of fields it is advised to apply NPK (20 : 40 : 20 kg per acre) in the field after the first rainfall.

The crop is generally free from 111 most of the pests and diseases. Sometimes water logging causes tuber rot which can be cured by stopping further irrigation. For the control of *illi* and other pests, it is advisable to spray Thyrum 5 ml/liter every month. This could be applied as a precautionary measure also.

Harvesting of the Crop

After 3-3.5 months of sowing (generally in the month of

October-November) the leaves of Musali start yellowing. Subsequently they become dry and fall off and get detached from the tuber/disc. Some research scientists are of the opinion that this is the maturity stage of the crop and once the crop attains this stage then it means that the crop cycle is complete, hence its tubers should be dug out.

But, people engaged in the cultivation of Musali in the most professional manner do not agree with it. They find that even after the drying up of the leaves tubers/disc lingers of Musali keep on maturing in the soil. So if at this stage the tubers are dug out they will not be fully mature and hence will lack some medicinal properties. Thus, these tubers should be left in the soil for some more time where they keep on maturing.

During this time also the moisture level in the soil should be maintained. By January-February (when this crop is 7-8 months old) the skin of the tubers matures and it turns to dark brown. This is the right time to dig out the tubers. Thus on attaining this stage, the tubers should be dug out.

After digging out the tubers they are properly washed. Of these tubers, the longer and healthy fingers are detached from the tubers, whereas the smaller ones are kept to be used as planting material for the next season. The longer and healthy fingers already detached from the tubers are then taken for processing. During processing the skin of the fingers are peeled out with a stainless steel knife and then they are kept in the sun for drying. Within 3-4 days these fingers get dried up. These are then packed in polythene bags and sent to the pharmaceutical industries.

The planting material obtained from the tubers is properly treated with fungicides and stored in sand bags maintaining a temperature of 28-31 degree celcius and a relative humidity of 50-65 per cent.

Production Per Acre

On an average this crop gives an yield of 20-22 quintal of wet (fresh) Musali per acre. After peeling and drying up nearly 20 per cent (4-4.5 quintals) dry Musali is finally obtained. In the indigenous market, the present rate of dry Musali ranges between Rs. 800 to Rs. 1,800 per kg, whereas it is beyond Rs. 3,000 per kg in the international market.

Economics of Musali Cultivation

The per acre expenditure and income from the cultivation of the crop of Musali could be seen from the table.

How to Reduce Initial Expenditure?

It may be seen from the economic aspects that the initial expenditure in respect of this crop specially on procuring planting material is quite high. This could be quite discouraging for the persons desirous to undertake the cultivation of this crop.

But one possible alternative to this state of affairs could be the procurement of planting material from the forests which could be procured in cheaper rates (Rs. 5 to 20/kg.). But, in this case it will take atleast 2 years for the crop to mature. In the initial year the yield may not be that of a commercial level.

Thus the procurement material from the forest could be a possible solution to avoid heavy initial expenditure on this crop.

A. Expenditure		
(*a*) Cost of planting material : (5 quintals of seed at the rate of Rs. 45,000/quintal)	=	Rs. 2,25,000
(*b*) Cost of manure (FYM)	=	Rs. 10,000
(*c*) Expenditure on the preparation of land, beds and irrigation facilities	=	Rs. 15,000
(*d*) Expenditure on weeding, other labour charges	=	Rs. 5,000
(*e*) Expenditure on tonics, fertilisers and pesticides	=	Rs. 5,000
(*f*) Expenditure on harvesting and processing	=	Rs. 25,000
(*g*) Miscellaneous expenses	=	Rs. 10,000
Total expenditure	**=**	**Rs. 2,95,000**
B. Income from the Crop		
(*a*) Sale of dry Musali (average 400 kg of dry Musali at the rate of Rs. 1,200 per kg.)	=	Rs. 4,80,000
(*b*) Planting material for the next crop	=	Rs. 1,50,000
Total income	**=**	**Rs. 6,30,000**
(*c*) Net income per acre	=	Rs. 3,35000

Training Facilities

Unfortunately no efforts have been made the government level to promote this valuable and high yielding medicinal crop.

Significant contribution to popularise the cultivation of Musali in the country is that of Mittal Musali Farm, Jalgaon Jamod (Maharashtra), where not only free training facilities for the cultivation of Musali are offered to the desirous persons, but, free services are also extended to those persons also who are interested in undertaking experiments and research in the field of the cultivation or processing of Musali.

One of the leading farms of our country which has undertaken the cultivation of Safed Musali in a highly professional manner in nearly 50 acres of land is Mittal Musali Farm which could be unhesitatingly contacted for any type of assistance or guidance in the cultivation of Musali starting from the supply of planting material, to the training and guidance in growing this high yielding crop.

CHAPTER 38

STEVIA

0% Calory, 100% Sweet, 100% Natural

All over the world there is increasing tendency towards consuming natural products and thus living a natural life and at the same time our life styles have changed so much over the last 4-5 decades that sweeteners (either high calorie natural or processed sugars or high potency and low calorie sweeteners such as Aspartame) have become integral part of our natural daily diet.

Due to sedentary life styles that we all tend to lead these days the incidence of obesity and diabetic conditions are increasing dramatically. In India number of diabetic people in the age group of 25-45 is about 15% and is increasing at an alarming pace.

In addition, we are largest consumers of sugar in the world. This single factor, we have come to understand, would greatly contribute to increasing the number of diabetic people and related problems.

Stevia, botanically known as *Stevia rebaudiana Bertoni* (Family-Asteraceae) is a sweet herb. A perennial herb, Stevia is a member of the daisy family. The leaves are mid green and intensely sweet. The compounds in the leaves are called stevioside and rubaudioside and they can be more than 200 times sweeter than sugar. The plant bears greenish cream flowers in autumn.

Although Stevia has been in use in Asia and Europe of years, it was only in the past couple of years that it really started to capture attention in the Indian market as a healthy alternative sweetener to sugar. Stevia has no calcium cyclamate, no saccharin, no aspartame, and no calories. It is safe for diabetics, as it does not affect blood sugar levels; it does not have the neurological or renal side effects associated with some of the artificial sweeteners.

Stevia is a new crop in that is gaining very high popularity amongst all type of sweetener users as most ideal substitute for sugar. Sugar is basically a chemical that has grown in market over last many years. But in this age of changing life styles and people becoming more conscious of their health, the worldwide sugar consumption is going down and is getting replaced by low calorie sweeteners. Many of these Sweeteners are complex chemicals or many times naturals as well.

Stevia is gaining more acceptance amongst countries of eastern block such as Japan and Korea. There are many advantages of using Stevia over conventional sweetners the main one being :

- Stevia leaves are 20-30 times sweetner than sugar.
- Stevia leaves can be dried and stored.
- Stevia can be used in raw form.

Stevia is short duration crop. It be harvested 3/4 times a year.

The yearly yields can be in the range of 3-4 tons.

Stevia can be sold @40–70 Rs. per kilo of dried leaves and is thus economically extremely beneficial to growers.

Uses of Stevia Leaves

Currently consumers have more inclination towards products that are claimed to be 'All Natural' and 'Low CHO'. Hence, the food industry could grab a major share in the market if Stevia, the natural sweetner, is used as sweetening agent in products like biscuits, jams, chocolates, ice creams, baked foods, soft drinks, soda, candies and also common beverages like dip tea, coffee and herbal tea that are targeted particularly at the diabetics and also the health conscious consumers.

Application in Bakery, Confectionery and Beverage Sector

All cooked and baked, food items like puddings; desserts can be sweetened with only very small quantities of Stevia leaf powder as compared to table sugar. Just 50 gms. of Stevia leaf can replace 1000 gms. of cane sugar. The sweetness of stevioside is non-fermenting and it does not display browning when cooked. This further widens its area of application in baking. This also enhances the quality and safety of usage with a longer shelf-life period. Breads made with Stevia as an ingredient (for diabetic patients) have been found to display improved texture, softness and an increased shelf life.

The confectionery industry is yet to reap the benefit of stevia, which has the potential to replace sugar as sweetening agent. The leaves can be used in chocolates and candies not only to meet the requirement of diabetic and health conscious consumers, but also to harvest the added advantage, that it does not encourage tooth decay. Stevia possesses an antimicrobial property and can be used in all the sweets that children are fond of. A mere fragment of the leaf is enough to sweeten the mouth for an hour. So Stevia can also be used in the manufacture of chewing gums, mints, mouth refreshers and even in *pan*.

The soft drink manufacturers have introduced several health drinks and many food supplementary beverages, especially for the diabetic patients. The addition of Stevia leaves (dried) as such or as powder in such products would not only aid in increasing the sweetness naturally but also helps in rejuvenating the pancreatic gland. Apart from this, Stevia is rich in nutrients, containing substantial amounts of protein, calcium and phosphorus.

The beverage industry can blend stevia leaves with tea and coffee, which will ensure value addition, simultaneously, easing the preparation of the drink. Stevia leaves can be made available in tea bags and ready to mix coffee to make tea/coffee more delicious for the diabetic and calorie conscious consumers.

Stevia consists of 10-12% of stevioside, which can be extracted as a liquid concentrate that can be used directly in soft drinks, beverages, chocolates etc.

Medicinal Properties of Stevia

Stevia is a versatile herb with incredible sweetness that can be safely used in herbal medicines; tonics for diabetic patients and also in the daily usage products. Stevia leaves can be used because of its anti-fungal and anti-bacterial property. Mild stevia leaf tea offers excellent relief for an upset stomach.

A wet Stevia leaf bag provides a cooling effect on eyes (similar to using cucumber). The leaves effectively tighten the skin and are good for wrinkles. Stevia has proved to give exceptional benefits when used regularly in skin care. It also has a healing effect on blemishes, wounds, cuts and scratches.

Stevia is helpful in weight and blood pressure management. It has also been reported that stevia lowers incidence of colds and flu.

Commercial Cultivation

You need not be a South American farmer to be a successful Stevia grower. A short day plant, it only reaches a height of 45 cm in three months. The concentration of Stevioside in the leaves of Stevia increases when the plants are grown under long day condition.

Stevia grows best in upland areas in sub-tropical climates. In other places it can be grown as an annual. The plant prefers a lightly textured, well-drained soil to which organic matter has been added. It needs ample water so that the soil is consistently moist, but not wet. In hot, sunny climates it will do best in semi-shade. Propagation is from seed sown in spring, but germination rates can be low – expect half the seeds sown not to germinate. Plant seedlings out once all danger of frost is over. Leaves are best harvested just before flowering. The plants will also grow from cuttings, which are best taken in late winter.

The concentration of stevioside in the leaves of Stevia increases when the plants are grown under long day condition. While, cultivating stevia on a large scale, it can be grown in well-drained red soil and sandy loam soil. The soil should be in the pH range of 6.5-7.5. Saline soils should be avoided to cultivate this plant.

Stevia can be successfully cultivated all around the year all over India except the areas, which receive snowfall, or temperatures go below 5 degree celcius in winter. The summer temperatures actually do not affect this plants if the high summer temperatures have already been factored in the cultivation practices.

Since seed germination rate is very poor, it is propagated vegetatively. Though stem cuttings are used for vegetative propagation, tissue culture plants have proved to be the best planting material for Stevia. Tissue culture plants of Stevia are genetically pure, free from pathogens and have excellent vigour. The tissue culture plants can be planted throughout the year, except during peak summer. An ideal planting density is 30,000 plants per acre with a spacing of 25 × 40 cm in a raised bed system. The soil can be enriched with a basal dressing of 25 tons of well rotten farmyard manure/hectare.

Land Preparation

Land should be ploughed initially with a disc plough or harrowed to break down the clods. Fine tilth is required. One to two times ploughing has to be done after harrowing. Around 25 MT of FYM/Ac has to be applied as a basal dressing during the last ploughing to incorporate the manure with the soil.

Soil Type

Stevia requires very good drainage any soils that retain the moisture for very long period of time are unsuitable for Stevia cultivation and should be religiously avoided. Similarly black cotton soils with very heavy clay content should also be avoided. It is mandatory that before planting, each and every soil sample should be properly tested and analyzed by an expert. Red soil and sandy loam with a 6-7 pH are best for the cultivation of Stevia.

Raised Bed Preparation

Forming raised beds is the most economical way to grow Stevia. The raised bed should be of 15 cm in height and 60 cm in width. The distance between two rows should be 40 cm and distance that between each plant 23 cm. This would give a plant population of around 30,000 per acre.

Planting

Stevia can be planted in many ways. The types of agronomic practices generally depend on type of soil and climatic conditions. Generally it is advisable to plant minimum of 40,000 plants per acre. Since the economical part is the leaves it is very important to achieve highest vegetative growth.

Planting Material

There are basically two options for multiplication. The first is the tissue culture and second the stem cutting. Tissue culture is the best option but many farmers are tempted to try the stem cutting method for multiplication. As per practical experience, stem cutting is sometimes more expensive to produce than the tissue culture since the success rate of the stem cuttings establishment is very low, it takes minimum of 25 weeks for the stem cutting to develop in proper feeding roots for transplantation (younger stem cuttings transplants have shown more than 50% mortality in first few weeks of transplantation in main field). It is also noticed that Steviocide content of the stem cutting plants is lower by at least 2 percentage points in Indian conditions.

Plant Varieties

There are about 90 varieties of *stevia rebaudiana* developed all around the world. Basically all these varieties have been developed for different climatic requirements. Many times these varieties perform strangely in different climatic conditions. At the end of the day, just like sugarcane, it is the steviocide and rebaudiocide content in the Stevia leaves that determine the price and marketability of Stevia leaves. In many cases in South India it was observed that steviocide content was as low as 3.5%, which was below the minimum market requirement of 9%. Hence it becomes imperative that the grower selects proper varieties with adequate guarantees from the planting material suppliers about minimum assured steviocide and rebaudiocide contents.

Irrigation

Stevia requires ample supply of good quality water all year around. As the plant cannot tolerate drought, frequent irrigation is required. Micro sprinklers are the best method of irrigating Stevia plants. Flood or canal irrigation would not supply the required amount of water at the right time. So through micro sprinklers, the water can be sprinkled once in a day in winter and two to four times in a day in summer depending upon the heat and relative humidity in the air. Watering frequency should be scheduled so that the plants do no wilt for want of water.

Fertilizer Application

The recommended dose of fertilizer is 110:45:45 NPK/ac. This requires 4½ bags of urea, 2 bags of DAP and 2 bags of Potash. The entire dose of DAP is applied as the basal dressing. The Nitrogen and Potash fertilizer can be split and applied as 10 doses in every month. Nitrogen application is a must for the production of dry matters.

Plant Protection

Organic gardeners in particular should find stevia an ideal addition to their yield. Though nontoxic, stevia plans have been found to have insect-repelling tendencies. Their very sweetness, in fact, may be a kind of natural defense mechanism against aphids and other bugs that find it not to their taste. Perhaps that's why crop-devouring grasshoppers have been reported to bypass stevia under cultivation. In case any disease symptoms are noticed, spraying of neem oil diluted in water in the best organic method.

Weeding

Removal of weeds can be done manually. Since the crop is grown in raised beds, intercultural operations are easier by manual labour.

Maintenance

Stevia plants do best in a rich, loamy soil–the same kind in which common garden variety plants thrive. Since the feeder roots tend to be quite near the surface, it is a good idea to add compost for extra nutrients if the soil in you area is sandy.

Besides being sensitive to cold during their developmental stage, the roots can also be adversely affected by excessively levels of moisture. So take care not to over-water them and to make sure the soil in which they are planted drains easily and isn't soggy or subject to flooding.

Frequent light watering is recommended during the summer months. Adding a layer of compost or your favourite mulch around each Stevia plant will help keep the shallow feeder roots from drying out.

Stevia plants respond well to fertilizers with lower nitrogen content than the fertilizer's phosphoric acid or potash content. Most organic fertilizers would work well, since they release nitrogen slowly.

Flowering of the plant should be avoided. Since Stevia has a significant apical dominance, the plant tends to grow tall and lanky. Pinching of the apical bud would enhance bushy growth of the plant with side branches.

Harvesting

Depending on climatic conditions one can achieve the yields of 2000-4000 kilos in three to six harvests annually. Another important aspect of harvesting is the timing of harvest. It should be noted that at no point of time plants should be allowed to flower since after flowering the Steviocide percentage goes down rapidly and leaves are rendered unmarketable. Leaves are harvested by plucking in a small quantity, or the entire plant with the side branches is cut leaving 10 to 15 cm from the base.

The first harvesting can be done four to five months after planting. Subsequent harvesting can be done every three months, for three consecutive years. The sweetener in the leaf is maximum till the plant flowers. Just before flowering, the plant should be cut completely leaving 10 cm from the ground. The new flush of leaves will sprout from here. The new plant will be ready for harvest again in three months. The plant yields around 3000 kg of dried leaves from an acre of plantation every year. Harvesting should be done as late as possible, since cool autumn temperatures and shorter days tend to intensify the sweetness of the plants as they evolve into a reproductive state.

Unlocking the Sweetness in your Harvest

Once all leaves have been harvested its required to dry them. This can be accomplished on a net. The drying process is not one that requires excessive heat; more important is good air circulation. On a moderately warm fall day, stevia crop can be quick dried in the full sun in about 12 hours. (Drying times longer than that will lower the stevioside content of the final product).

Crushing the dried leaves is the final step in releasing stevia's sweetening power. The dried leaves are powdered, sieved and the fine powder is stored in containers. This can be done either by hand or, for greater effect, in a coffee grinder or in a special blender for herbs.

Fresh Stevia Leaves

This form of stevia is the herb in its most natural, unrefined state. A leaf picked from a stevia plant and chewed will impart an extremely sweet taste sensation reminiscent of licorice that lasts for quite a while. For stevia to have a more practical application as a tea or sweetner, the leaves must be dried or put through an extraction process, which makes the sweet taste even more potent.

Dried Leaves

For more of the flavour and sweet constituents of the stevia leaf to be released, drying and crushing is necessary. A dried leaf is considerably sweeter than a fresh one, and is the form of stevia used in brewing herbal tea.

Stevia Extracts

The form in which Stevia is primarily used as a sweetener in Japan is that of a white powdered extract. In this form it is approximately 200 to 300 times sweetener than sugar (by weight). This white powder is an extract of the sweet glycosides (natural sweetening agents) in the Stevia leaf.

Not all Stevia extract powders are the same. The taste, sweetness and cost of the various white stevia powders will likely depend on their degree of refinement and the quality of the stevia plant used. Since extracted stevia powder is so intensely sweet, it is always recommend that it be used by the pinch (or drop if diluted in water). Once mixed, this solution should be stored in the refrigerator.

Liquid Concentrates

These come in several forms. There's a syrupy black liquid (that results from boiling the leaves in water), which can enhance the flavour of many foods. Steeping stevia leaves in distilled water or a mix of water and grain alcohol makes another type.

Future of Honey Leaf (*Madhu-Patraa*) Stevia

The leaves of this splendid plant are 30 times sweeter than sugar; with zero calories where as pure extract is 300 times sweeter than sugar. This sweet-honey-leaf herb is likely to become the major source of high potency sweetener for the growing natural food market, in the years to come.

Stevia finds its use as a natural sweetener, replacing the chemical sweeteners and even table sugar, the sweetness in leaf is due to the presence of an intensive-sweetening agent called stevioside and the leaf by itself is about 20 to 30 times sweeter than sugar. The leaf has stevioside of 10-12% or dry weight basis.

Stevia is a new promising renewable raw material for the food market. The market potential for this natural sweetener is still untapped. It is estimated that about 30 million Indians are presently suffering from diabetes and it is estimated that by 2025 India's contribution to the diabetic global population would be a whopping 80 million. With such a huge share of the population being diabetic, the new ventures in the food industry are focused entirely on them.

The soft drink manufactures in India are yet to exploit the sweetness of this herb by its addition in their product. Though many soft drinks are introduced in the market with the prefix 'dia' connoting that it is meant for the diabetics, the usage of stevia in such products would fetch a greater demand than for the one with artificial sweeteners.

The beverage industries like tea and coffee manufacturers have just started introducing new products for the diabetic patients, realizing the major share held by them in the consumer market. As tea and coffee has been the non-replaceable and the best beverage for every Indian, this sector has enormous potential to come up when the natural sweetener Stevia is used in their products.

All those 'dia' prefixed products in the market at present are sweetened with artificial sweeteners that is, of course, equally sweet, but with undetectable side effects in due course. Stevia is fully plant-based, natural sweetener can be used to replace the artificial sweeteners completely. As stevia leaf powder with no processing is highly safe to use, calorie-free and moreover around

20-30 times sweeter than cane-sugar, it can replace canesugar too. The process of manufacturing Stevia leaf powder is quite simple, when compared to the tedious steps involved in cane sugar manufacture.

Market Potential

India being largest consumer of cane-sugar along with largest diabetic population in the world. Stevia is ideally poised to make significant contribution in satisfying the Indian demand of natural low calorie sweetener.

Today *Stevia rebaudiana* extract accounts for 40% of the sweetener market in countries such as Japan, Korea and Malaysia.

Stevia is relatively new to Indian market. Hence there is plenty of confusion with regard to marketing of this unique product. There are some farmers insist on buy back arrangement.

There are reputed business houses those offer buy back @ of Rs. 60-70 per kilo. This price is on lower side, on an average the farmer should get a price of Rs. 100 to break even in first year.

International Market

There is growing international market for Stevia as well. There are offers to buy container loads of Stevia leaves at a price of 25 dollars per kilo. As mentioned earlier international demand is only for high quality Stevia leaves having minimum 9% Steviocide content. There is no buyer for Stevia leaves having less than 8% Steviocide content.

Stevia cultivation can be a successful venture provided you are able to produce high quality leaves have minimum Steviocide content of 9% at the same time produce optimum quantity of marketable leaves per acre to keep the production cost below Rs. 25 per kilo on five years basis.

Conclusion

The demand for high potency sweeteners is expected to increase Worldwide. The increase in the number of diabetic patients and health conscious individuals would push forward the need for alternatives to sugar.

Stevia is a potential alternative source for replacing artificial sweeteners like saccharin, aspartame, asulfam, etc. Unlike many low calorie sweeteners, stevioside is stable at high temperature and over a pH range of 3-9. Steps need to be properly aligned to exploit the natural sweetness to Stevia.

Food industrialists have to start launching new products utilizing stevia. This would obviously provoke the need to grow more and finally result in more area under stevia cultivation.

In India several important and necessary steps have to be taken up for its propagation. Development of seedlings suitable to India would be the first requirement. A crop production system, providing information on optimized crop inputs, weed and disease control, harvesting and handling methods would have to be detailed out. Awareness has to be created about the natural herb and the products manufactured out of it by the industries.

It's time to streamline necessary forces to have access to Stevia. Initiative needs to be undertaken to promote this natural sweetener and create product awareness. This would be the right approach to unlock the sweetness of this herb in our day-to-day life.

CHAPTER 39

INDUSTRIAL UTILIZATION OF MEDICINAL PLANTS

Introduction

Most developing countries are endowed with vast resources of medicinal and aromatic plants. These plants have been used over the millennia for human welfare in the promotion of health and as drugs and fragrance materials. This close relationship between man and his environment continues even today as a large proportion of people in developing countries still live in rural areas. Furthermore, these people are precluded from the luxury of access to modern therapy, mainly for economic reasons.

About 80% of the population of many developing countries still use traditional medicines for their health care. Modern pharmacopoeia still contain at least 25% drugs derived from plants and many others which are synthetic analogues built on prototype compounds isolated from plants. China, India, Sri Lanka and a few other countries have officially recognized the use of traditional medicines in their health care delivery systems.

A major factor impeding the development of the medicinal plant based industries in developing countries has been the lack of information on the social and economic benefits that could be derived from the industrial utilization of medicinal plants. Except for the use of these plants for local health care needs, not much information has been available on their market potential and trading possibilities. As a result, the real potential of these plants has not been exploited by the governments or entrepreneurs.

The demands of the majority of the people in developing countries for medicinal plants have been met by indiscriminate harvesting of spontaneous flora including those in forests. As a result many plant species have become extinct and some are endangered. It is therefore necessary that systematic cultivation of medicinal plants be introduced in order to conserve biodiversity and protect threatened species. Systematic cultivation of these plants could only be initiated if there is a continuous demand for the raw materials. It is therefore necessary to establish processing facilities preferably in the vicinity of cultivations in order to create a demand and assure farmers of the sale of raw material. Thus cultivation and processing should be started simultaneously. In cases where parts of mature plants from forests are to be harvested, sustainable harvesting techniques are needed to protect these plants.

The promotion and development of processing of plant-based products have been given a fresh impetus due to certain ground realities:

- Green consumerism and the current resurgence of interest in the use of "Naturals" in developed countries.
- Free market economy bringing in more openness and expanding markets and demand for new resources, materials and products.
- A growing acceptability of the social responsibility of minimizing socio-economic inequalities in favour of rural people resulting in creating additional job and income opportunities for poor people.
- Poor economic conditions in developing countries restricting import, thereby placing increased reliance on medicines using local plant resources.
- Increasing awareness regarding biodiversity conservation and the sustainable and protective use of plant resources.
- Search for new phytopharmaceuticals for the prevention and cure of deadly diseases such as cancer and AIDS.

In 1981, the United Nations Industrial Development Organization (UNIDO) established a programme for the systematic utilization of this renewable natural resource for the benefit of the populace of developing countries. The programme aims for initiating development projects for increasing the industrial output of developing countries in the field of locally-used herbal drugs and phytopharmaceuticals which affect the economies of both developing and developed countries.

Some of the constraints associated with the processing of medicinal plants which may result in reducing their competitiveness in global markets and which have to be remedied are :

- Poor agricultural practices.
- Poor harvesting (indiscriminate) and post-harvest treatment practices.
- Lack of research on development of high-yielding varieties, domestication etc..
- Poor propagation methods.
- Inefficient processing techniques leading to low yields and poor quality products.
- Poor quality control procedures.
- High energy losses during processing.
- Lack of current good manufacturing practices.
- Lack of R & D on product and process development.
- Difficulties in marketing.
- Lack of local market for primary processed products.
- Lack of trained personnel and equipment.
- Lack of facilities to fabricate equipment locally.
- Lack of access to latest technological and market information.

Systematic cultivation of many medicinal plants needs specific cultural practices and agronomical requirements. These are species-specific and are not only dependent on soil, water and climatic conditions. Hence research and development work has to be done to formulate Good Agricultural Practices which will include proper selection and identification, propagation methods, cultivation techniques, harvesting, step-wise quality control of raw material up to processing stage, post-harvest treatment, storage and safety. These aspects have to be incorporated into protocols for the cultivation of medicinal plants.

Organic farming is another practice that is gaining wide acceptance as world demand particularly in developed countries for organically grown crops is rapidly on the increase. Farmers have to be trained in all aspects of organic farming of medicinal plants and herbs including obtaining certit. ation from associations that do the monitoring starting from cultivation to final harvesting. Organic farming which is labour-intensive gives the developing countries the comparative advantage to be competitive.

Process Technology

Traditional Medicines – Modern Technology

The medicines for internal use prepared in the traditional manner involve simple methods such as hot- or cold-water extraction, expression of juice after crushing, powdering of dried material, formulation of powder into pastes via such a vehicle as water, oil or honey, and even fermentation after adding a sugar source. The range of products that could be obtained from medicinal plants is given in Figure 1.

Traditional herbal medicines were produced using age old methods by the practitioner him/herself who was able to identify the correct plant species. This practice of the traditional practitioner dispensing his own medicines is being gradually shifted to herbal drug stores which are profit-oriented. As a result, there is no guarantee of the authenticity and quantity of plant material used in the preparations. The quality of traditional medicines so produced vary widely and may not even be effective. Therefore, there is a need to select proper and appropriate technologies for the industrial production of traditional medicines such that the effectiveness of the preparation is maintained. Traditional methods used have many disadvantages which could be corrected by selecting the suitable technologies. It has to be stated that the traditional methods were dependent on the status of technology that was available at that time. It therefore follows that these can be modified and improved using the technologies available today to make them more effective, stable, reproducible, controlled and in dosage forms that can easily be transported or taken to office.

Hence the introduction of appropriate, simple and low-cost technologies should be encouraged maintaining as much as possible the labour-intensive nature of such activities, conservation of biodiversity through small-scale production and preservation of cultural knowledge. Use of sophisticated modern technolo-gy will alienate the traditional practitioner as he has no control over such production methods. Even in the use of appropriate technologies, the practitioner who produces these drugs has to be educated about the advantages of using such production and quality control methods.

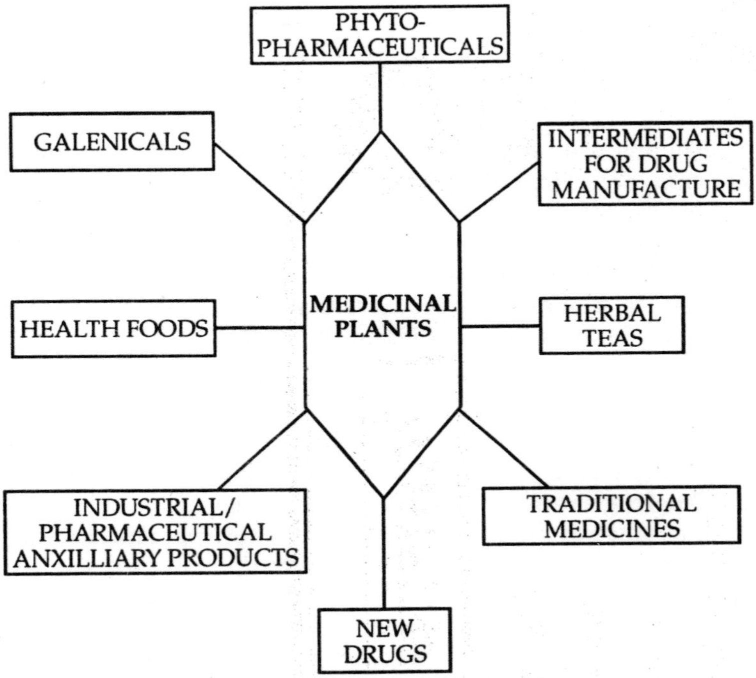

Fig. 1 : Industrial uses of Medicinal Plants

One major concern in introducing modern technology for the production of traditional medicines is whether the final preparation will be acceptable to the practitioner who has sole faith in extemporaneous preparations. This problem has to be overcome by a process of education, whereby the disadvantages of the old methods and the advantages of the new methods can be imparted.

Value Added Products

The value of medicinal plants as a source of foreign exchange for developing countries depends on the use of those plants as raw materials in the pharmaceutical industry. These raw materials are used to:

- Isolate pure active compounds for formulation into drugs (quinine, reserpine, digoxin, etc.)
- Isolate intermediates for the production of semi-synthetic drugs
- Prepare standardised galenicals (extracts, powders, tinctures, etc.)

If one is to produce known pure phytopharmaceuticals used in modern medicine, more processing stages and more sophisticated machinery are required. Furthermore, safety and pollution aspects have to be considered. Most of these processes and formulations are patent-protected. Even transferring technology through contractual agreements and payment will not be of much help unless there is a large local demand for these drugs. Often the drugs so produced are more expensive than world market prices owing to the limitations of the economy of scale of production.

Certain plants are rich sources of intermediates used in the production of drugs. The primary processing of parts of plants containing the intermediates could be carried out in the country of origin thus retaining some value of the resource material. For example, diosgenin (from *Dioscorea* sp.) and hecogenin (from sisal) used in the production of steroids can be commercially produced in the countries of origin where there are steady supplies of sufficient raw materials.

Processed products (galenicals) from plants could be standardised fluid/solid extracts or powders or tinctures. Standardised extracts of many plants (e.g. *Aloe* species, *Atropa belladonna, Cassia angustifolia, Capsicum annuum, Centella asiatica, Cephaelis ipecacuanha, Digitalis* species, *Commiphora mukul, Panax ginseng)* are widely used in health care. Some of these have to be formulated for incorporation into modem dosage forms. New formulations require some development work, particularly on account of the nature of the processed products. Plant extracts are difficult to granulate, sensitive to moisture and prone to microbial contamination. Hence the types of excipients to be used and the processing parameters have to be determined.

Downstream processing activities leading to different medicinal plant based products are indicated in Figure 2.

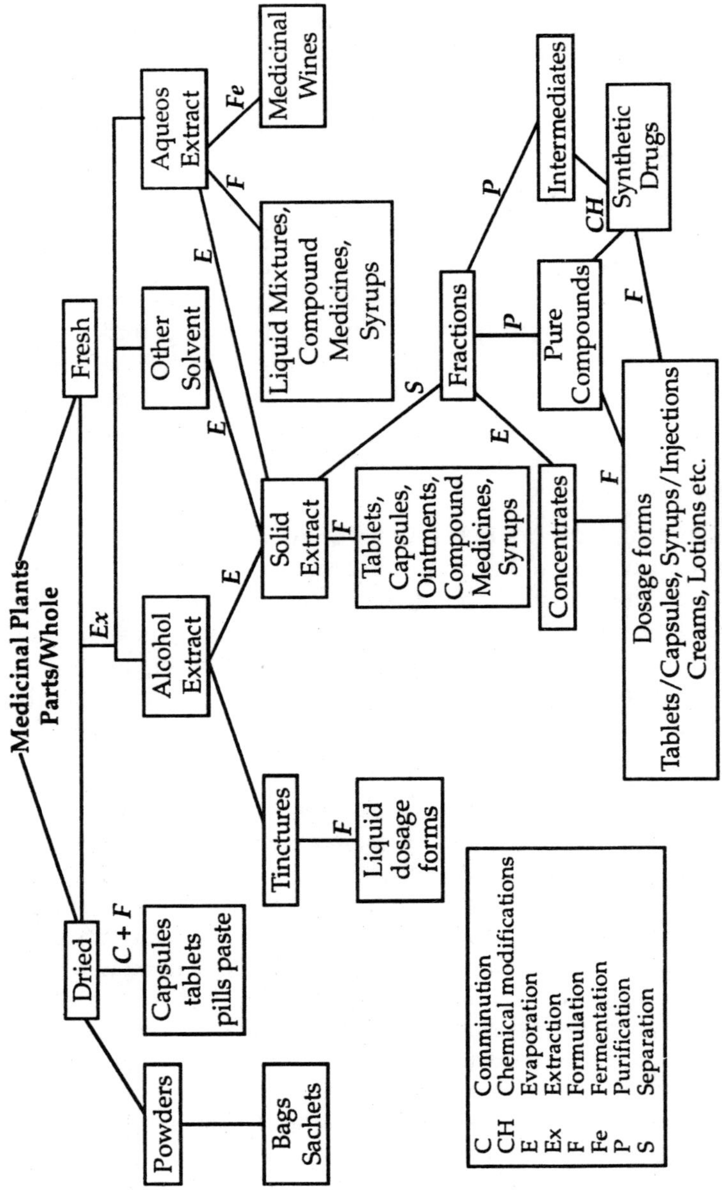

Fig. 2 : Downstream processes.

Most of the activities of UNIDO in the industrial utilization of medicinal plants have involved: the production of standardised traditional medicines, galenicals and extracts, the formulation and development of dosage forms, the development of new preparations based on the traditional pharmacopoeia, research and development in processing and formulation and basic chemical and pharmacological studies. Processing using clean and pollution free technologies have been introduced.

Pilot Scale Production

Development of process parameters has to be carried out at a pilot scale in order to be upscaled to industrial production. Many R&D institutions and universities in developing countries do not have such facilities and are therefore unable to pass on their R&D findings to the industry. UNIDO has played a significant role in achieving university-industry interaction by designing a polyvalent pilot plant which could be used for many unit operations needed for the production of plant-based products. The pilot plant has enabled the R&D work to be demonstrated to produce final products which could be market tested. Universities in turn have been able to realize the hidden potential of industrialization of their R&D work. Any institution which wants to initiate downstream processing of medicinal plants and other non-wood forest products should possess facilities for pilot scale production.

New Technologies

Improved methods for the processing of medicinal and aromatic plants and new techniques for quality assessment are being developed rapidly and continuously and they have to be introduced to developing countries if they are to forge ahead to keep up with recent developments and new international requirements. One such method of recent development is Supercritical Fluid Extraction of natural products as an alternative for solvent extraction. It is now used both for processing of phytopharmaceuticals and other plant products and for the removal of trace amounts of contaminant substances such as pesticides, toxins and surfactants.

Quality Assurance

The control of the quality of the raw materials, finished

products and of processes is an absolute necessity, if one is to produce goods for world markets and human consumption.

International Standard Specifications exist for some processed products and some countries and buyers have their own requirements. The quality requirements for medicinal plant preparations are stringent in terms of content of active principles and toxic materials. Whereas the production of traditional medicines for local use does not require such stringent standards, what is produced will be a much more improved version of the already produced medicines using traditional methods.

Quality has to be built into the whole process beginning from the selection of propagation material to the final product reaching the consumer. It is therefore a management system where all steps involved in the industrial utilization process have to be properly and strictly controlled to produce the desired quality products. All elements of Total Quality Management (TQM) have to be introduced in any industrial project. The requirements for ISO 9000 certification and Good Manufacturing Practices (GMP) have to be introduced and the personnel trained so that enterprises could introduce the proper systems needed for certification. Furthermore eco-audit procedures (ISO 14000) leading to eco-labelling will be required for safeguarding environmental damage.

Human Resource Development

Many developing countries have a core of trained personnel in the fields of chemistry, biology, agriculture, pharmacology and pharmacy. They lack resources in such fields as chemical engineering and technology. This can be considered as a major constraint upon industrial development. UNIDO therefore has developed the human resources required for this specific area of industrial development by conducting training in industrial practice including quality assurance, management and marketing.

Marketing

Marketing is an unsurmountable problem besetting the development of the plant-based industry in developing countries and marketability will be a crucial factor in determining the failure or success of these industries. The market outlets can be for local use and for export. As for local use, some products could reach the consumer directly while others have to be either further processed or used as secondary components in other industrial products. Hence user industries have to be promoted so that locally produced extracts can be used to save foreign exchange needed for importation of such additives.

Further processing to yield value added products will be limited by the local demand situation unless they could be produced at prices to be competitive in the world market. Even if the cost of production is low and quality of the products is good, substantial market promotion has to be undertaken in order to penetrate the world market.

Market tie-ups with progressive entrepreneurs from the developed world would be a convenient and a realistic option for securing markets for the finished products. Joint ventures or trade agreements could be initiated with companies that are expanding their production to meet the ever increasing demand for green products.

Research and Development

Industrial development requires parallel research and development. Research in chemistry and bioactive components of indigenous flora of developing countries has been ongoing for quite some time, funded by multi-/bi-lateral aid or non-governmental donor organisations. A systematic and a concerted approach to this activity has not been maintained for want of sophisticated equipment and high-cost chemicals. Much of the research has been mainly academic. The concept of applied research in the industrial use of plants has not received much attention.

Research in support of industrial development encompasses all activities ranging from the development of superior propagation materials, agrotechnology, low cost and efficient processing technologies to improve quality and yield, new formulations to new products and the marketing of finished products.

Registration and Property Rights

Many developing countries do not have procedures to register medicinal plant preparations although they are widely used for the health care needs of the majority of people. The

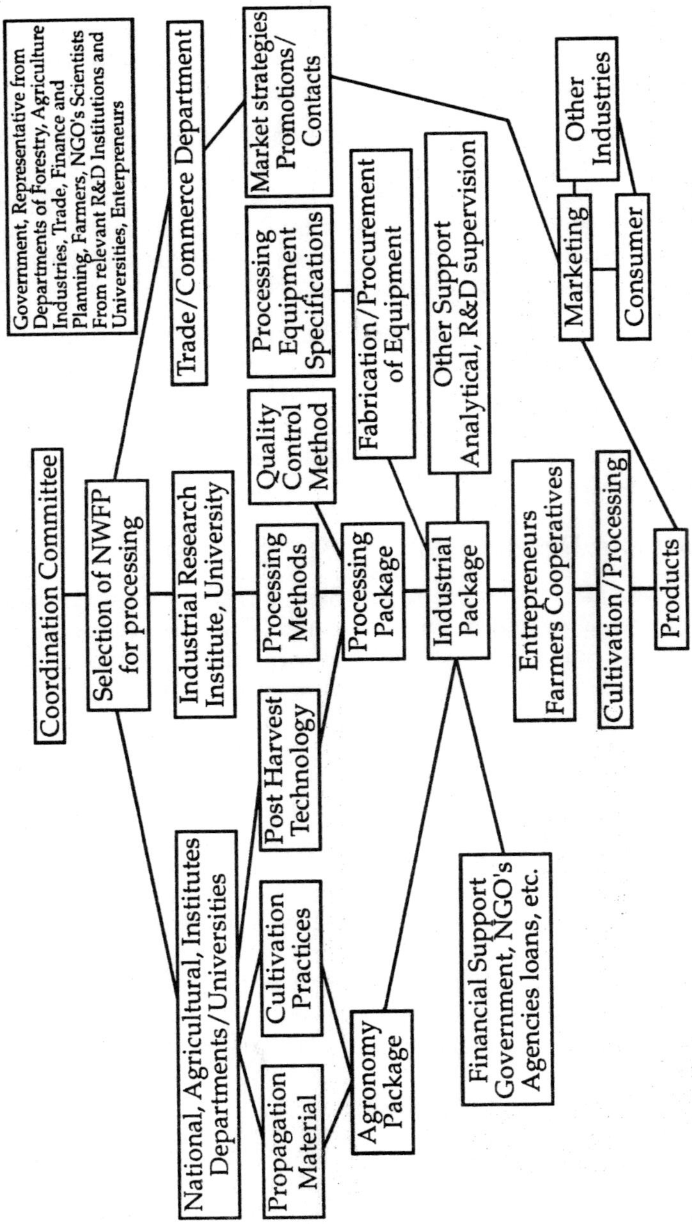

Fig. 3 : Scheme for Development of Processing of NEFP

regulations if any, are very stringent requiring the same standards expected of modern medicines. WHO has recently published guidelines for the assessment of herbal medicines taking into account the long and extensive usage of them (WHO, 1991). These guidelines should encourage developing countries to relax some of the current regulations to be realistic in recognizing the role of traditional medicines in the health care delivery of their countries.

The vital question of property rights to developing countries for the use of know-how and genetic resources in the development of modern drugs in developed countries has been discussed in many forms but without a final solution.

Technical Information

As the use of plant-based drugs has declined with the introduction of synthetic drugs especially in developing countries, the need to collate all available knowledge on medicinal plants and their uses has become imperative particularly because of the recent revival of interest in the use of "natural" medicines. In view of the greater demand for information on traditional uses and proper identification, and the enormous volume of information being generated, a user-friendly information storage and retrieval system will be of considerable importance. As such a computer database would be of invaluable assistance to developing countries in R&D work on industrial utilization of local plant resources.

Data on the medicinal plants are also available in international journals and a number of databases, the premier being Napralert established at the University of Illinois in Chicago with WHO assistance. Many developing countries lack the resources to subscribe to research journals or acquire access to these databases. In fact, the data required by the scientific personnel in developing countries with respect to technologies and methods used for processing and formulation of medicinal plants are not readily available in the literature nor in the databases as some of these are patent-protected.

Conclusions

A scheme for the establishment of processing industries based on plants is given in Figure 3. It must be emphasised that the proper coordination of the multidisciplinary activities needed for processing is vital for the success of industries venturing into this field.

The development of any industrial processing activity has to be linked to the specific needs, socio-cultural background, resource potential and the technological capabilities of each country. Consequently, any integrated development approach adopted has to exploit the full potential of this natural resource by providing a proper scientific, technological, economic and an industrial base. Above all, developing countries' research and development capabilities, including human resources, have to be greatly enhanced so as to ensure the growth of industries based on medicinal and aromatic plants.

Sustainable development of industries based on plants requires multi-disciplinary activities and close collaboration between scientists, government officials, NGOs and international organizations. Such efforts have not received much attention resulting in the slow development of the economic uses of renewable resources. Hence it is important that more action-oriented plans are initiated to exploit the full potential of these resources, bearing in mind the conservation issues such that rural household incomes and national economies could be augmented.

CHAPTER 40

BIOCULTURAL HERITAGE OF INDIA'S MEDICINAL PLANTS

FRLHT is engaged in one of the most comprehensive efforts being implemented in India to 'conserve' medicinal plants in their natural habitats. Its work is supported through a bilateral aid agreement between the Government of India and DANIDA. FRLHT is concerned not only with the question of conserving the medicinal plant diversity but also its associated cultural diversity. India has one of the richest and oldest, unbroken bio-cultural heritage related to medicinal plants.

Introduction

It cannot be denied that the initiative for the biodiversity convention was a northern initiative, one inspired by two compelling needs: First, a sense of insecurity due to the precarious nature of the planet's ecology (caused, in the first place, almost entirely by the western model of unsustainable development); and second, a desire to have access to the South's abundant biogenetic resources.

The South has reacted to the North's agenda by asking for financial compensations to conserve its tropical forests and has demanded transfer of the North's biotechnology on favourable terms. The North has accepted some of the South's demands after hard bargaining and negotiations are still going on.

Amidst these negotiations on biodiversity conservation and its high-tech utilisation, we in the South seem to have paid insufficient attention to the fact that millions of our people have traditionally been using biodiversity for a wide variety of purposes. Many of these traditional uses have relevance both to our present way of life and to our future well being. As they have a time-tested competence, they are likely to represent the most ecologically sound, safe and sustainable ways of using our bioresources.

India's (and other southern countries') own agenda in biodiversity conservation should not be merely to see on what financial terms access to our biogenetic raw materials should be granted but more importantly to engage creatively in the revitalisation of traditional sciences and technologies, and to use their fruits for the benefit of our own people and to share them on fair terms with the rest of the world.

India's Legacy of Medicinal Plants

The ethos of conserving biodiversity is deeply ingrained in ancient Indian culture. Traditionally, patches of pristine forests were set aside as sacred groves; planting of shade and fruit bearing trees had religious sanction, as also prescriptions to regulate hunting, *e.g.*, ban on killing a pregnant female animal and designating certain parts of the year as closed season. This serious concern for ecology continued to thrive side by side with developments related to economic progress, international trade, and science and technology. In fact the Indian health system which goes back to 1500 BC, arid is mainly based on plants and animals, symbolises how deep-rooted was this concern for biodiversity conservation and its sustainable use.

The Indian people know a great deal about medicinal plants. Studies reveal that the largest proportion of the biodiversity in all our ecosystems is used by village communities for human and veterinary health care (See Table 1).

Table 1 : An example of uses of biodiversity by "ecosystem people": Some of the medicinal and other plants used by the Mahadev Koli Tribals.*

Purpose	*Number of Species*
Medicinal uses	202
Veterinary uses	109
As fish poisons	23
For pest control	51
For water purification	3
As wild edible plants	87
As fodder plants	65
For fuel	30
Hunting purposes	3
Cultural and religious purposes	38

**Source : D.K. Kulkarni Agharkar Institute, Poona.*

Over 7,500 species of plants are estimated (AICEP, 1994) to be used by "the ecosystem people" (See Figure 1) who belong to some 4635 ethnic communities. India probably has the oldest, richest and most diverse, cultural traditions in the use of medicinal plants.

Medicinal Plants used in Folk & Tribal, Ayurveda, Siddha, Unani, Amchi (Tibetan) and Allopathic Health Care

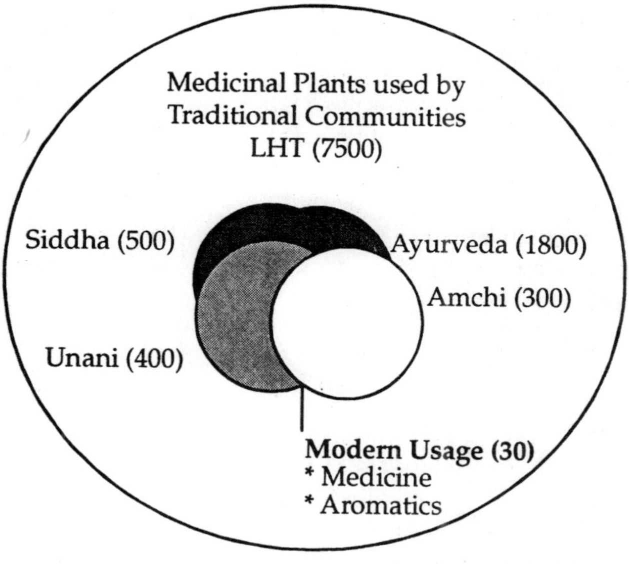

NON-INSTITUTION USERS	INSTITUTIONAL USERS
Physicians	Traditional Pharmacies
Veterinary Doctors	Pharmaceuticals
Farmers/Collectors	Perfumery Industry
Village Communities	Exporters
	Botanical Gardens
	Medical Institutes

Fig. 1 : Resource base of traditional medicine.

Medicinal plants continue to provide health security to rural people in primary health care (Table 2). According to WHO estimates over 80% of people in developing countries depend on traditional medicines for their primary health needs. In India the coverage of rural population by the modern health system varies between different regions from three to thirty percent. Thus, for some 4-5 hundred million people, traditional medicine is the only alternative. This is also borne out by the fact that there still exist over one million traditional, village-based carriers of the herbal medicine traditions in the country.

Table 2 : Richness of folk medicine: Examples of ethno-medicinal plants with ten or more uses reported across ethnic communities in South India.*

Plant Species	*Reported Number of Uses*
Centella asiatica	33
Pergularia daemia	23
Aristoloclhia indica	22
Ichnocarpus frutescens	22
Alstonia scholaris	19
Holarrhena antidysenterica	18
Trachyspermum ammi	16
Hygrophila auriculiculata	15
Trianthema portulacastrum	15
Semecarpus anacardium	15
Hemidesmus indicus	15
Catharanthus roseus	14
Apama siliquosa	13
Anacardium occidentale	12
Costus speciosus	12
Justicia gendarulssa	11
Perguiaria extensa	10

**Source : FRLHT Research Department.*

Today these biodiversity-dependent rural communities are facing a serious resource threat because of the rapid loss of natural habitats, and the over-exploitation of medicinal plants from the wild (Figs. 2, 3 and 4, and Tables 3 and 4). To meet the demands of the Indian herbal industry which has annual turnover of about US$ 300 million tons of medicinal plants are being harvested every year from some 165,000 ha of forests (estimates based on extrapolation of regional trade figures by FRLHT Research Department).

The biodiversity loss is not only a threat to ecology of the planet but a more immediate threat to the livelihood security of rural communities.

Table 3 : List of species banned from export by the Ministry of Commerce (vide notification no. 47 (PN)/92-97 dated 30 March 1994.

Aconitum sp.	*Gnetum* sp.
Acorus sp.	*Gynocardia odorata* (Chaulmogri)
Angiopteris sp.	*Hydnocarpus* sp.
Aristolochia sp.	*Hyoscyamus niger* (Black henbane)
Artemisia sp.	*Iphigenia indica*
Arundinaria jaunsarensis	*Meconopsis betonicifolia*
Atropa sp.	*Nardostachys* sp. (Jatamansi)
Balanophor sp.	*Nepenthes khasiana* (Pitcher plant)
Berberis aristata (Indian barberry; Rasvat)	*Osmunda* sp.
Colchicum luteum (Hirantutya)	*Paphiopedilium* sp. (Ladies' slipper orchid)
Commiphora wightii	*Physochlana praealta* (Bajarbang)
Coptis sp.	*Podophyllum hexandrum*
Coscinium fenestratum (Calumba wood)	*Pratia serpumlia*
Costus specious (Keu, Kust)	*Rauvolfia* sp. (Serpgandha)
Cyathea gigantea	*Renanthera imschootiana* (Red vanda)
Cycas beddomei (Beddomes cycad)	*Rhem emodi* (Dolu)
Didymocarpus pedicellata	*Rhododendron* sp.
Dioscorea deltoidea	*Saussurea lappa* (Kuth)
Dolomiaea pedicellata	*Strychnos potatorum* (Nirmali)
Drosera sp.	*Swertia chirata* (Charayatah)
Exphedra sp.	*Taxxus baccata* (Yewu, Birm)
Gentiana Kurroo (Kuru, Kutki)	*Urginea* sp.
Gloriosa superba	*Vanda caerulea* (Blue vanda)

Table 4 : A first red data list of South India Medicinal Plants (based on CAMP report; IUCN version 2.2)

Species	Status	Species	Status
Acorus calamus	VU/N & CR/R	*Lamprachaenium microcephalum*	EN
Adenia hondala	VU	*Madhuca diplostemon*	E
Adhatoda beddomei	CR	*Madhuca insignis*	EX
Aegle marmelos	VU	*Michelia champaca*	VU
Aerva wightii	EX	*Moringa concanensis*	VU
Amorphophallus paeonifolius	VU	*Myristica malabarica*	E
Ampelocissus araneosa	VU	*Nervilia aragoana*	EN
Ampelocissus indica	EN	*Nilgirianthus ciliatus*	EN
Andrographis paniculata	LR	*Nothapodytes nimmoniana*	VU
Aristolochia bracteata	LR	*Ochreinauclea missionis*	VU
Aristolochia tagala	VU	*Operculina turpethum*	LR
Artemisia nilagirica	LR	*Oroxylum indicum*	VU
Asparagus rattleri	EX	*Paphiopedilum druryi*	CR
Balanites aegyptiaca	LR	*Phoenix pusilla*	LR
Buchanania lanzan	LR	*Piper barberi*	CR
Cayratia pedata	CR	*Piper longum*	LR
Cleome burmanni	DD	*Piper mullesua*	VU
Commiphora mukul	VU/R	*Piper nigrum*	VU
Coscinium fenestratum	CR	*Plectranthus vetiveroides*	EW
Cycas circinalis	VU & CR/R	*Pterocarpus santalinus*	E
Cyclea fissicalyx	E	*Pseudarthria viscida*	LR
Drosera indica	LR	*Puereria tuberosa*	LR
Drosera peltata	VU	*Rauvolfia serpentina*	E
Elaeagnus conferta	LR	*Saraca asoca*	VU E/R
Embelia ribes	LR	*Schrebera swietenioides*	VU
Garcinia indica	VU	*Symplocos cochinchinensis laurina*	VU
Garcinia morella	VU	*Symplocos racemosa*	LR
Gardenia gummifera	LR	*Syzygium travancoricum*	CR
Glycomis macrocarpa	LR	*Tinospora sinensis*	VU
Gloriosa superba	LR	*Tragia bicolor*	VU
Hedychium coronarium	LR	*Trichopus zeylanicus*	CR
Heliotropium keralense	EN	*Utleria salicifolia*	CR
Holostemma annulare	VU	*Vateria indica*	LR
Hydnocarpus macrocarpa	VU	*Vateria macrocarpa*	CR
Janakia arayalpathra	CR	*Vernonia anthelmintica*	LR
Kaempferia galanga	CR/R	*Woodfordia fruticosa*	LR
Kingiodendron pinnatum	E		

* CR = Critically endangered; E = Endangered; EX = Extinct; VU = Vulnerable; LR = Low risk; DD = Data-deficient; EW = Extinct in wild; R = Regional

Source : FRLHT Research Department

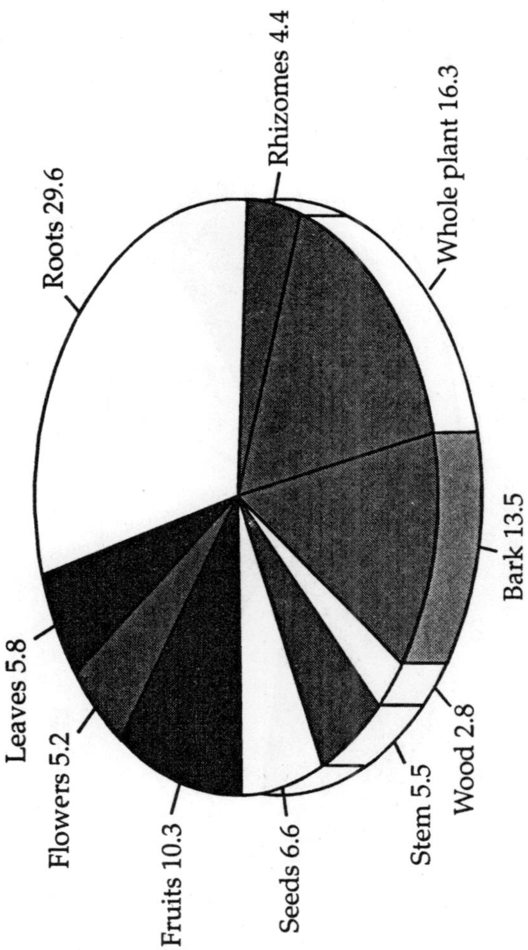

Fig. 2 : Destructive collections : distribution of medicinal plants by parts used (72% destructive and 28% non-destructive)

Source : FRLHT Research Department

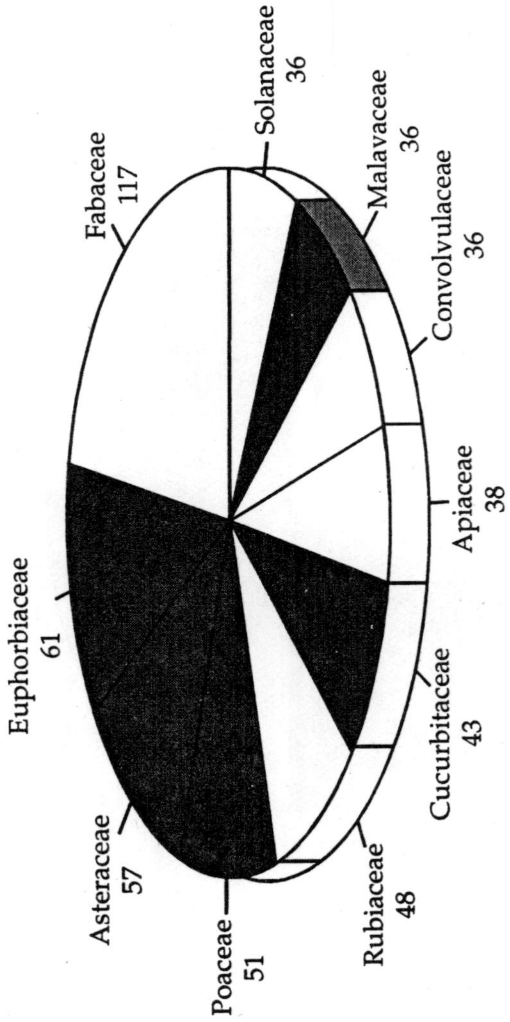

Fig. 3 : Distribution of Ayurvedic medicinal plants among the ten most represented families

Source : FRLHT Research Department

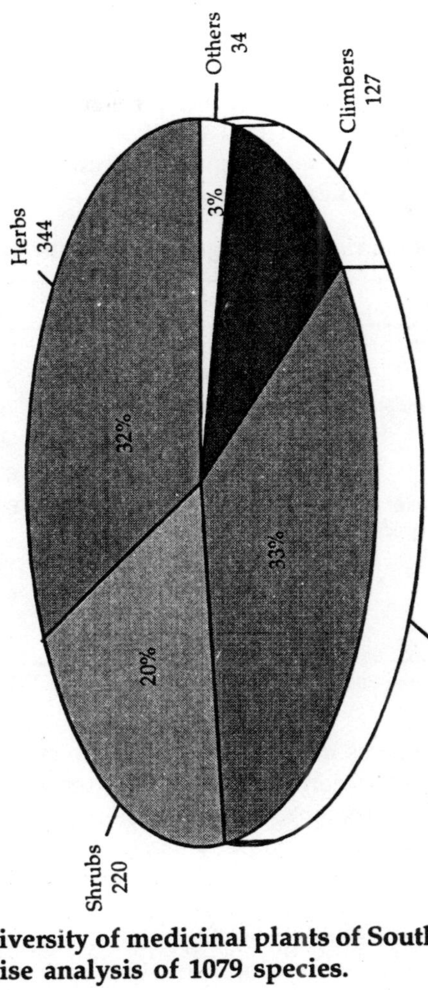

Fig. 4 : Diversity of medicinal plants of South India : Habitat-wise analysis of 1079 species.

Source : FRLHT Research Department

Nature of Traditional Medicine in India

The biodiversity of medicinal plants is associated with a very rich cultural diversity related to India's traditional systems of medicine. Traditional Medicine as practised in India consists of two streams, *viz.* folk medicine and the codified systems of medicine.

Folk Medicine

This is a diverse stream which is ecosystem and ethnic community specific. It is an oral tradition purely empirical in nature that exists in all rural communities throughout the length and breadth of India (Table 5).

Table 5 : Types of Carriers of village-based health traditions*

Traditional Carrier	*Conditions Treated*	*Numbers*
Housewives and elder	Home remedies, food and nutrition-related	millions
Traditional birth attendants	Normal deliveries	700,000
Herbal healers	Common ailments	300,000
Bone setters	Orthopaedics	60,000
Visha vaidhyas (snake, scorpion, dog bite specialists)	Natural poisons	60,000
Other specialists	Eyes, skin, respiratory, dental, arthiritis, mental, diseases, gastro-intestinal, wounds, fistula, piles	1,000 in each area

* Figures bases on extrapolations from micro-studies by FRLHT Research Department.

One comes across many examples of the great depth and range of the folk tradition in unpublished reports on medical practices in different regions. For instance, in 1793, two medical officers of the East India Company–James Finlay and Thomas Cruso reported on the practice of rhinoplasty by a potter's community in Pune district in the Madras Gazette (and later in 1794 in the London Gentleman's Magazine). It was this technical report that led to further developments in Britain of plastic surgery of the nose. To cite a current example, it is well known in south Karnataka that certain paralytic conditions can effectively be treated using 'Ankola' oil (In a place called Ankola in southern Karnataka, this treatment is administered by a tribal practitioner). Ankola is the name of a village; this particular herbal oil is part of its folk medical legacy. There is also the case, in Tamil Nadu state, of the so-called 'Coimbatore orthopaedic treatment' for straightening out a club foot by the combined use of a special herbal oil which softens the bony tissues and traditional forms of traction (Telungu Palayam Original Hospital in Coimbatore, Tamil Nadu, is a traditional orthopaedic centre where club foot and other outstanding orthopaedic procedures are practised). Yet another example is the use of the bitter aqueous extract of *Alstonia scholaris* bark by many rural communities at the start of monsoons, as a protection against malarial fevers.

Codified Traditional Medicine

Systems like Ayurveda, Unani, Siddha and the Tibetan system are expressions of this stream. The 'codified' stream consists of medical knowledge with sophisticated theoretical foundations expressed in thousands of regional manuscripts covering treatises on all branches of medicine and surgery. However, of an estimated 100,000 medical manuscripts lying in oriental libraries and private collections in India and abroad, less than one percent are available and in current use by students and teachers in Indian medical schools. The earliest Ayurvedic texts, the Susruta Samhita and Caraka Samhita, are believed to have been written between 1500 and 1000 BC. The main branches of Ayurveda are Kayacikitsa (General Medicine), Balacikitsa (paediatrics), Grahacikitsa (psychiatry), Salakya Tantra (Ophthalmology and ENT), Salya Cikitsa (Surgery), Visa Cikista (Toxicology), Rasayana (Rejuvenation) and Vajikarana (reproductive health). Besides these, there are specialised treatises on a range of subjects including Pharmacy (Bhesaja Kalpana), Pharmacopoeia (Nighantu), Diagnostics (Nidana), Special diagnostic techniques like Pulse diagnosis (Nadi Cikista), Iatrochemistry (Rasasastra), Dietetics (Pathyapathya), Pharmacology (Dravyaguna) and Positive health and preventive medicine (Svastha Vrtta).

Today there are over 400,000 licensed registered practitioners of the codified stream practising in the towns and cities of India (Ministry of Health, 1991). They offer a wide range of treatments as shown in the examples listed in Table 6.

Unfortunately, despite the large presence of a "living biohealth culture" in villages throughout India, it receives marginalised policy and financial support from national and international agencies, either for conserving its bioresources or for maintaining its indigenous knowledge base. This is evidently due to the western ethnocentric bias in health policies around the globe.

Challenge Facing the Cultural Heritage

Whereas the reasons for loss of plant diversity are well

known, the reasons for loss of cultural diversity are much less understood. These are briefly outlined here as a detailed treatment of the issue is outside the scope of this article.

Table 6 : Examples of strength of codified stream

Discipline	*Nature of Treatment or Advice Offered*
Prenatal care	Detailed diet and promotive herbs; behaviorial advice for healthy progeny and for all stages of foetal development
Obstetrics	Reliable advice which can help in case of delivery
Post-natal care food & nutrition	Herbs that can raise the general immunity of the mother & child advice on seasonal diets suited to different constitutions; advice on specific (therapeutic) diets for various ailments/disease stages; food values provided on a range of relevant parameters different from modern nutrition; advice on incompatible foods
Gynaecology	Safe herbal treatments for all typical ailments of women; better management of suspected infertility conditions
Respiratory disorders	Safe herbal treatments for entire range of respiratory disorders including asthamatic conditions & allergies
Skin disorders	Safe herbal treatments for wide range of skin conditions including diseases like Psoariasis and Erysipelas
GIT disorders	Safe herbal treatments for hyperacidity, ulcers & metabolic disorders
Cardiac diseases	herbal drugs for lowering chlorestrol, chronic angina and congenital heart diseases
Ophthalmology	Safe herbal treatments for intra-occular haemorrhage, diabetic retinopathy, paediatric myopia and a range of common eye diseases
Orthopaedics	Management of compound factures with open wounds; management of polio; management of osteal deformities
Nervous & muscular disorders	Herbal drugs for degenerative diseases in their early stages; hemiplegia, paraplegia, cereberal palsy etc.
Mental health	Safe herbal drugs for anxiety neurosis, obsessions, hysteric manifestations, epilepsy etc.
Specific diseases	Herbal treatments for non-insulin-dependent diabetes, arthritis, hepatitis, rheumatic fevers, gall & kidney stone, ano-fistulas & haemorrhoids, promotive care of cancer and AIDS cases

In the domain of knowledge, non-western medical systems like Ayurveda have so far failed to stake their claims to originality. Their epistemological foundations lie unexplored outside of their cultural worlds. The universal attributes of the indigenous knowledge systems remain unrecognised due to the marginalised political status of the southern societies and more mundanely due to the limitations of language and ethnicity.

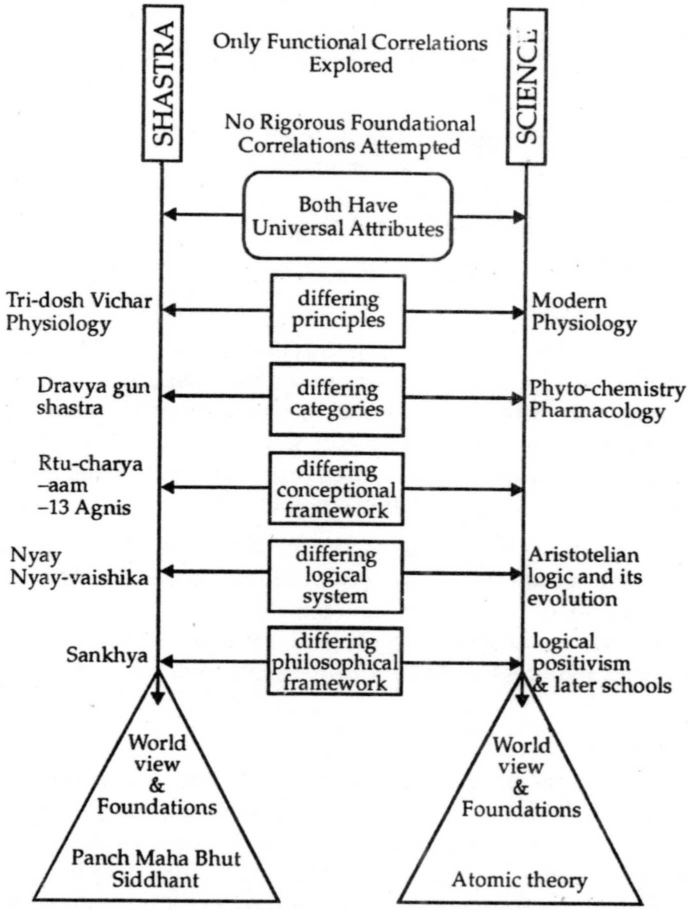

Fig. 5 : Epistemological foundations of Ayurveda and modern science.

A major problem that non-western societies have to contend with in any serious evaluation of their own indigenous knowledge systems, is the common claim of all western scientists and philosophers that, after all, science is one, universal and uniquely expressed in the western scientific paradigms. Thus, while it may be possible to conceive of alternative methodologies, theories and practices in other domains such as music, linguistics, logic, art and politics, there is no such possibility conceded with regard to alternative sciences.

The fact, however, is that different cultures have developed fundamentally different ways of perceiving and viewing nature and this in turn has given rise to different traditions of knowledge (See Figure 5).

The chart in Figure 5 depicts the differences in foundations, concepts and categories of Ayurveda and modem medicine. These differences should cause no surprise to anyone who is willing to accept the inevitable plurality of cultures.

Can traditional systems of medicine be explained in terms of modern medicine?

The comparative understanding between Ayurveda and modern medicine in fact has not progressed sufficiently to correlate diagnosis of the two systems of medicine or translate Ayurvedic pharmacology (Dravya gun shastra) in terms of modern pharmacological parameters or reduce Ayurvedic lines of treatment to satisfy Allopathic therapeutic logic.

Efficacy of disease management by Ayurvedic means and methods based on its own theories and monitored using modern parameters could form the basis for a dialogue between the two systems where the outcome would be in the form of *prima facie* evidence of success or failure in management but not a one-to-one correlation in diagnosis or a modern explanation of how and why of the traditional line of treatment.

The scope of this kind of comparative research designs from the modern medicine viewpoint will inevitably be limited. It will provide modern medicine only with some evidence of the efficacy or otherwise of Ayurvedic management. If modern medicine wants to apply the success of Ayurvedic treatment more widely based on the *prima facie* results, it will have to take the trouble to learn Ayurvedic principles and its diagnostic, pharmacological and treatment theories and procedures. The kind of comparative research outlined above will not, for instance, lead to any 'short-cut' ways to directly co-opt Ayurvedic drugs into the modem Materia Medica. This should be seen as a short-term loss. A pursuance of genuine dialogue with the traditional systems of medicine may pave the way for more meaningful long-term advances in world medicine that may broaden the scope of modern understanding of health and disease at more fundamental levels.

Conclusion

Conservation of medicinal plants in its biocultural perspective not only implies conservation of biodiversity but also places an equal emphasis on conservation of cultural diversity. The debilitating ecological consequences of monoculture in biological life are well known. The effects of promoting a monoculture in the civilisational context is only now being recognised.

CHAPTER 41

LIST OF GRASSES AND BAMBOOS OF INDIA

The forest grasses and bamboos are mostly used as food, fodder, house building materials, paper pulps, thatch materials, soil binders, for rope, mat, broom making as indicators, fishing rod, musical instruments, bow and arrows, hunting implements, agricultural implements etc. To avoid the bulkiness of the book the individual taxa of the grasses and bamboos have not been dealt with like other forest taxa. But the distribution, phenology has been provided though some species do occur flowers after a long period.

Bamboos

Arundinaria callosa	Munro (Assam, Meghalaya, Arunachal Pradesh, Nagaland) **Fl & Fr.** : Season taken from Herbarium Sheets CNH (CAL): March–May.
Arundinaria falcorneri	Benth. et Hook. f. (Himachal Pradesh, Uttar Pradesh, Sikkim and West Bengal) **Fl. & Fr.** : May–October.
Arundinaria khasiana	Munro (Assam and Meghalaya) **Fl. & Fr.** : August–November.
Arundinaria mannii	Gamble (Meghalaya, Assam, Arunachal Pradesh) **Fl. & Fr.** : April–October. Clayton and Renvoize in 1986 placed the taxon under the genus

Racemobambos	Holttum but R.B. Majumdar in 1989 transferred this species under the genus ***Neomicrocalamus*** Keng f.
Arundinaria racemosa	Munro (Sikkim, Arunachal Pradesh and West Bengal) **Fl. & Fr.** : May–October.
Arundinaria suberecta	Munro (Khasi hills) **Fl. & Fr.** : Not seen.
Bambusa arundinacea	(Retz.) Willd. (Uttar Pradesh, Jammu-Kashmir, Maharashtra, West Bengal, Meghalaya, Assam, Orissa, Bihar, Andhra Pradesh and Tamil Nadu) **Fl. & Fr.** : March–August.
Bambusa balcooa	Roxb. (Assam and West Bengal) **Fl. & Fr.** : June–December. In Assam it is known as *Balkua Bans.* In eastern part of Bengal it is known as Teli Bans.
Bambusa khasiana	Munro [Manipur and Meghalaya (Khasi hills)] **Fl. & Fr.** : March–December.
Bambusa mastersii	Munro (Assam) **Fl. & Fr.** : Not seen. In Assam it is known as *Beti Bans.* Dr. R.B. Majumdar in 1989 transferred the species in the genus *Melocalamus.*
Bambusa nana	Roxb. (West Bengal, Himachal Pradesh, Assam) **Fl. & Fr.** : December–February. This bamboo used as hedge plant in tea gardens.
Bambusa nutans	Wall. ex Munro (Uttar Pradesh, Sikkim, Assam and Orissa) **Fl. & Fr.** : May–December.

Bambusa pallida	Munro (Sikkim, Meghalaya, Assam, Arunachal Pradesh, Nagaland, Tripura) **Fl. & Fr.** : October–January. In Assam it is known as "Makal Bans".
Bambusa polymorpha	Munro (Meghalaya) **Fl. & Fr.** : August–January. In Bangladesh it is known as "Jama bans".
Bambusa teres	Ham. ex Wall. (West Bengal and Assam) **Fl. & Fr.** : January.
Bambusa tulda	Roxb. (Madhya Pradesh, Meghalaya, Assam) **Fl. & Fr.** : December–June. Locally known as "Mulya Bans".
Bambusa vulgaris	Schroder (Uttar Pradesh, Madhya Pradesh, Sikkim, West Bengal and Andaman) **Fl. & Fr.** : April–December.
Bambusa jaintiana	R. B. Majumdar (Meghalaya) **Fl. & Fr.** : Not seen.
Chimonobambusa callosa	Nakai (Meghalaya, Manipur and Mizoram) **Fl. & Fr.** : Not seen.
Chimonbambusa grifFlthiana	Nakai (Meghalaya and Nagaland) **Fl. & Fr.** : Not seen.
Chimonobambusa hookeriana	Nakai (Arunachal Pradesh) **Fl. & Fr.** : Not seen.
Chimonobambusa intermedia	Nakai (Arunachal Pradesh) **Fl. & Fr.** : Not seen.
Chimonobambusa jainiana	Das et Pal (Sikkim) **Fl. & Fr.** : December-May.
Dendrocalamus giganteus	Munro (Malay peninsula) Fl. & Fr. : Not seen.

Dendrocalamus hamiltonii	Nees et Am. ex Munro (Uttar Pradesh, Assam, Meghalaya and West Bengal, Sikkim) **Fl. & Fr.** : Not seen.
Dendrocalamus hookeri	Munro (Uttar Pradesh, Assam, Meghalaya) **Fl. & Fr.** : March–September.
Dendrocalamus longispathus	Kurz. (West Bengal) **Fl. & Fr.** : Not seen.
Dendrocalamus patellaris	Gamble (West Bengal, Assam and Nagaland) **Fl. & Fr.** : Not seen.
Dendrocalamus sikkimensis	Gamble ex Oliver (Sikkim, Meghalaya, Naga hills) **Fl. & Fr.** : Not seen.
Dendrocalamus strictus	Ness (Throughout India) **Fl. & Fr.** : March–October
Dinochloa compactiflora	(Kurz) Mc Clure (West Bengal) **Fl. & Fr.** : Not seen.
Dinochloa nicobariana	R.Majumdar (North Nicobar island) **Fl. & Fr.** : Not seen. Culm used in house building.
Melocalmus compactiflorus	Benth. & Hook.f. (Assam, Meghalaya) **Fl. & Fr.** : March–December
Melocalmus gracilis	R.Majumdar (Meghalaya) **Fl. & Fr.** : April–December
Melocanna baccifera	(Roxb.) Kurz (West Bengal and Meghalaya) **Fl. & Fr.** : Not seen. Culm used for thatching.
Ochlandra beddomei	Gamble (South India) **Fl. & Fr.** : Not seen.
Ochlandpa brandisii	Gamble (South India) **Fl. & Fr.** : May–August.

Ochlandra ebracteata	Raizada et Chatterjee (Kerala) **Fl. & Fr.** : June–October.
Ochlandra scriptoria	(Dennst.) C.E.C. Fischer (Nilgiri hills, Kerala) **Fl. & Fr.** : March–June.
Ochlandra Setigera	Gamble (Tamil Nadu) **Fl. & Fr.** : Not seen.
Ochlandra travancoria	Benth (South India) **Fl. & Fr.** : Not seen.
Oxytenanthera nigrociliata	Munro (Garo hills, Orissa, Andaman) **Fl. & Fr.** : March–June.
Phyllostachys assamica	Gamble ex Brandi (Assam and West Bengal) **Fl. & Fr.** : June–September.
Phyllostachys mannii	Gamble (Assam, Meghalaya, Arunachal Pradesh, Nagaland) **Fl. & Fr.** : June–September.
Racemobambos prainii	(Gamble) keng f. et Wen (Assam, Naga Hills) **Fl. & Fr.** : Not seen.
Schizostachyum capitatum	(Munro) R. Majumdar (Eastern Himalayas) **Fl. & Fr.** : Not seen.
Schizostachyum dullooa	(Gamble) R. Majumdar (Eastern Himalayas) **Fl. & Fr.** : Not seen.
Schizostachyum grifilthii	(Munro) R. Majumdar (Eastern Himalayas) **Fl. & Fr.** : Not seen.
Schizostachyum helferi	(Munro) R. Majumdar (Valleys of North East Himalayas) **Fl. & Fr.** : Not seen.
Schizostachyum kurzii	(Munro) R. Majumdar (South Andaman) **Fl. & Fr.** : Not seen.

Semiarundinaria pantlingii	Nakai (Arunachal Pradesh) **Fl. & Fr.** : Not seen.
Sinarundinaria griffithiana	(Munro) Chao et Renv. (Meghalaya, Khasia and Jaintia hills, Darjeeling, Assam, Mizoram) **Fl. & Fr.** : Not seen
Sinarundinaria longispiculata	Chao et Renvoize (Assam) **Fl. & Fr.** : March–June
Sinarundinaria densifolia	(Munro) Chao et Renv. (Kerala) **Fl. & Fr.** : April–November
Sinarundinaria elegans	(Kurz) Chao et Renv. (Nagaland, Sikkim) **Fl. & Fr.** : March–November
Sinarundinaria falcata	(Nees) Chao et Renv (Himachal Pradesh, Uttar Pradesh, Darjeeling, Nilgiri hills) **Fl. & Fr.** : April–June
Sinarundinaria hirsute	(Munro) Chao et Renv. (Manipur, Khasi hill, Naga hills) **Fl. & Fr.** : Not seen.
Sinarundinaria hookeriana	(Munro) Chao et Renv. (Sikkim, Darjeeling, Arunachal Pradesh) **Fl. & Fr.** : Not seen.
Sinarundinaria maling	(Gamble) Chao et Renv. (Darjeeling, Sikkim, Meghalaya) **Fl. & Fr.** : May–August.
Sinarundinaria pautlingii	(Gamble) Chao et Renv. (Darjeeling, Sikkim) **Fl. & Fr.** : Not seen.
Sinarundinaria polystachya	(Kurz ex Gamble) Chao et Renv. (Darjeeling, Khasi hills), Uttar Pradesh) **Fl. & Fr.** : July–September.
Sinarundinaria rolloana	(Gamble) Chao et Renv. (Naga hills) **Fl. & Fr.** : May–October.

Sinarundinaria walkeriana	(Munro) Chao et Renv. (Travancore) **Fl. & Fr.** : Not seen.
Sinarundinaria wightiana	(Nees) Chao et Renv. (Nilgiri hills) **Fl. & Fr.** : Not seen.
Sinobambusa elegans	Nakai (Nagaland) **Fl. & Fr.** : Not seen.
Thamnocalamus aristatus	(Gamble) E.G. Camus (Sikkim and Arunachal Pradesh) **Fl. & Fr.** : Not seen.
Thamnocalamus falconeri	Hook. f. apud Munro (Uttar Pradesh, Sikkim and Arunachal Pradesh) **Fl. & Fr.** : March–September.
Thamnocalamus prijnii	(Gamble) E.G. Camus (Meghalaya and Nagaland) **Fl. & Fr.** : Not seen.
Thamnocalamus spathiflorus	(Trin.) Munro (Arunachal Pradesh and Uttar Pradesh) **Fl. & Fr.** : October–May in 1994 Stapleton described one subspecies ***Occidentalis stepleton*** from Himachal Pradesh and Uttar Pradesh.
Thyrsostachys siamensis	Gamble (Arunachal Pradesh) **Fl. & Fr.** : March–June.

GRASSES

The grass species that are commonly found in the forests of different parts of India has been discussed here.

Acrocra munroanum	Hem. (Assam, Meghalaya, Kerala) **Fl. & Fr.** : October–November.
Acroceras tonkinense	C.E. Hubb.ex Bor. (Assam) **Fl. & Fr.** : September–November.
Acroceras zizanioides	Dandy. (West Bengal, Assam, Nagaland) **Fl. & Fr.** : April–January.

Aristida adscensionis	L. (Throughout India) **Fl. & Fr. :** August–December.
Aristida setacea	Retz (Throughout India) **Fl. & Fr. :** June–February.
Aristida redacta	Stapf. (Madhya Pradesh, Maharashtra, Rajasthan, West Bengal, Karnataka, Tamil Nadu) **Fl. & Fr. :** August–December.
Arundo donax	L. (Throughout India) **Fl. & Fr. :** April–December.
Arundinaria Khasiana	Munro (Assam and Meghalaya) **Fl. & Fr. :** August–November.
Bothriochloa intermedia	(R.Br.) A. Camus (West Bengal, Bihar, Orissa, Assam) **Fl. & Fr. :** May–December.
Capilipedium assimile	(Steud) A. Camus (Throughout India) **Fl. & Fr. :** March–December.
Capilipedium parviflorum	(R. Br.) Stapf, (Northwest and Eastern India) **Fl. & Fr. :** August–November.
Cenchrus ciliaris	L. (Drier parts of India) **Fl. & Fr. :** July–December.
Centotheca lappacea	*Desv.* (Throughout India in forest margin) **Fl. & Fr. :** May–September.
Chloris dolichostachya	Lagasca (South India) **Fl. & Fr. :** October–December.
Chrysopogon verticillatus	(Roxb.) Trin ex Steud (West Bengal, Orissa, Bihar, Tamil Nadu) **Fl. & Fr. :** February–March.
Cymbopogon flexuosus	(Nees) Wats. (South India) **Fl. & Fr. :** August–November.
Cymbopogon nardus	(L.) Rendle (Central and Eastern India)

	Fl. & Fr. : May–February.
Cyrtococcum oxyphyllum	Stapf. (Throughout Indian in damp slops in forest) **Fl. & Fr.** : February–April.
Cyrtococcum trigonum	A. Camus (West Bengal, Kerala, Tamil Nadu, Andhra Pradesh) **Fl. & Fr.** : September–March.
Eragrostiella bifaria	(Vahl) Bar. (Madhya Pradesh, Uttar Pradesh, West Bengal, Bihar, Orissa, Maharashtra, Karnataka, Andhra Pradesh, Tamil Nadu, Kerala) **Fl. & Fr.** : July–December.
Erianthus ravennae	(L.) P. Beauv. (Uttar Pradesh, West Bengal, Orissa, Bihar, Assam, Meghalaya) **Fl. & Fr.** : March–December.
Eulalia leschenaultiana	(Decne) Ohwi (Uttar Pradesh, West Bengal, Bihar, Karnataka)
Hackelochloa granularis	(L.) O. Ktze. (Deciduous forest) **Fl. & Fr.** : May–January
Hystrix duthiei	Bar (Uttar Pradesh) **Fl. & Fr.** : March–September.
Leptaspis urocolata	R. Br. (Kerala) **Fl. & Fr.** : October–January.
Lophatherum gracile	Brongn (Assam) **Fl. & Fr.** : October–January.
Microstegium ciliatum	A. Camus (Uttar Pradesh, Madhya Pradesh, Maharashtra, Bihar, West Bengal, Assam, Meghalaya, Andhra Pradesh, Nagaland, Kerala) **Fl. & Fr.** : October–February.
Microstegium petiolara	W.D. Clayton (North West Himalaya) **Fl. & Fr.** : October.

Myriostachya wightiana Hook.f., Sporobolus tremulus (Willd.) Kunth, Porteresia coarctata (Roxb.) Tateckao, *Spinifex littoreus* (Burm.f.) Merr. **Zoysia matrella** (L.) Merr., are the principal grasses in mangrove forests. The flowering months of these grass are between March–November.

Narenga porphyrocoma (Hance) Bor. (Uttar Pradesh, West Bengal, Madhya Pradesh, Bihar, Sikkim, Assam, Meghalaya)
Fl. & Fr. : November–February.

Oplismenus compositus P. Beauv. (Throughout India)
Fl. & Fr. : May–February.

Oplismenus undulatifolius P. Beauv. (Punjab and Uttar Pradesh)
Fl. & Fr. : May–October.

Ottochloa nodosa Dandy (Assam, Lushai hills)
Fl. & Fr. : October–February.

***Panicum brevifolium* L.** (Sikkim, Assam, Meghalaya, Arunachal Pradesh, Nagaland, Orissa, Andhra Pradesh, Tamil Nadu, Kerala)
Fl. & Fr. : August–February.

Panicum montanum Roxb. (Throughout India)
Fl. & Fr. : October–January.

Panicum sarmentosum Roxb. (Maharashtra, Bihar, West Bengal, Sikkim, Assam, Meghalaya, Arunachal Pradesh, Tripura).
Fl. & Fr. : August–November.

Pennisetum pedicellatum Trin. (Throughout India)
Fl. & Fr. : September–December.

Pennisetum purpureum Schumach. (Gujrat, West Bengal, Andhra Pradesh, Maharashtra, Kerala)
Fl. & Fr. : August–December.

Pennisetum Setosum	L.C. Rich. (Bihar, Madhya Pradesh, Nagaland, Orissa, Andhra Pradesh, Kerala) **Fl. & Fr.** : May–December.
Phragmites karka	(Retz.) Trin. (Throughout India) **Fl. & Fr.** : March–December. Are the principal grasses of deciduous forest with special reference to *Sal* forests.
Pogonatherum contortum	Brongn (Hill slopes) **Fl. & Fr.** : March–December.
Pseudechinolaena polystachya	Stapf. (Eastern Himalayas–Khasi hills, Nagaland, Tamil Nadu) **Fl. & Fr.** : May–November.
Schizachyrium exile	(Hochst.) Staff. (Uttar Pradesh, Madhya Pradesh, West Bengal, Bihar, Meghalaya, Tamil Nadu, Kerala) **Fl. & Fr.** : August–December.
Setaria palmifolia	Stapf. (West Bengal, Assam, Meghalaya, Arunachal Pradesh, Andhra Pradesh, Kerala) **Fl. & Fr.** : August–November.
Sorghum halepense	(L.) Pers. (Different parts of India) **Fl. & Fr.** : May–September.
Streptogyna crinita	P. Beauv. (Kerala) **Fl. & Fr.** : Not seen.
Stipa sibirica	Lamk (Kashmir) **Fl. & Fr.** : August–December It is the principal grass in pine forest.
Themeda anathera	Hack. (Kashmir, Uttar Pradesh) **Fl. & Fr.** : June–September. Principal undergrowth of Chir pine forest.
Thysanolaena maxima	(Roxb.) O. Ktze. (Throughout India) **Fl. & Fr.** : April–January.

Themeda quadrivalvis	(L.) O. Ktze. (Uttar Pradesh, Maharashtra, Rajasthan, Bihar, West Bengal, Sikkim, Nagaland, Manipur) **Fl. & Fr. :** April–November.
Themeda triandra	Forssk. (Throughout India) **Fl. & Fr.** : March–October.
Themeda villosa	(Poir) A.Camus (Eastern India) **Fl. & Fr.** : August–November.

CHAPTER 42

BAMBOO

Technology Innovations Towards Value-Added Applications

Activated Carbon from Bamboo

Bamboo when processed for slivers, slats, sticks for *agarbattis,* chopsticks, toothpicks, matchsticks etc. or for value-added products such as flooring tiles, reconstituted wooden block, chip and wafer boards etc., a lot of waste is generated.

The waste could be in the form of nodes, upper and lower portion of bamboo etc. This could be effectively converted into other value-added products such as activated carbon.

Even during the ancient Egyptian culture, the beneficial properties of charred coal were known for improving the quality of drinking water. The modern successor, relying on the same principle and being used in numerous applications, is activated carbon. Activated carbon is a carbonaceous adsorbent with high porosity, and hence a large surface area. Commercial activated carbon grades have an internal surface area of 500 to 1500 m^2/gms. Related to the type of application, two major product groups exist :

- Powdered activated carbon; particle size 1-150 micron.
- Granular activated carbon (granulated or extruded), particle size in the range of 0.50-4.00 mm.

Activated carbon mainly consists of elementary carbon in graphite like structure. It can be produced by heat treatment, or "activation" of raw materials such as wood, coal, peat, coconut shell, bamboo and other plantation wood. During the activation process, the unique internal pore structure is created and it is this pore structure, which provides activated carbon its outstanding adsorptive properties. There are over **150** different activated carbon grades available in the market for suiting different application.

Pores in Activated Carbon

A properly activated carbon has unique characteristics *i.e.,* large internal surface area, dedicated (surface) chemical properties and good accessibility of internal pores. According to IUPAC definitions three groups of pores are distinguished :

- macro pores (above 50 nm diameter)
- mesopores (2-50 nm diameter)
- micropores (under 2 nm diameter)

Micropores generally contribute to the major part of the internal surface area. Macro- and mesopores can generally be regarded as the highways into the carbon particle, and are crucial for kinetics. The pore size distribution is highly important for the practical application; the best fit depends on the matrix (gas, liquid) and treatment conditions. The desired pore structure of an activated carbon product is attained by combining the right raw material and activation conditions.

Activated Carbon : A Myriad of Applications

Activated carbon has hundreds of different applications. In fact, virtually every product manufactured today has been improved upon at least once by the use of activated carbon. In its numerous applications activated carbon represents a number of different functionalities as hereunder :

- **Adsorption** : The most well-known mechanism, through physical adsorption (Van-der Waal's forces) or chemisorption
- **Reduction :** *e.g.* removal of chlorine from water is based on chemical reduction reactions
- **Catalysis :** activated carbon can catalyze a number of chemical conversions, or can be a carrier of catalytic agents (*e.g.* precious metals)
- **Carrier of biomass** : support material in biological filters
- **Carrier of chemicals** : slow release application of colourant.

Types of Activated Carbon

Apart from the activated carbon product to be selected, a key issue to address is the type of technology to apply.

Typically, ***powdered activated carbon*** is dosed into the process stream (gas or liquid) and, after a certain contact time, separated by filtration or settling. Some of the issues involved : required contact time, dosing system, single or multi-stage dosing, carbon separation and safety measures.

Granular activated carbon is mostly used in fixed filter beds, or alternatively in (pseudo-) moving filter beds. Some of the issues involved : required contact time (alternatively : hydraulic space velocity), permanent or mobile filter vessels, filling and emptying facilities, safety measures. Further, a crucial consideration regarding granular activated carbon refers to possible regeneration, in situ or off site.

Methodology to Produce Activated Carbon from Bamboo

(*a*) Activation and Carbonization

1. *Size reduction* : Cutting down to small pieces and then grinding to different mesh sizes (*e.g.* 60-80, 100-120 and below 200).
2. Acidify with dilute acetic/phosphoric/hydrochloric/ sulfuric acid.
3. Drying in oven up to 200°C overnight.
4. Carbonization at 800°C for 4 hours in the presence of nitrogen flow.
5. Gradual cooling down to ambient temperature in nitrogen flow.
6. Char sample for characterization.
7. Activation at 800'C for 4 hours using steam/carbon dioxide.
8. Cooling down to ambient temperature under nitrogen flow.

(*b*) Characterization

1. *Pore structure* : Porosity, average pore size, pore size distribution, Langmuir and total surface area by nitrogen atmosphere using BET apparatus and porosimeter.

2. Density measurement by usual density bottle method.
3. Iodine adsorption.
4. Testing for mechanical and thermal strength.

(*c*) Performance Studies

1. *Liquid mixtures* : Removal of phenol from aqueous waste; removal of color impurities from waste.
2. **Gaseous mixtures :** Removal of sulfur dioxide, chlorine, VOCs.

To produce activated carbon from bamboo, TIFAC has undertaken an investigative study to bring out a complete *business opportunity package* comprising of the process optimization of carbon conversion for bamboo, characterization of activated carbon on the basis of properties such as adsorption, surface area generation, porosity etc., required specifications of plant and equipment/machinery, process economics etc. Under the study, the following activities are being carried out :

- Detailed feasibility study and evaluation on possible methodologies of converting waste bamboo (D. strictus, B. bamboos and P. stocksii) to usable activated carbon in an economic manner.
- To culminate it into a commercialisable process to be scaled up in partnership with an entrepreneur after evaluation of merits and demerits of different methods.
- Carrying out the experimentation of identified methods producing activated carbon from bamboo.
- Characterization of activated carbon establishing all its properties and its category.
- Comparison of properties of different types of activated carbon obtained from petroleum coke, coconut shell and forest wood (hard wood or plantation wood).
- Suggestion on the potential application/use of different types of activated carbon generated.

The experimental study would establish the technology package for entrepreneurs in setting up production units for bamboo based activated carbon.

Processing of Bamboo Shoots

Bamboo shoot is the young culm of a bamboo harvested at the time, or shortly after its appearance above the soil surface. It is rich in vitamins, cellulose, amino acids and trace elements and has the same nutritional value as an onion and is a good source of fiber.

Bamboo shoot comprises 90% water. Edible content of a newly harvested shoot is usually 40%-50% , with smaller shoots yielding a lower percentage edible content.

Bamboo shoots can be consumed fresh or processed. Fresh bamboo shoot has a crisp and sweet flavour. Bamboo shoot is used in oriental cooking as an extender because it takes on the flavor of the ingredients it is cooked with.

It is reported that bamboo shoot has the effect in cancer prevention and is very popular in the South-east Asian countries. It is also used to increase the appetite and decrease blood pressure and cholesterol. As a widely consumed vegetable, bamboo shoot can be labeled as a heart protective vegetable and its component phytosterols may be suitable as nutraceuticals.

Nutrition Chart

Sl. No.	*Item*	*Value*
1.	Calories	14.00
2.	Total fat (gms.)	0.30
3.	Saturated fat (gms.)	0.10
4.	Mono-saturated fat (gms.)	Nil
5.	Polyunsaturated fat (gms.)	0.10
6.	Dietary fibre (gms.)	1.20
7.	Protein (gms.)	1.80
8.	Carbohydrates (gms.)	2.00
9.	Cholesterol (mg.)	Nil
10.	Sodium (mg.)	5.00
11.	Potassium (mg.)	640.00

The nutritional values for 1 cup (100 gms.) of bamboo shoots are as follows :

FDA laid down the following specifications for fresh and canned bamboo shoots :

- **Nutritional Data for Raw Bamboo Shoots (100 gms) :** 8.3% cals from fat, 32.7% cals from protein, 58.9% cals from carbohydrates.
- **Nutritional Data for Bamboo Canned Shoots (100 gms) :** 16% calories from fat, 31.3% cals from protein, 52.7% cals from carbohydrates.

Species Suitable for Bamboo Shoots

Most bamboo species produce edible shoots. The shoots vary in size and degree of bitterness and most commercially marketed shoots are derived from a small number of chosen species. Bamboo species are split into two groups : the clumping types with short rhizomes (*i.e.* underground stems) and the running types with long rhizomes. The Indian bamboo species suitable for producing shoots are as follows :

- ***Bambusa polymorpha*** found in Nagaland, Meghalaya, Assam, West Bengal, Bihar and Uttaranchal.
- ***Dendrocalamus brandisii*** grown in Jirbam area of Manipur and Andamans.
- ***D. giganteus*** widely distributed in Arunachal Pradesh, Assam, Manipur, Nagaland and West Bengal.
- ***D. hamiltonii*** distributed in North West Himalayas, Sikkim, Arunachal Pradesh, Assam, Manipur, Meghalaya, Mizoram, Nagaland and Tripura.
- ***D. strictus*** grown in UP, MP, Orissa and Western Ghats, semi-dry and dry zones along with plains and hilly tracts.
- ***Meloccana balmbusoides*** cultivated in Assam, Manipur, Meghalaya, Mizoram, Tripura, W. Bengal, Maharashtra and Karnataka.
- *Others* such as ***phyllostachus bambusoides, tulda/nutan/ pallida.***

The above species are a generic recommendation covering most commonly available species.

Harvesting of Bamboo Shoots

Quantum of shoot production varies with the species and locality. The clumping type produce shoots after mid-summer, while the runners produce shoots in spring.

Bamboo shoots are available, depending on the species, for a limited period of time *i.e.* approx. 5 months in a year.

Most species provide shoots in the period March–September and the typical shooting season for a species rarely exceeds two months.

The shooting period can be enhanced if other species are also utilized and/or the shooting season is extended by controlled cultivation methods and management practices.

Value Addition to Raw Bamboo Shoots

Bamboo shoots could be hygienically processed and packaged for a good value-addition. At present, bamboo shoots with limited shelf life, harvested under hot and humid seasons, are largely sold in unprocessed form. Some quantities are also sold dried and preserved with simple additives like salt and lemon. In addition to fresh bamboo shoots, items such as shredded shoots and pickled shoots are also available.

In the bamboo food processing sector, the focus is on improving the availability, supply chain and shelf life of shoots, processing and packaging under hygienic conditions, marketing etc.

With appropriate technology and market support, it should be possible to value add bamboo shoots into preferred products which provides employment and income to cultivators, processors and intermediates engaged at different stages. With minor additions/modifications to the existing equipment, mushroom and other vegetable processing units can add bamboo shoots, to their portfolio.

The integration of the growers into the manufacturing set up or the development of cooperatives at the village level is the only mechanism, which would ensure the success of the processing units. Initiatives are being taken to encourage the commercial scale plantations of bamboo shoot and also to ensure a ready market for the same.

Quality standards for the product as well as the process would also have to be introduced by the concerned authorities in addition to obtaining the certificate of approval from the FDA. At full production (80% capacity) every year a processing unit can operate as follows :

- Requires 5300 tonnes of raw bamboo shoots
- 960 tonnes of processed shoots
- Generates employment
- Generates over Rs. 50 million foreign exchange with a net profit over Rs. 10 million.

Possible technologies for the value addition of bamboo shoots are as follows :

- **Preservation of harvested bamboo shoots** : Peeling • Rinsing • Boiling • Dehydration • Sterilization • Packing (polythene bags or containers), adding salt between shoot layers, compressing and sealing.
- **Boiled shoots** : The technical process for producing *boiled shoots* is as follows :

  Cleaning raw shoots. • Boiling • Cooling • Sorting • Rinsing • Canning • Adding juice • Can sealing • Sterilizing • Cooling • Storage.
- **Dried Bamboo Shoots** : This technology develops tasty open-and-eat dried shoots, multi-taste dried shoots and local flavour dried shoots, all of which were vegetable-type leisure food products.

  For a plant to be commercially viable, it requires a higher input of bamboo shoots, good processing machinery and processing technologies ensuring quality norms.

  Peeling • Rinsing • Boiling, dehydration, Sterilizing • Compress de-watering • Bake de-watering • Squeeze de-watering • Re-bake de-watering • Semi product storage • Sorting, packing • Market.
- **Canning of Bamboo Shoots** : Raw Bamboo Shoots • Washing • Preparation • Blanching • Filling in can with brine solution • Sealing • Sterilizing in retorts at 12-15 psi for 30 mins • Cooling to ambient temperature in cooling tanks

- Cleaning and stacking cans • Labeling cans and cartons
- Packing cans in cartons • Market

The bamboo shoot industry, using shoots from both natural forests and bamboo plantations, makes a substantial contribution to the economy and development of rural communities, in which people use bamboo shoot as food and also earn substantial income from bamboo shoot harvesting, preserving, transporting, processing, marketing etc.

It provides rural people with income during the lean rainy season when no other major agricultural crops can be produced as bamboo shoot cultivation is profitable when compared with paddy. One hectare of bamboo plantation yields 4.5 tonnes of bamboo shoots compared with 2.0 tonnes of rice per one hectare.

Market Scenario

In India, despite the availability of six varieties of shoots in abundant quantities the commercial potential has not yet been exploited. A 5% penetration in the export market would yield a market size of Rs. 200 crores for the Indian bamboo shoot processing units to start with. The international market has been increasing rapidly at almost 25% per year on higher acceptance of bamboo shoot as a delicacy and also because of its high nutrition content.

At present, consumption of bamboo shoot is largely restricted to local tastes and availability, mostly in the North-East region. Though a small proportion does find its way into the market, the commercial potential of bamboo shoot remains untapped. The total value of the bamboo shoot sold commercially is currently (2001) estimated at mere Rs. 4.8 crore in India. There is a high potential for increasing the domestic market size for bamboo shoot through creating awareness and regular supply.

Chinese populations worldwide consume bamboo shoot, demands exist in Singapore, Malaysia, Vietnam, the USA, Canada and Europe for canned (and in the future fresh) shoots. Shoots of D. asper and T. siamensis could be the main export items from India; shoots of other species are sold in the domestic market.

More than 90 per cent of the overseas market consists of steamed shoots, while the rest is composed of deep-frozen and dry shoots. Japan is the main market for steamed shoots of

D. asper, accounting for more than 70 per cent of the total export volume. Other markets include USA, Germany, Saudi Arabia and Hong Kong.

In view of such an impact of bamboo shoot production and processing to rural economy, TIFAC has launched a major *business opportunity* study with the following objectives in this area :

- Identification of **2-3** Indian bamboo shoot species based on quantum of their availability in terms of commercial exploitation; the identification should take into account of a speciemix for their availability over a considerable part of the year.
- Detailed study and development of recommended practices for harvesting methodologies.
- Analysis of detailed nutritional values for the identified species.
- Experimental investigation to establish the detailed processing technique with the relevant process parameters for bamboo shoots.
- Experimental evaluation of commercially viable packaging methods for the processed shoots based on process economics *vis-a-vis* shelf life of packaged products.
- Preparation of a detailed *'business opportunity report'* indicating optimal requirement of processing and packaging equipment (specs. and sizes), processing hygiene regime, utility and effluent system requirements, operational costs, consumables, manpower etc.

  The report should bring out raw material availability and pricing, investment size, production capacities, ROI analysis and market scenario to be established from the following step.
- A survey to ascertain high-end market for processed bamboo shoots in India presently of imported origin; organoleptic product evaluation of 5-7 dishes prepared from bamboo shoots suitable for Indian palate.

The aforesaid exercise would contribute towards evaluating the market acceptance and potential for bamboo shoots.

The *Business Opportunity Report* would be ready for dissemination among potential entrepreneurs by early February 2003. The investment under the National Mission on Bamboo Applications and the consequent value addition will lead to substantial employment, directly in the production units, and in the linked processes of cultivation, processing and marketing.

Several products will function as intermediates; for instance bamboo board sections as material, for furniture. Several processes and products entail the forging of linkages to corresponding mutual benefit between the organised and unorganised sectors, between household, cottage, tiny, small and medium scale enterprise. There are substantial environmental and ecological benefits to be gained from this investment.

Many products, especially in the boards and composites segment, and activated carbon will substitute materials and products currently imported, thereby leading to a saving of foreign exchange for the economy. Other products carry the promise and potential of exports as well, particularly high value laminates, flooring boards, and bamboo shoots.

New tools and techniques would be introduced to enhance the manner in which the bamboo resources are used in the country leading to greater efficiencies and a sensitive use of new materials.

The opportunities created by these interventions will help build suitable models for sustainable livelihood and also augment the income of farmers and artisans who are dependent on this resource.

The shift from timber to bamboo and from steel to bamboo would help reduce the pressures on our environment and ecology.

INDEX

A

B

C

F

G

N

O

P